ICTs AND INDIAN SOCIAL CHANGE

ICTs AND INDIAN SOCIAL CHANGE

Diffusion, Poverty, Governance

Editors

Ashwani Saith
M. Vijayabaskar
V. Gayathri

Institute for Human Development
Institute of Social Studies

SAGE
Los Angeles • London • New Delhi • Singapore
www.sagepublications.com

First published in 2008 by

SAGE Publications India Pvt Ltd
B1/I-1 Mohan Cooperative Industrial Area
Mathura Road
New Delhi 110044, India
www.sagepub.in

SAGE Publications Inc
2455 Teller Road
Thousand Oaks, California 91320, USA

SAGE Publications Ltd
1 Oliver's Yard, 55 City Road
London EC1Y 1SP, United Kingdom

SAGE Publications Asia-Pacific Pte Ltd
33 Pekin Street
#02-01 Far East Square
Singapore 048763

Published by Vivek Mehra for SAGE Publications India Pvt Ltd, typeset in 10/12 pt Galliard BT by Star Compugraphics Private Limited, Delhi and printed at Chaman Enterprises, New Delhi.

Library of Congress Cataloging-in-Publication Data

ICTs and Indian social change: diffusion, poverty, governance/editors Ashwani Saith, M. Vijayabaskar and V. Gayathri.
p. cm.
Includes bibliographical references and index.
1. Information technology—India. 2. Telecommunication—India. 3. India—Social conditions—1947. I. Saith, Ashwani. II. Vijayabaskar, M. III. Gayathri, V.

HD30.2.I28 303.48'330954—dc22 2008 2008006728

ISBN: 978-0-7619-3612-1 (Hb) 978-81-7829-776-7 (India-Hb)

The SAGE Team: Sugata Ghosh, Vaijayantee Bhattacharya, Mathew P.J.
Jacket design by Kamal P. Jammual

Contents

List of Tables

List of Figures

Preface

The services sector has arguably been the driver of India's growth process in recent years. The contribution of IT related services to this phenomenon is widely acknowledged and well documented. In an earlier volume, we analyzed the economic implications of this sector's growth, looking at both the macro and micro-level impacts. Apart from its emergence as a major global player in the software services and ITES markets, India is also home to one of the largest set of civil society and government initiatives to use ICTs to empower and enhance their transformative capabilities and to extend the range of services to the marginalized at reduced costs. India, therefore, suggests itself as a most appropriate site for investigating the potential of ICTs to bring about social transformation as desired and anticipated in many quarters. The important question is, ***can new technologies transform the lives of the poor and socially marginalized?*** There is much hype and many loud claims on these issues raising much dust that tend to cloud the reflective gaze. This volume seeks to provide some clarity of vision in this vital contested arena of public policy.

These issue have been extensively debated in the context of advanced market economies. Their significance in low-income economies like India cannot be overstated, but has been seriously understudied. The pace and forms of utilization of available technologies are crucially dependent on the institutional dimensions of the economy and society in question. Mapping the process of institutional mediation from the stage of technology diffusion to their downstream social outcomes in India offers rich lessons for low-income countries seeking to 'develop' through adoption of new technologies. This volume, which though independent, forms a sequel to our earlier book focussing on ICTs and Indian economic development[1], is also largely based on a set of extensively revised papers originally discussed at a conference on a similar theme held at Bangalore in December 2002. The conference was funded by the Indo-Dutch Programme for Alternatives in Development (IDPAD). IDPAD's support extended further into providing us with a small grant for work required to put this and its preceding companion volume together. We are extremely thankful to IDPAD for this, Sanchita Dutta (IDPAD India) and Cora Govers (IDPAD The Netherlands) in particular, and also to IDPAD's partner in India, the Indian Council for Social Science Research (ICSSR).

The large international conference, which brought together economists and other social scientists, technologists and philosophers of science, policy makers and activists, civil servants and private sector actors, together in one intellectual forum was held in Bangalore jointly by the Institute of Social Studies (ISS), the Hague, and Institute for Human Development (IHD), New Delhi. Support from members of both the institutes was crucial for the conference. The coordination and logistic support undertaken by IHD for the conference was critical to its success and we are especially grateful to Professor Alakh N. Sharma, Mr Prem Chandra and their indefatigable team at IHD for seamlessly mounting such a complex event. The moral and financial support provided by our own 'home' institutions, Institute of Social Studies, The Hague, Madras Institute of Development Studies, Chennai, and Indian Institute of Information Technology, Bangalore, is also happily acknowledged, as are the efforts of the rapporteurs and discussants of the various papers that have often played an important role in the revisions leading up to this book. SAGE has provided much appreciated encouragement and much useful support during the editing and production process, and thanks go in this regard to Sugata Ghosh and especially to our highly efficient editor, Vaijayantee Bhattacharya.

Note

1. Saith, Ashwani and M. Vijayabaskar (eds). 2005. *ICTs and Indian Economic Development: Economy, Work, Regulation*. New Delhi. Sage Publications.

1

Introduction: ICTs and Indian Social Change—An Agenda of Concerns

Ashwani Saith
M. Vijayabaskar

The Force of Technology: Optimism, Reductionism, Realism

We seem to live increasingly in a world and era of projected and attempted quick fixes; most fail—occasionally there is a much-celebrated exception to the rule. Technology is one, perhaps the most prominent, source of ready-made, instant, solutions. Technology and technologists usually carry with them spectacular promises of problem-solving, of overcoming resource constraints, of generating growth and of delivering development. Technological optimism, if not its companion technological reductionism, is stamped on the technologist's *weltanschauung*. 'Prometheus Unbound'—indeed, it is expected to sweep aside all that is before it. But what about the problems created by unbridled technological change itself in the process of modern industrialization? The technologist has no truck with such concerns of the historian, ecologist or sociologist: problems created by technological change are expected in time to induce technological solutions to themselves. It is here that uncontrolled optimism slides into reductionism.

Who would wish to challenge the power of technological change? Successive waves of Schumpeterian creative destruction, Kuznetsian cycles of technologically defined leading sectors, and the concomitant Kuhnian paradigmatic revolutions, all testify to its transformative capacity. It cannot be gainsaid that technological change has been a primary driver and generator of long-term growth and structural change for the past two centuries. But equally, the stoical peasant or the emaciated rural labourer as the intended beneficiaries of the green, white or biogenetic revolutions, or the daily-threatened migrant slum-dweller, living under the motorway arch or in the shadow of urban skyscrapers, might wish to testify that somehow such technology-driven growth seems to have passed them by. Far from technological change being a problem-solver, their perception might often be that of it being a problem-creator.

The optimism of the technologist has found an enthusiastic soul-mate in the market-driven, neo-liberal, globalizing economist, who betrays no residual sign of any legacy of economics being a dismal science. Far from the world of stationary states, of diminishing returns, of falling rates of profit, the gung-ho neo-liberal economists are not far behind the technologists in offering their map of the way to the promised land—which lies just on the other side of that hill called globalization. Since such economists tend, at present, to dominate decision-making structures in most developing economies, it is no surprise that most governments also tend to subscribe to the same book of recipes, and carry the same road map and navigation system. The long-suffering peasant and the slum-dweller have 'been there, done that', too.

This school of technologists, economists and policy makers boasts of similar mind sets and approaches: there is 'a solution'; they have 'it'; the world should adapt as necessary to make 'it' happen in order to obtain the hypothesized benefits. If the results deviate from those predicted, it is because this adaptation of behaviour and policies was inadequate. Ensuring such adaptation for enabling the solution then defines the framework of policy. In an era of government roll-back, public–private partnerships and good governance, enabling such adaptive responses to the needs of policy both defines and induces the role of civil society and development agencies, and of NGOs. Thus, the fundamental structural issues of inequality and exclusion are bypassed and technological solutions become a substitute for deep structural reform—rather as in the case of rural micro-finance programmes. The key policy instrument becomes how to link, say, poor rural people to information and communication technologies (ICTs), not directly to land, nor directly to employment; access to ICTs is implicitly assumed to enable access to sustainable livelihoods.

But is technology really neutral and therefore sufficiently pliable to be shaped at will, regardless of the socio-economic nature of the system, in ways that would orient it to human needs rather than to market demand? Pulled by inexorable gravitational force, the use of technologies follows the shortest line to sustainable profits. The World Trade Organization (WTO) and Trade-Related Aspects of Intellectual Property Rights (TRIPS) regimes seek to ensure the profitability of the patented innovator, but they also simultaneously raise the barriers to cheap diffusion and access for the followers, providing a prime example of various kinds of protectionist measures (mis)used by developed countries to 'kick away the ladder' in order to avoid competition from follower countries. More than in perhaps any other sphere, the potential does not translate into the real in the case of projected benefits of technological change, especially when viewed from the vantage point of the followers, or the poor; this applies with full force to the case of ICTs.

While it might well be appropriate, now and again, to interrogate and recalibrate the excessive assertions of technological optimism and reductionism, it would be churlish, if not self-defeating, not to recognize the rich and worthy seam of technological idealism that seeks to adapt and shape the processes and outcomes of scientific and technological discourse and practice, with a view to making it more socially relevant and responsive. The reference here is less to the quick-fix approach, as for instance visible in the Millennium Project, *a la* Sachs, with regard to instrumental technological solutions for African village development, and more explicitly in the direction of the dedicated efforts of socially minded scientists in designing and producing technological products directly answering to the needs, and also the capacities and constraints, of the poor. Equally, it would be self-defeating not to identify and exploit available spaces for influencing, even if not governing, this process of technological change.

The distinguishing feature of ICTs is the important role that users play in innovations. Its history is replete with instances of users finding new ways of deploying technologies. The rise of the Internet is the most-cited illustration of this process. This characteristic of the new technologies allows for their diffusion across various realms of production and governance processes, and also seemingly vests users with an equally important role in innovation. The possibility of such 'creative consumption' has opened up spaces for articulation of alternate, politically empowering uses. Further, this possibility, 'cyber-feminists' argue, can potentially redefine the relationship between technology and women. The traditional feminist discourse tends to pit technologies against women. Produced by men, they partake in perpetuating relations of dominance, with women relegated to being passive consumers.

New ICTs, on the other hand, subvert the authority of the producer. With users assigned a primary role in the technology's growth and innovation process, traditional barriers between the 'active male producer' and the 'passive female consumer' are undermined. Women can redefine the direction of the growth of this technology to undermine existing gender relations of power. Likewise, low-income economies can innovate new ways of using ICTs and hope to chart out alternate technological trajectories. Developmental interventions can overcome institutional barriers to information flows through appropriate use of ICTs. ICTs, therefore, can constitute an important component of the developmental toolkit. Such possibilities, of course, quickly run against the reality of a growing divide in access to these technologies. New technologies increasingly are found to aggravate inequities, since the skill and capital resources required to access them are predominantly confined to high-income groups and regions.

The open-ended nature of ICTs has however spawned a range of initiatives to address, even if not really redress, such inequities. India is home to probably the largest number of such initiatives. Taking stock and comprehending the implications of such interventions is therefore important. In this volume, we seek to map some of the important initiatives underway, the theoretical underpinnings of such initiatives, limitations and possible lessons for technology-driven/enabled developmental interventions.

Technology-enabled reforms have particularly assumed importance under current regimes of neo-liberal governance. Governance issues constitute a core component of neo-liberal reforms and increasingly the reforms are anchored around deploying the new technologies to overcome institutional rigidities that retard the reform process. This technocratic process under the rubric of e-governance has been diffused and even financially supported by several multilateral agencies. This is not to imply a collision of interests between technocrats and the neo-liberal agenda. Rather, a liberal conception of technology as essentially neutral can rest quite easily with governance initiatives directed at making markets more efficient.

An Agenda of Concerns: The Chapters Contextualized

The capacity of technological change, in particular ICTs, to radically transform and reshape economy, culture and society in a condensed period of time is unquestionable. While a distant view confirms this generalization, a closer scrutiny would reveal a multi-stranded web of interlinked processes, each calling for specific definition and analysis. This sets up a wider agenda of concerns and issues, in which the more specific question of the impact on the poor and excluded on development and democracy is embedded.

This book brings together a set of papers that speak to a growing body of literature on the relationship between new technologies and socio-economic divides, and the deployment of new technologies in the realm of governance and other public policy initiatives to overcome such divisions. In doing so, it juxtaposes two sets of views on the nature of the relationship. On the one hand is the vision of the technocrat. In this vision, essentially a liberal one, technology is a neutral tool and the impacts it effects are essentially conditioned by the kind of uses to which it is put. Those who are in a position to decide upon its uses, therefore, can produce the required impacts. The liberal view assumes special importance in the case of ICTs. Unlike many technologies that carried the promise of socio-economic transformation in the past, ICTs appear to have certain characteristics that make it a prime candidate. The second vision is a social constructionist view that sees technology as being constructed by and in turn constitutive of social relations. This view seldom finds its way into the domain of the policy maker and the planner. By bringing in these divergent voices together, we hope to stimulate a dialogue between two perspectives of technology. The themes and argumentation of the essays are briefly profiled.

Ways of Seeing

Chapter 2 exemplifies and elaborates this constructionist view of technology. The author points out how an instrumentalist view of technology can simply not account for the complex interactions that both technologies and cultures undergo when technologies are transplanted. Despite such cultural mediations producing varied outcomes, the dominant perspective continues to vest it with a dynamic outside the realm of the cultural because of its association with science, and hence with claims to truth. Its extreme functionality and historical association with perpetuating and defining relations of dominance have only served to reinforce such claims. Using illustrations from history and the present, the author highlights the lack of correspondence between technology and science, and hence the absence of any specific relationship with truth claims. Technology transfer implies not merely a transfer of objects, but also meanings embodied in such technological practices. He thus says that to understand 'technology' we need to pay close attention to the meanings that people associate it with. Moving on to the specific case of ICTs, he discusses its deployment in governance and argues that the increasingly technicist understanding of governance is driven by the rise of these technologies. As these technologies have not been developed for improving governance, their potential for stimulating democratic governance can too easily be overstated.

Commercially Viable Technological Idealism?

Commercially viable technological idealism? That might sound like an oxymoron to some, as a necessary check for sustainability to others and as an opportunity for an expeditious public–private partnership to yet others. The development sector is awash with schemes and initiatives where idealistic cosmetics dress hard commercial interests of corporate benefactors and philanthropists. For these development entrepreneurs, the so-called bankability of the poor testifies to them constituting the final hitherto un-penetrated market. But not all initiatives are driven top-down from commercial or sectional interests; others, indeed, derive their energy and resources from the radical, idealistic imagination which attempts to go against the gravitational force of the market. However, are viable and sustainable projects of diffusion and use possible below the radar visibility of the commercial market?

For instance, is there room for technologically viable, serviceable ICT platforms which have been overtaken by the enforced acceleration of technological obsolescence in the ICT sector, but which are perfectly capable of providing all relevant services to users who can do without the latest models and systems? Is such a dualistic system, relying on hand-me-downs, possible as an instrument for bridging the various digital divides and what form it might take?

Socially driven, idealist technologists have been more direct and ambitious in addressing the issue of the digital divide. They have focussed their energies on lowering effective threshold levels of the basic socially useful ICT package through technological interventions that would simultaneously bypass physical infrastructural, human capital and financial constraints and thereby include the vast mass of poorer, socio-economically excluded constituencies within the ambit of new generation ICTs. The celebrated case of the Simputer, created and launched by a group of so-motivated Indian scientists, provides an immediate example; another distinct example takes the form of the corDect innovation at the Indian Institute of Technology, Chennai. The chequered history, lengthy gestation and teething problems of the Simputer reveal both the potential and limitations of the creative model through which it arose: where university scientists are encouraged to convert their breakthroughs from blueprints into commercially viable products and services. The diverse economic, industrial, institutional and legal complexities and hurdles involved in successfully traversing this gauntlet explains the lengthy lags, and also points to the inherent difficulties of overcoming the structural constraints of developing advanced technological packages in a developing economy, even such a mature one as India, in the cut-throat open market competitive environment. It is not surprising that other competitor products, similar in orientation to the Simputer, have been effectively developed and marketed from developed economy locations in much shorter time spans.

Continuing with the theme of socio-economic impacts, two chapters explore the ability of ICTs to overcome the digital divide through appropriate ICTs and packages (Jain et al., and Jhunjhunwala and Ramachandran). If technologies are deployed without factoring in the needs of the poor and marginalized, they are more likely to aggravate social inequities. Innovative solutions based on existing technologies of information and communication, however, can be devised to counter the dequalizing effects of ICTs. Use of open-source software and cheap infrastructure-based wireless local loops technologies to provide low-cost connectivity to rural India are some of the experiments underway in India.

Jhunjhunwala and Ramachandran locate their chapter in the context of the rapid diffusion of telephony in India in the last decade or so. Their question is whether the new technologies can provide solutions to bridge the inequities in access to critical resources. They use the case study of N-Logue, an initiative of a team of engineers and scientists from the Indian Institute of Technology, Madras. Despite the rapid growth of telephony, telephone density in India is still low at 9 per cent and importantly, skewed heavily against the rural sector with a density of only 1.5 per cent. Further, private operators have relied more on developed regions, exaggerating disparities. N-Logue strives to create new telecom technologies that enable provision of resources like healthcare, education and market information to this segment. Importantly, it has created a business model that renders providing services to poorer rural regions attractive to private capital. Of course, it does require state support in terms of providing infrastructure and incentives, but the authors argue that multimedia kiosks on the lines of public telephone booths promoted by the Indian government would be ideal to spread information-based services to rural areas. The corDect technology developed by this team has found markets even in other low-income countries.

Jain et al. highlight two domains where ICTs can be very useful in reducing what they refer to as income and non-income poverty. Lack of access to primary and secondary education is one manifestation of non-income poverty and in rural areas, whereas improving market access for agricultural and non-agricultural products can reduce income poverty. They contend that a combined provision of these two services can be commercially viable and enable rural citizens to benefit from new market opportunities thrown up by globalization. Primary education is a public good with little incentive for private investment. Investment that enhances incomes, however, is attractive to investors. The authors call for an effort that combines the two initiatives. In doing so, they also critique current modes of pushing development through creation of pilot models, and so on, which have problems of both sustainability and 'scaling-up'. Instead, they stress the need for 'disruptive' solutions that clearly depart from incremental efforts that currently dominate the 'ICT for D' landscape.

Their model once again involves formation of kiosk-like structures in rural areas, called TeleInfoCentres, based on four technology building blocks; thin clients, server-centric computing, open-source software and WiFi. Server-centric computing and the use of recycled computers as spares overcome problems of maintenance in rural areas. These centres can be used to provide both content for education as well as for income augmenting activities like agricultural extension services and financial intermediation. Even the government can use them for IT-enabled services like data-inputting.

Diffusion across the Eternal Divides: Digital and Social

The recognized intractability of the deep socio-economic and structural factors underlying the digital divide have induced the spawning of a plethora of civil society and NGO initiatives designed to provide various bridging mechanisms. This is a densely populated field of activity, with a rapid turnover, fast learning curves and constant challenge and innovation. As with other NGOs, there is a critical need for systematic and impartial studies of this family of initiatives, to sift the serious from the self-serving, the purely commercial from the socially oriented, the really effective from the self-projecting experiences. There is enough evidence from some of the branded programmes that there could be another chiasmic divide between the aspiration and the achievement, between the professionals and the poor, between the hope and the hype surrounding these interventions which assert the power of ICTs to overcome deep rural poverty.

While Jhunjhunwala and Ramachandran, and Jain are enthusiastic about the potential of such techno-economic interventions, Saith underlines the need to exercise caution in stating that provision of cheaper ICTs to the poor can alleviate poverty. Mere access is insufficient if it is not backed by a strong agency capable of directing its use to desired ends. While diffusion of ICTs in rural areas can improve certain kinds of access to services and information, impacts can be improved through mediation by civil society organizations working towards poverty alleviation.

ICTs and Civil Society

The ways and means by which civil society organizations can use ICT-based tools towards their objectives has been an important subject of discussion the world over, especially among multilateral developmental agencies. Civil society initiatives using ICTs in India are aplenty and quite a few of them are held as models to be emulated. But they are seldom the subject of academic scrutiny. More often, the organizations themselves make such claims with little evidence of serious analyses of its pitfalls. Sreekumar, rooting himself in a framework that recognizes the important role that institutions have in shaping outcomes, examines a few of the more recognized and acclaimed initiatives in different parts of India. He particularly looks into claims about transformation of gender relations through IT deployment by civil society organizations in rural India. Not only does he observe a low participation by women, he also captures the age-based barriers to access IT-enabled service provision in rural areas. Even when there are women participants as in the case of the Embalam project run by MSSRF, their participation is indicative of an unwillingness of men to participate in unremunerative tasks rather than a sign of re-ordering of gender relations. He finds that though ICTs have led to progressive changes in both the economic and social spheres, the impact of such initiatives are limited by a lack of understanding of the complex ways in which technological use is played out in society. He further distinguishes between what he terms as social movements that are strengthened by use of ICTs and civil society initiatives that can be termed better as social enterprises contesting within the realm of the civil society.

The Significance of Regional Initiative

Notions of digital and social divides need to be defined and analyzed along various axes of dis-aggregation if they are to be operationally addressed. There are various relevant binaries here: young and old, male and female, highly qualified and educationally excluded, rural and urban, small town and megapolis, and rich and poor. An intermediate level is provided by the region, where many of these intersecting factors enter a common policy arena. The role of structural and policy as well as serendipitous factors in influencing the outcomes at the regional level remains a highly pertinent concern that needs to be addressed much more systematically.

As mentioned earlier, several studies address the possibility of accentuated inequities with the onset of new technologies. Regions with the ability to take advantage of the availability of such technologies tend to be those endowed with better infrastructure and skill sets. An agency that can play a crucial role in reducing digital divisions is the state. As has been exemplified by the experience of South East Asia's newly industrializing countries, state policies have a determining role in reducing existing technological gaps between nations. Within India, a similar role is envisaged for state governments, especially with lower levels of income and infrastructure, to ensure that the benefits of ICTs do not accrue only to the high-income states. Apart from policies at the national level, the 1990s saw the rise to prominence of state-level policies in promoting IT-related investments and growth. Interestingly, Nirvikar Singh points out in Chapter 7 that lack of appropriate moves even within high-income states can lead to failures, as in the case of Punjab. Despite relatively high-income levels and high levels of traditional infrastructure, the state has failed to take advantage of opportunities brought by ICTs in India. Nirvikar Singh attributes policy failure to its focus on incremental changes when what was required was a radical leap to bridge the ICT chasm. A key role that the state can play in the promotion of ICT sectors is through its use of ICTs for better governance.

Decentralized Democratic Communication: The Community Radio, A Forgotten Friend?

The dominance of new technological media is overwhelming and irresistible; rather like a swirling dust storm, it sweeps aside all ahead of it and drags much behind it. It beams the imagery of the new world into the remotest of locations, thrusts it into the most excluded of lives and superimposes it unchallenged into the most gullible of eyes and minds. Yet, there remains an audible murmur that questions its appropriateness for addressing the digital and social divides in meaningful ways. What about the old friend, the radio, especially in the hands of a community-driven and -oriented service and all the more when it is linked indirectly to the power of the Internet? Is local community radio not more accommodative of local voices and sustained conversations which can never reach the expensive and ephemeral glamour channels of high media with its severe infliction of attention-deficit-disorder syndrome?

New technologies, as they diffuse through stratified social formations, are likely to produce differential impacts. At the same time, they also have the potential to subvert the basis of such stratifications by enabling more egalitarian and decentralized communication networks. Such networks can add critical capabilities to new social movements to spread and build collective resistance. Importantly, it is a potential tool in strengthening democratic institutions. It is in this context that Pavarala asks in Chapter 8, 'How can ICTs be used to stimulate a kind of publicness, which is neither part of the state nor wholly dependent on the processes of the market?' He uses case studies of initiatives to push radio into uses that militate against the dominant mode of consumerism to address this question. Community radio has been an important movement that seeks to provide a means to dialogue among decentralized networks and has the potential to resist the dominant forms and content of communication. Premised also on the ability of community radio to actively constitute a counter-public sphere, Pavarala discusses the politics and potential of community radio towards this in India. Radio, being a low-cost and highly flexible technology, according to Pavarala, constitutes an important mode of communication that can have local content and meaning, transforming listeners from being passive consumers to active producers of information. Its potential has been recognized by several high-income countries and countries like Australia and the US have a long history of promoting it. Among the low-income countries, policies formulated by post-apartheid South Africa have been one of the most progressive. In discussing the various initiatives in South Asia, particularly India, he stresses the indifference of the state in this domain and the scattered initiatives of a few civil society organizations. Community radio is therefore a potential yet to be fully realized in India.

Government and Governance

So much is made of the role of ICTs in the provision of public services that the distinction between government and governance is all too often blurred, if not lost. Clearly, the role of ICTs in the provision of government and local public services is greatly valuable and has the potential capacity to enormously simplify the life of the citizen across a wide arc of the daily interface with government; from paying bills, to processing taxes, passports, using land records, accessing information, available services, and so on. What effective institutional models are available in this domain?

But this is distinct from the issues of governance, which goes well beyond the one-way provision of instrumental information. How can ICTs be harnessed to monitor the quality of public service and of civil rights? Can such governance structures be effective in monitoring, if not governing, the government in its dealings with the citizenry and in its exercise of its powers? Can it assist in exposing, curbing and penalizing corruption and crime, especially within government structures themselves? Or will those that govern be more successful in using the power of ICTs to monitor or snoop into the lives of citizens, especially when they are thought not to subscribe to authorized mainstream views and positions? Does this undermine the democratic rights of legitimate oppositional projects? What are the guarantees, say with regard to electronic eavesdropping in an era of global securitization in an overcharged, and perhaps overdone, atmosphere of threat and fear? When will creativity and debate and challenge be construed as a subversive security risk and by whom? Will excluded social constituencies be able to affiliate, network and organize collectively and effectively on a pan-country basis via ICTs? Will the news of and responses to both good and bad practice in panchayats be shared and collectively learnt from? Will global and national structures and instruments of information control, censorship and surveillance be open to scrutiny? Will the monopolistic structures that govern the global information regime, especially with regard to the Internet, be reformed and democratized, or will each new generation of ICTs simply continue and maintain the dependency of the south on the gatekeepers and monitors of the north? These dark facets and potentially negative outcomes of ICTs on the quality of citizen rights and governance demand serious reflection. Given the weakened capabilities of the state and permeation of the market in more and more realms of socio-economic space, civil society is expected to mediate and ensure that the voices against dominant exclusionary paradigms of development are heard and addressed. ICTs can play a powerful role in giving voice to such concerns and civil society initiatives can enable the state to fulfil its obligations to its citizens better. Gopakumar documents such an initiative, Transparency

International, a global initiative against corruption in low-income countries. Citing the successes of such initiatives, he hints at the immense scope for democratizing governance processes and rendering them transparent.

E-governance is believed to improve governance through transparency and better access to information about public services. To the extent that efficient provisioning of public services and goods hinges upon fast exchange of information between citizens and government agencies, ICT tools can go a long way in aiding this process. Democratic institutions can also be strengthened as information flows can enhance awareness among citizens of their rights apart from enabling their voices to be heard better. Further, such initiatives can also serve as seedbeds of technological learning for firms providing technical support to such initiatives. While it is true that use of ICT tools may improve such processes, the very act of use of ICTs by itself is very unlikely to create improvements if it is not accompanied by appropriate institutional innovations, argue Singh and Mitra.

Singh and Mitra also point out the differences in imperatives of e-governance between 'developed' and 'less developed' regions. While initiatives in the former countries are directed towards targeting marginal sections to improve their access to information resources, in low-income countries, the primary aim is to enhance the quality of service delivery. Since e-governance involves re-engineering of processes, bureaucracy is likely to resist such initiatives. As India has emerged as an important destination for outsourced software development including software meant for governance processes, the authors argue a strong case for interaction between the software sector and the Indian government in establishing healthy e-governance practices. This is especially possible when public organizations increasingly look to adapt private sector practices to enhance performance. A number of e-business models like virtual storefronts, marketplace aggregator, information broker, online service provider, reverse auction, content provider, and so on, developed in the private sector, can be applicable to public service delivery, a combination of information brokering and content provision, in particular.

The success of e-governance initiatives hinges more on institutional changes than on the mere introduction of ICTs; this is stressed by Shirin Madon as well. Reiterating the social constructivist perspective, Madon stresses the need to analyze the trajectories of such initiatives to see to what extent agency plays a role in conditioning outcomes.

Her chapter has to be seen in the context of several studies raising questions about the long-term sustainability of governance initiatives. The Government of India hopes that at least 25 per cent of government services will be delivered electronically by 2005. Adopting Sen's capabilities approach, she explores the relationship between the capabilities of end-users and the ways in which implementation of e-governance gets shaped. In this, she highlights the agency of mediating institutions like local government organizations, civil society organizations and private bodies. One key argument that she makes is that institutionalist accounts of constraints on IT innovation projects do not help us understand how and why these constraints change over time. Based on three case studies of e-governance projects in South India, she makes a strong case for shifting current emphasis in e-governance initiatives on infrastructure provision to building of capability among end-users.

Gender Values and Value-chains: Profit, Work, Alienation

Technological change is not, never has been or can be gender neutral. The impact of ICTs on the lives and livelihoods, well being and rights of women is a vast area that calls for substantial sustained research. Accumulated experience throws up many insights and also raises questions for follow up. Computerization of industrial manufacturing processes transformed the workplace and reconfigured the skilling of the workforce, with major implications for the location, exploitation and opportunities of working women in the labour force. There are analogous questions with regard to the transformation of the service sector through the absorption of ICTs. The phenomenon of the call centre and its mixed impact on women's work and domestic well-being have been extensively commented upon; yet systematic studies are few. Recently, the conditions prevailing in such workplaces were likened to those prevailing on slave galleys! While this might have been overly dramatic, the risks and vulnerabilities associated with night work have, unfortunately, been all too frequently realized and are part of the public imagination. Such workers might indeed be well paid, but the attendant personal non-monetary costs are not negligible, including the psychological damage of sustained forms of alienation where the worker has to take on an alien virtual identity for the purpose of trying to sell the product of some foreign company to some potential foreign customer, often with gratuitous insults and verbal abuse from the other end of the line. Claims of empowerment through ICTs are also ubiquitous, especially when emanating from the NGOs and organizations that run related ICT projects. Does the Grameen Bank's village payphone project really enrich and empower poor, excluded women? Does this marriage of two quick-fix panaceas, micro-finance and ICTs, highlight new potent avenues to pro-poor, gender-sensitive development? Early pointers are sobering: evidence from the payphone scheme itself reveals that, as in the case of the micro-credit, the formal beneficiary is the woman, but the real operator of the facility is an adult male of the family. And for each female that owns a phone, there could be hundreds that neither own nor use one, most often since they declare that they have no one to call, unlike the better-off families with overseas, remitting, migrant members.

An important locus of intersection between gender and new ICTs is in the domain of the labour market. Again, ICTs tend to disrupt existing gender relations in the labour market as they reduce the need for traditional 'masculine' skills even as they create new skill requirements that favour women's participation in the labour market. Two chapters explore this theme in this volume, based on fieldwork among workers in the Indian call centre industry. Gayathri and Piyush Antony examine the factors that draw women into the Indian call centres' workforce to evaluate the implications for their 'empowerment'. They distinguish 'push' factors like aspirations of women workers and changing family norms that favour women seeking paid employment from 'pull' factors that make the sector attractive to this set of women, which revolves around the nature of skills, work and employment, salary levels, and so on. The chapter innovatively uses excerpts from interviews to point out that women's employment in this sector is essentially framed within existing gendered norms and are hardly indicative of any transformation anticipated. In fact, the very 'over-representation of women' in this sector, a low value-adding segment of IT-based services, is indicative of persistence of traditional gender relations.

While Gayathri and Antony concentrate more on household-level factors, Vijayabaskar focuses on how capital and labour construct specific meanings of work and employment in the industry. By locating the characteristics of the workforce in a comparative perspective, he highlights how specific meanings of work here in India allow for intensive labour use practices without much resistance from labour. Labour is drawn from fairly elite families whose incomes constitute only a supplementary income to the household. To most employees, work here does not hold career prospects but offers a lucrative avenue for short-term employment. Resistance is individualized and primarily through exit, with little interest among workers in collective initiatives like formation of trade unions. Further, the chapter points to the fluidity of gendering processes and how masculinist constructions of work in this sector can create new meanings of work. Both the chapters also highlight how work in this sector is tied to the life cycle of women workers, with very few employees hopeful of continuing their employment after marriage or child birth.

The Diasporas and Transnationals: Virtuality and Cultural Reincarnations

Technologies, apart from socio-economic impacts, also foster changes in the realm of culture through creation of new work and consumption identities. When such sectors diffuse in a society that is characterized by technological backwardness in most sectors, such impacts may be much larger, as identities generated in use and involvement in production of such technologies can wield much more power than in societies that are already permeated by a high level of technological advancement. Such power, in turn, can be used to reinforce ideological positions that seek to maintain their status. Attempts to see in ICTs the best solution to socio-economic problems may be one such mode of reinforcing their locational importance. Simultaneously, formation of such identities is also conditioned by factors like migration, religious affinities, and so on.

The impact of ICTs on the cultural configuration and reproduction of diasporas has been nothing short of profound. It has instigated regurgitations and kaleidoscopic realignments of the relationships between sections of the Diaspora and their 'home'-related identities and interventions. Persisting forms of social exclusion in their new locations, regardless of the visible economic success in their new lives, have often placed firm ceilings in their *sanskritization* aspirations and strategies in their new homes. This tendency, possibly processed by a sense of catharsis, has often intensified the contacts with the country of origin while also making these linkages far more volatile.

New-found techno-religiosity has combined with the power of ICTs and the Internet to create new channels for virtual expression and also for ideological, institutional and financial interventions on a disproportionately significant scale in the ongoing cultural reproduction processes 'at home', namely, in the country of origin. This volatile and dynamic interface, with its mutual rivalries and relations of affection and rejection, or of imitation and distancing, is in obvious need of extensive study.

There are diasporas and diasporas. The rural Bangladeshi migrant to the Gulf economies might indeed find the village payphone scheme very handy in making economical and quick remittances to kin back home, quite apart from its welfare-enhancing effect through facilitating affordable, regular family contact. This is but a sideline in the large mosaic of relationships between ICTs and global economic flows, whether of various forms of capital, or of different categories of labour, ranging from the raw and unskilled workers to the creative and highly qualified engineers and entrepreneurs. Embedded in, and driving the growth process in the emerging global accumulation regimes, are new value ICT chains involving body shopping and cyber coolies; the subsidized production of high skilled, low cost, pliable, disenfranchized, ICT workforces at the dependent periphery, contributing, in highly regulated and regimented ways, to accumulation processes in the centre.

Xiang Biao's chapter shifts the focus on technology and its impacts to the practices of one group who are involved in the production of these technologies, the Indian software professionals. Despite the sector being seen as critically dependent on human capital, few studies pay attention to the modes of production and appropriation of this human capital at the global level. He puts forth the category of the 'ethnic transnational middle class' to explain the role of immigrant software Indian professionals in the US and other high-income countries. They are transnationals at one level, but straddle that identity with their ethnicity to create networks that act as a conduit both for global capital to source labour from low-income countries like India, as well as enable this segment of labour to accumulate capital through its global pursuits. His study is based on fieldwork among software professionals from Andhra Pradesh, which includes participant observation through living with some of these professionals in Sydney, Australia.

He highlights the nexus between current regimes of capital accumulation at the global level and the increased visibility of the IT sectors.

Electronically linked finance capital constitutes the major source for IT investments. His primary contention is that flexible labour is critical to accumulation under conditions of economic volatility and body-shopping, driven largely by ethnicization, has contributed to the production of such a flexible global workforce. He maps this process of ethnicization both by state and capital agencies as well as by the professionals themselves. Extrapolating this exercise on to the debates on the link between globalization and socio-economic development in low-income regions, he suggests that rather than resist globalization, it may be worthwhile to pursue a strategy that allows for adequate articulation between this growth process and other sectors in the economy through investments in primary education and promotion of IT-enabled services.

Such immigrant communities, interestingly, have historically been at the forefront of formation of nationalist identities driven by their marginalized existence in the host countries (van der Veer). The Indian software diasporas, despite being located in high-tech modern environments and in high-income brackets, also tend to identify themselves with such movements and identities. Van der Veer juxtaposes the rise of nationalist sentiments among Indian indentured labour in the late 19th century with the rise of anti-Muslim Hindu nationalist feelings among the software professionals in the US. Modernity, rather than obliterate such 'pre-modern' identities, is accompanied by the reinforcement or formation of new identities.

He also points out how the hopes built around the software sector's ability to fuel India's prospects is not merely reflective of the collective dream of the Indian middle class. It is also critical to anchor the developmental trajectory of the state around this sector. He brings into relief the role of the nation–state in the formation of a transnational workforce via its immigrant regulations. An interplay of immigrant regulation laws and generation of low-skilled, cheap labour in the home country ensures production of a 'flexible' labour with also flexible citizenship rights that seem to work to the benefit of capital and upper-income sections among this transnational workforce. He further argues that in the Indian context, despite the greater economic significance of migration to the Gulf, it is the software migrant population that has attracted greater state attention and favourable legislation giving concessions to non-resident indians and People of Indian Origin). This, he attributes to the dominant Hindu middle class origins of this particular migrant segment with its resonance with the political mobilization of upper caste Hindu by the BJP and other parties in recent years. Their religio-nationalist imagining is critical to the furtherance of such transnational networks. The opposition of religion and science, he concludes, has been only a marginal phenomenon and increasingly we witness the combined pursuit of techno-science and religious nationalist dreams.

Ravi Sundaram too works with the category of transnational networks, but uses it in the domain of cityscapes to point to the new sources of threat to the global elite through threats to intellectual property regimes. Using the notion of 'splintered urbanism', he explores the changes in the urban landscape of Delhi in recent years, especially those brought about by the techno-elite like malls, private townships, and so on. The rise of this brand of consumerism is accompanied and in fact supported by a class of products that can be labelled 'pirated', including CDs, VCDs, MP3s, assembled grey market computers, and so on, the production and circulation of which lie quite outside the realm of state regulation. The vibrant market for music and films brought about distinct localities within Delhi for different segments of this market and linked with a transnational network of suppliers and sellers in many parts of Asia. He thus points to the impossibility of a human–non-human distinction that tends to inform studies of urbanism. On the contrary, the 'urban' cannot be understood without the 'technological'.

Increasingly, more than control over technology of production and reproduction, it is the control over branding and fashion creation that the capital relies on for value creation. When production networks are diffused across the globe, the possibility of reproducing technologies becomes that much more accessible, and hence the possibility of piracy as well. This is particularly true in the case of non-material goods like software, music, and so on. He goes on to outline the various components of the enforcement regime prevailing in India, the act of the 'raid' in particular. The 'copy' therefore may turn out to be a key marker of resistance in contemporary urban forms.

It is with this range of concerns that the chapters in this volume[1] variously engage. There is no claim to the collection being comprehensive in its coverage, or seamless in its integration. That is perhaps inherent in the nature of researching a fast, and often unpredictably, moving subject area. It is to be hoped, nevertheless, that these findings and reflections will further debate, contention and counter-contention in the understanding of this critically important phenomenon, not just in India, but globally.

Note

1. This collection of essays is ideally read in conjunction with its independent companion volume: Ashwani Saith and M. Vijayabaskar (eds), *ICTs and Indian Economic Development: Economy, Work, Regulation*. New Delhi: Sage Publications 2005. Both volumes contain recently updated versions of papers presented originally in a major international conference on ICTs and Indian Development held in Bangalore. This volume also includes a few additional papers that address important facets of the theme.

2

Culture of Technology and ICTs

Sundar Sarukkai

Introduction

There are diverse views held on the use of the phrase 'culture of technology'. In general, we can identify three broad positions that explicate the relationship between technology and human societies. One is that technology is culture-neutral; that is, the domain of technology is independent of the culture into which it is placed. Second is an intermediate position that believes technology exhibits different characteristics across different cultures but yet is independent of the influence of culture and human subject. Third is a form of cultural determinism, which essentially states that technology is as much a cultural activity as other human activities.

There are two major problems which I would like to briefly address before further discussion. Both of them have to do with the confusion inherent in the terms 'culture' and 'technology'. Since the aim of this essay is to understand the specific Indian experience with technology and the character of 'Information Technologies (ITs)', I shall only briefly touch upon these larger questions.

In what follows, culture will stand for the domain of human activities of various communities. This includes their eating habits, dress codes, moral and ethical norms, religious and social rituals, private and family values, and so on. All these human activities are mediated by specific historical, traditional, philosophical and cultural world-views. The claim that some activity is culture-neutral implies that the specificity of the world-view of a community does not play an essential role in a culturally neutral activity. Similarly, to say that some activities are responsive to culture means that these activities express themselves in different cultures in different ways. It is commonly believed that only culture-neutral activities can be transplanted across cultures, as manifested in arguments for culture neutrality of technology based on the claim that technology seems to be easily transplantable across widely differing cultures. But transplantation across cultures does not *necessarily* imply culture neutrality. I think this view is misguided for the simple reason that many cultural activities get transferred, appropriated and modified across societies. Among these are cuisine, dress, music, literature and so on. In technology transfer, there is not only a movement of objects of technology, but also 'objects' of the cultural discourse. These objects of discourse can (and perhaps need to) modify the new culture in order that the objects of technology 'function' in the same way as they used to in the original culture. The fact that technology can be transferred into different communities is not proof of the culturally neutral aspect of it. It is because technology is not culturally neutral that transference of technology never works the same way across societies.

It is also quite difficult to clearly categorize as to what we mean by technology. Technology stands for a plethora of 'things': machines of various sizes and shapes, products of machines which themselves perform various functions and consumer products which carry a validation by being a product of something 'technological'. It is also generally believed that any material product associated with and probably arising from a scientific enterprise is technological, thus making the link between science and technology explicit. The belief that technology is culture-neutral means that all these which go to constitute technology is independent of the world-view of the community in which it is placed and does not have to change its 'form' in order to accommodate the cultural world of that community.

A useful way to understand the notion of 'culture of technology' is to first acknowledge that both culture and technology are defined in terms of the meanings which accrue to them. Technology is the domain of meanings we give to it and implied in the way we talk about it. Our response to technology and our relationship with it is largely the domain of meanings we ascribe to it and which is reflected in the ways by which we understand and use technology. Thus, to understand what technology means is to look at the narratives we form about technology. These narratives create the culture of technology. The view that technology is applied science is one such narrative.

Why is there a reaction against understanding technology as a cultural activity? What is lost if we claim culture specificity to technology? Viewing technology as a cultural activity will not imply that it is the same as activities like eating habits and dress codes. I believe that technology can continue to retain a technological narrative even as it allows for an intrinsic 'culturality'.

Assuming that the meaning of technology lies in the narratives we hold about it, what is the narrative which makes us hold the position that technology is culture-neutral? I would like to consider three clear arguments within this narrative.

First is the tendency to equate technology with its functionality. The view that technology is culturally neutral follows from the position which equates technology with functionality and nothing more. Consider this simple example of washing machines in India. Functionally, a washing machine washes clothes. But the set of meanings associated with an object called washing machine is more than its function. It carries within it a sphere of meaning such as cultural notions of washing, labour distribution of washing within a household, link with resource-specific constraints of water and electricity, private and public spaces for drying clothes, and so on. The work pattern of a household changes with the introduction of a washing machine. This work pattern in India is very different when compared to other societies.

There are also many constraints under which the machine is asked to perform. For example, where water supply is not assured through the day, automatic washing machines can only be used at times when there is water flow. In India, 'semi-automatic' machines are very popular because they allow the possibility of using these machines even when there is no water flow in the taps, by manually pouring water into the machine with the help of buckets. These are also popular because cultural habits of drying clothes are still very prevalent. Thus, if we collapse technology into mere functionality and then argue for its culture neutrality, we are missing the point altogether.

The second dominant element in the narrative of culture neutrality is technology's link with science. Through this link, technology establishes a claim to truth and following science assumes these 'truths' as being independent of culture. In taking a stand of cultural neutrality, technology is ***appropriating*** to itself the epistemological status of science.

There are problems with this link. It is not clear, even in the case of 'truths' in science, that such a culture-independent realm of reality and truth can be accessed as claimed. Even granting this claim to scientific truth, how does this follow in the case of technology? The activity of technology is not to generate truth claims about the world; rather it is most importantly an activity which allows us a handle to 'operate' in the world. Operating on and negotiating with the world are very different activities compared to describing and theorizing about it. This confusion between science and technology is further aggravated by mistakenly ***equating*** technology with 'applied science'. Science claims to discover objective, acultural truths and the title 'applied science' tends to carry over these claims also.[1]

The third narrative element is the creation of an image of technology, exhibited most prominently in the colonial discourse as an 'objective' parameter to compare different cultures. Such comparison was useful in the project of colonization and even today is paramount to the creation of hierarchies of societies based on technological development. Thus, developed countries are those which are far more advanced technologically in comparison to lesser developed countries. For technology to function in this role, it has to be culture-neutral. For, it is by being outside the ambit and influence of culture that it can be used to compare and rank different cultures. This kind of 'objectivization' of technology has played an important role in making technology seemingly neutral to culture.[2]

I believe that what has been discussed captures the important narrative elements of the view that technology is culture-neutral. But it is not enough to establish the intrinsic link between culture and technology. One has to adduce more reasons, along the lines described earlier, in order to thematize the culture of technology. I am particularly interested in doing this by looking at the Indian experience with technology. I believe that this interplay between technology and culture, as it occurs in the rich and complex cultures in India, has interesting insights into technology and is perhaps the best example of exhibiting how technology is inherently cultural. The example with the washing machine, as given earlier, reflects this complexity. The Indian experience with technology richly illustrates the need for a discourse of technology which goes beyond mere functionality and instrumentality.

Any discussion of technology in the Indian context must take account of three different aspects related to it. The first is to understand technology in ancient India in a proper perspective; second, to exemplify the connection between modern technology and the discourse of colonial India; and finally, to study the cultural elements intrinsic to technology. I will look at the specific characteristics of information technologies against this background.

Technology in ancient India is a very good example of how 'technologies' with no explicit link to modern science have and can flourish, thus questioning the facile link between science and technology. It is a paradigm for technologies outside the gamut of modern science even as it says something about the essential nature of technology itself. The discourse of technology in colonial India is an important factor in understanding why culture neutrality has been so vehemently claimed on behalf of technology. This has very little to do with technology per se, and has more to do with creating 'objective' parameters for comparing civilizations and thus validating hierarchies of societies. The example of offering prayers to machines ties the arguments together by explicating how the set of meanings, which goes to create an image of technology, gets enlarged through cultural interactions and appropriated into the cultural realm. This appropriation is possible only because technology is intrinsically cultural in character. Through this, I enter into the postmodern discourse and argue that the idea of technology should be ideally seen as belonging to the realm of the 'sublime'.

The Discourse on 'Indian' Technology

Increasingly, there is a strong need to understand technology over and beyond its genesis in the 'modern' West with its concomitant link with science. In looking for the genesis of technology in the industrial revolution or within the praxis of science, there is a negation of technologies which do not belong to or arise from this historical and scientific tradition. Ancient Indian technology is a very good example of this negation. Indian technology here refers to technological progress made in ancient India without being based on an epistemology arising from modern science. Examples of this are the synthesis of steel, zinc distillation, manufacturing of dyes, chemical technology and a whole host of other such processes and products (Alvares, 1991;

Dharampal, 2000). The presence of these technologies in ancient India, especially in areas like metallurgy, chemistry, textiles, and so on, highlights the problem in accepting a simple relation between technology and science. It also suggests that it is possible to have a domain of technology which can arise outside the boundaries of modern science and not validated by it.

The example of Indian technology is used here merely to ask the following question: What does this say about the character of technology such that it could flourish in a system which was totally divorced from modern science?

Without going into an elaborate discussion, let me state that this character of Indian technology illustrates something about modern technology also. There is a commonly held belief that there was no methodology or theory underlying Indian technology. There are two responses to this view. One is to do with the link between method/theory and technology: is there always a theory behind any technology? This view cannot be sustained when we look closely at modern technology also. The link is quite confusing most of the time. The other is to do with the truth value of theories upon which some technologies are based. That is, if technologies arise out of some theory, then does this prove the veracity of that theory? For the most part, I believe that the link between theory and technology is a tenuous one. It is not enough to say that Indian technology has no validity as technology because the theories of modern science were not the basis of it. There is the possibility that this ancient technology could have been purely 'empirical', but the complexity of technological processes do not seem to imply a simple rule-following picture. The neat picture of simple correspondence between technology and theory (as also exemplified in calling technology 'applied science') is largely to find an epistemological validation for what is a complex, interactive phenomenon.

The lesson from this is that technology should be seen from a much larger perspective and not be restricted to some specific links with modern science. Indian technology in this sense should be seen as a technological enterprise similar to modern technology. The character of technology that both embody is the way in which technology functions so as to allow us to get a handle on and operate in the world. By doing this, both arise from and respond to the cultural situation they are situated in.

Technology and Colonial India: Historical Lessons

Historical lessons are filled with difficulties and open to many misinterpretations. Fully aware of this, I isolate some elements of the history of the colonial rule in India, not to make any points about colonialism, but because these historical pointers allow one to understand how technology was used to establish the cultural superiority of the British. In particular, the view of technology as culture-neutral is needed to establish the colonial superiority and thus is part of colonial ideology.[3]

I use history here not as an overarching discourse on colonialism. Historical 'facts' are chosen selectively in order to consolidate my position that a powerful and influential image of India was created for the Europeans essentially by a few individuals with very clear agendas of their own. This image strongly influenced the subsequent image of technology in India and their legacy continues to be seen in the ways we respond and react to technology even today. I believe that it is important not to underestimate the power of these image-makers and planners.

As is well known, up to at least the late 17th century, the response of the European adventurers was tempered with awe of the material and technological progress of Indians (Adas, 1989; Alvares, 1991). A perceptible shift in this image of Indians happened, subtly at first but dramatically later, primarily influenced by the industrialization of Britain. What is of fundamental importance to this discussion is the way in which technology began to be used as a norm to compare and create hierarchy of civilizations. Technology became a validating mechanism used to 'prove' the superiority of Britain and Europe over Asia and Africa.

As far as India is concerned, two historical processes stand out, both of which happened contemporaneously. One was the steady disintegration of the Mughal empire, proliferation of wars, famine and disease in the sub-continent and two, the onset of the 'industrial revolution' in Britain. It is in this conjunction that one can begin to understand early industrialization in India and also its lessons for understanding technology today. These two 'events' are important markers, for it is in their dynamics that one begins to see the formation of an image of India as a land which had to be redeemed by the British. It is in the disintegration of India and the establishment of technology—as power and mastery over nature—that the first voices for the superiority of the British was articulated. Technology and science began to be used as 'objective' tools of comparison between societies.

Such an approach is clear at the end of the 18th and the early 19th centuries. Increasingly, technology and science were used as validating mechanisms to reiterate the superiority of the English. There had also been claims of the English being a superior civilization based on their arts, literature, religion, and so on, but none of these activities could establish a comparison of civilizations as 'objectively' and as powerfully as technology could. Even religious superiority was tempered by this acknowledgement of technology. David Livingstone looked upon railroads and telegraphs as God's gift to the chosen Christian few. He also regarded these inventions as 'breaking down barriers to Christian conversion' (Adas, 1989: 206). The impact of such beliefs should not be understated. Adas points to the enormous influence which a few writers, who wrote extensively on India, exerted on British administrators and others. One can, in a sense, trace back the incoherent understanding of technology in India today to the skewed view of those powerful voices.

Probably the most decisive influence in this matter was James Mill. His book *History of British India*, published in 1817, was one of the strongest critiques of Indian society. It was a book which selectively used information in order to propagate the utilitarian views on political economy, among other things. Science and technology are repeatedly emphasized in this book in order to reassert the backwardness of Indian civilization. Technology was essentially used as the best measure of a civilization. Mill's book had a tremendous impact—it was a must reading for those who were on their way to India. It strongly influenced the British administrators in India. Mill was probably the most influential image-maker of India in Europe, of his time (ibid.: 166).

Mill was also one of the earliest to link material achievement to the notion of civilization. His ideology, based on utilitarian tenets, stressed education, legal reform and free trade. Adas makes the important comment that Mill (along with a few more influential writers) were self-made men who had worked their way up from 'families of modest means' and whose views were shaped by their personal histories. He points out that their 'individual struggle to grow' suggested to them that change was the most important characteristic of any civilization.

The tension between change and stasis had been repeatedly used to present a picture of Indians as being 'inferior' to the British. Colonialism was strongly validated by the need to change a society which supposedly was mired in superstition and tradition. The emphasis on change, I believe, was most strongly established by the technological developments in Europe rather than any other single factor.

If we accept that writers like Mill were the strongest influence on British administrators, then we can situate the ideology inherent in Mill and others as the underlying ideology of Indian industrialization and consequent understanding of technology within India. This means that the reasons for industrialization were based on an imposed image created to uphold colonialism, based on religious and social convictions of a powerful few, as also on the psychological and historical forces which shaped these individuals. In the case of industrialization, this has been clearly enunciated by those British administrators who were instrumental in the industrialization process in India.

Two Governor-Generals of India in the 19th century exemplify this. In the 1830s, Lord William Bentinck looked upon technology as the medium through which Indians would be civilized. The emphasis on power and subjugation as part of colonization shifted to a self-imposed responsibility towards the ruled. This responsibility, founded perhaps in a theological doctrine of grace and 'saving', consisted in 'pulling-up' the unfortunate to the level of the more fortunate. And since the clearest indicator of this gap was exhibited through technology and science, industrialization became a means of saving the people of India and bringing them on par with the West. Bentinck, for example, viewed the steamboat as 'a great engine of moral improvement' (Adas, 1989). His successors too 'shared his faith in the power of technology to effect a 'complete moral revolution' (ibid.: 225).

The other prominent Governor-General, Dalhousie, in the 1850s, is better known for propagating such a view. He was the man behind the introduction of railways and telegraph in India. Other than various economic and political reasons which apparently motivated the introduction of railways, the ideological reason of saving India and the belief that technology was the prime motivator of global equality should also be noted. Dalhousie believed that only this imposed technology could 'shake India from its lethargy and alleviate the poverty and backwardness of its masses' (ibid.). Social revolution such as in breaking down of caste barriers was also seen as an added inducement for such processes.

This very sketchy historical account is given in order to understand the motives behind the early industrialization in India. Industrialization introduced into India by the British shows all these outlined characteristics of domination, means towards 'uplift', hopes of facilitating religious conversion, personal ideologies on change and economy, simplistic meanings of tradition, and so on. All these were to a large extent dependent on the construction of a belief that technology was indeed the indicator of civilization, of growth and progress. This in turn was based on the belief that science spoke about absolute 'truths' and was culture-independent and that mastery over nature (and other humans) as exemplified by technology constituted a clear proof of superior cultures. In this sense, science and technology were used in the late 19th and the early 20th centuries to give 'unbiased proofs' of the superiority of the Western world.

These conclusions are important in understanding the technological milieu in India. The validation of science and technology continues to rest on the belief that they are culturally neutral. I shall not even attempt to enter this contentious debate regarding science, but in the case of technology this is questionable. In this context, it may be useful to look at how technology in colonial India was intrinsically linked with cultural categories. Adas points out that a dominant reason for the invention of watches (and other precision instruments, some as a consequence) had to do with the monastic culture in Europe (Adas, 1989: 60). Control over time was established essentially through watches. Accuracy became an obsession not just with respect to time, but with measurements in general. This perceived control over time was mistaken for control of a primordial element of nature. More interestingly, the lack of such accurate control among other societies like Asians and Africans was argued to be decisive proof of the primitive status of these people. Records of the European descriptions of Asia and Africa continuously refer to the lack of a consciousness of clock-time, which supposedly proved the backwardness of these people.

It was similarly argued in the case of space also. Control and mastery over space, one of the reasons why railways were looked upon with such awe, were also seen to prove the superiority of the Europeans. It is interesting to note the comments of G.O. Trevelyan, a civil servant, who 'observed that in India all signs of civilization disappeared beyond one hundred yards on either side of the railway track' (ibid.: 228).

Even today, accuracy is seen as the hallmark of science and increasing accuracy as pointers of 'higher' civilizations. This world-view, of domination over space and time, arose out of a particular ethos of a society at a particular time. This also modified the culture of the societies into which it was imposed. As Adas remarks, once time was commodified and use-value ascribed to it, 'bells and clocks changed the European work patterns' (Adas, 1989: 250). They also helped in expanding the domain of conceptual categories like 'lazy' in relation to this particular world-view. Such categories were used as legitimization for colonization and subjugation of other communities.

A particular view of industrialization, which consisted in importing technology to uplift the natives to a predefined western conception of progress and development, continues to inform the 'catching-up' arguments for introduction of new technologies today.[4] Any imposition of an industrial model brings with it not just a set of machines but also the image of the maker. With the historical distance, this is clear in the case of early industrialization in India, but a similar mindset continues to prevail in policy with new technologies, including the example of ICTs. Moreover, science and technology continue to be viewed as the central parameters of 'development' at the expense of other human activities, which are often dismissed as subjective and therefore irrational and arbitrary.

The Culture of Technology: Offering Prayers to Machines

Technology expresses itself through culture. These expressions say something about technology as well as the particular culture. In one sense, it is this self-expression of technology through the medium of culture which constitutes what we may call as culture of technology. This also encapsulates the idea of a subject entering into a relationship with technology, an expression so forcibly used by Heidegger (1977).

Technology enters a culture with a restricted, given set of meanings. Placed in a cultural context, open to the gaze of its subjects, the domain of meanings keep getting enlarged. This enlargement of meanings manifest in the different ways we handle, refer to and talk about technology. This is what it means to enter into a relationship with it (Sarukkai, 1995).

Technology reflects, even as it sometimes distorts or amplifies, certain cultural traits. Culture of technology is not a 'mixture' of technology and culture; rather it is a co-habitation of the worlds (and cultures) of technology and society. If we continue to believe that they are separate, it is only because we privilege a particular narrative of technology as being culture-neutral and culture-independent.

In the Indian context, the formation of this culture of technology can be seen in the set of meanings that accrue to technology. I have given the example of a washing machine and how it derives a whole plethora of new meanings which essentially 'create' a washing machine which is quite different from the domain of meanings associated with a washing machine in another culture. The Indian experience with technology is replete with such examples. Rather than list these examples, I shall discuss one activity which captures the spirit, to a large extent, of these different examples. This activity also constructs a different narrative for technology; and it is within this narrative that the 'culturality' of technology gets clearly expressed. I refer to the phenomenon of offering 'prayers' to machines.

On a particular day sometime in September/October, a festival is celebrated all over India. This festival is *Ayudha Pooja* and is celebrated both in homes and in factories and offices. On this day, all material objects in the house including fans, vehicles, kitchen gadgets, and so on are cleaned and anointed with sandalwood, and prayer is offered in a manner similar to the worship of gods. In the factories, similar worship is offered to the machines, including computers, phones, fax and vehicles. One can see on that day buses, motorcycles and autos cleaned and being driven around with flowers decked all over them. Interestingly, all the new technological machines are most often the primary 'targets' for this worship. (Although this is primarily a festival for the Hindus, such religious interaction with technology occurs in other Indian communities also.)

Government offices, public and private factories and most other institutions are closed on this day, usually after the pooja is offered. Ironically, even in some premier scientific institutes, the staff follows this ritual towards laboratory equipments, computers, and the like.[5]

This kind of ritualization of technology does not happen on this day alone, although it is the one day of the year which is officially kept aside for this event. Rituals related to offering of prayers to machines, especially vehicles, have spread quite rapidly in the Indian society. I believe that this practice of offering worship to vehicles and machines is getting more prevalent. In fact, in many temples, there is a fair amount of professionalization associated with this ritual now, including charts of pooja rates which mention different amounts for scooters, autos, cars and buses! In some temples, the major activity now is this worship of vehicles.[6]

Technology as a 'Sublime Sentiment'

How do we understand this? The simplest answer is to say that Indians are by and large religious and they transplant this religiosity into the domain of machines. Some people, mostly the rationalists and scientists, have seen this as another manifestation of superstition inherent and flowing out of religion and tradition. Some others might claim that the penchant for finding gods everywhere is a part of the Hindu tradition as well illustrated in its myths and stories. Thus, they might claim that offering prayers to machines is but offering prayer to god, because god resides in machines and other material goods. I do not believe that this explanation is a sufficient one because it does not take account of, among other things, the treatment of the machines in a cavalier way (including various kinds of 'pollution' acts which occur within the vehicle).

There may be various other reasons to explain this. These explanations may place this act within religious and cultural domain even as they leave technology out of it. One explanation may be that one offers prayers to vehicles in the hope that the driver/passenger is safeguarded in his or her travels. But none of these explanations are complete, consistent and coherent. In the case of the explanation given earlier, one can ask, why is it not enough to offer prayers to the idol rather than go through the whole mimetic process in front of the vehicle? And why are they done to kitchen gadgets like grinders and so on? Why are not people 'worshipped' in similar manner especially if being safeguarded is a primary motivation? There are various other questions arising from this, but it will take me too far away from the points I wish to make.

It may be argued that this particular activity tells us more about the religious habits of the Indian (not just Hindu) society than about technology. I think we will be making a mistake if we try and remove technology from this act and place the act within the practice of religion. As I mentioned earlier, technology derives its meanings by expressing itself through culture (and that, especially in the Indian context, also implies religion). The practice of offering prayers is just this expression of technology through culture. Offering prayers to machines establishes a different kind of relationship with machines, *basically enlarging the domain of beliefs we hold about technology in order to enable us to deal with it in a manner suited to us*.

The modernist discourse of technology does not allow us avenues needed to reflect upon technology which take into account the diverse ways through which we relate to it. I have already argued that using technology as a parameter to compare civilizations, its link to science and 'objective' truth, rationality, and so on, are typical of the modernist discourse on technology. When the modernist discourse comes across the diverse ways by which technology expresses itself through culture, it removes them from the realm of technology because they do not fit into its view of technology. Therefore, we have to go beyond the modernist discourse to find meaningful ways to understand this complexity of technology. Postmodernism has done this effectively. It has forcefully argued for looking at technology not only through categories of modernity, but also through its intrinsic connection to culture, media, language, and so on.[7]

Activities like offering prayers to machines signify something important about technology. Lyotard (1984) tries to articulate the spirit of postmodernism by drawing upon the idea of the 'Kantian theme of the sublime'. His concern is not technology but modern art and literature. But his exercise is actually useful in leading me to believe that this idea of the sublime captures the complexity of technology as experienced in the Indian context. This is what Lyotard writes about the nature of the sublime (ibid.: 78):

> The sublime is a different sentiment. It takes place, on the contrary, when the imagination fails to present an object which might, if only in principle, come to match a concept. We have the Idea of the world (the totality of what is), but we do not have the capacity to show an example of it. We have the Idea of the simple (that which cannot be broken down, decomposed) but we cannot illustrate it with a sensible object which would be a 'case' of it. We can conceive the infinitely great, the infinitely powerful, but every presentation of an object destined to 'make visible' this absolute greatness or power appears to us painfully inadequate. Those are Ideas of which no presentation is possible... They can be said to be unpresentable.

The ambiguity in dealing with technology, especially as in the case of offering prayers to machines, arises because of the simple fact that we are unable to match our conceptualization about technology to the objects (representatives) of technology. It is here that one finds a strong link between art and technology. One of the important concerns of art has been to present the unpresentable and to present the fact that the unpresentable exists (Lyotard, 1984: 78). Technology is also a presentation like art, a product as well as a process. Every technological 'object' is an object akin to objects of art. In this presentation of technology, there is the background of the unpresentable which makes the presentation itself possible.

It is true that we have an idea of technology. When asked to point out to objects corresponding to our conceptualization of it, we are able to point at some representatives of technology. But there is no possibility of subsuming this conceptualization and the presentation of technology. There is always an inherent gap in this presentation. Although Lyotard says very little about his examples, one can infer the direction of his arguments. The Idea of the world and the Idea of the simple are not purely abstract terms; if they were so, then every 'Idea' would correspond to this sublime sentiment. The important point is to note that we can show corresponding 'objects' which stand in relation to these concepts, but the match is 'painfully inadequate'. This is what distinguishes these Ideas from abstract and artificial ones like Unicorns or ghosts. In our relationship with technology, we are always confronted with this conflict.

I believe that it is the response to such a conflict that essentially drives the creation of a culture of technology which is manifested in its complex interplay with other cultural activities like religion. Offering prayers to machines is a search to present the unpresentable in technology; a way to find forms of expression for the inadequacy of presentation of technology and the process by which one bridges the gap between 'product' and 'concept', thereby taking it into the realm of the sublime. It is not surprising that the uneasiness which is caused by this conflict is eased by taking recourse to a 'form' which is associated with solace—namely, rituals and simulated religiosity. This is yet another point which is close to Lyotard's remarks about art:

> Modern aesthetics is an aesthetic of the sublime, though a nostalgic one. It allows the unpresentable to be put forward only as the missing contents; but the form, because of its recognizable consistency, continues to offer to the reader or viewer matter for solace and pleasure (ibid.: 81).

Thus, offering prayers to machines must be seen as an activity which reinforces the view that technology should be recognized and understood as a sublime sentiment. Linking 'religion' to technology only emphasizes the recourse to 'solace' when faced with the nature of the unpresentable. The kind of 'solace' one looks for depends on the culture one is embedded in. I do not believe that this conflict about technology is present only in the Indian culture. What really differs is the response to this conflict in different cultures. As far as Western culture is concerned, technology has been ritualized in very specific cultural ways. The postmodern categories of simulation, seduction, performativity, the role of the mediascape in the context of technology, and so on, capture the complex character of technology in a postmodern society.

Kroker (1992), for example, situates the individual as no longer the possessive one who is 'consumer par excellence', but rather now a 'possessed individual as itself an object of consumption'. No longer is technology satisfied in its role of conquering nature, but is fast enveloped in a desire to conquer the social world too. Appropriate, therefore, as Kroker remarks, the will to technique, in this postmodern world, 'achieves its aestheticized point of excess'. This excess gathers technology, not as an alien outsider, but as a 'sign of the possession of the body and mind'. The individual becomes a 'technologically constituted self which is both a condition for the preservation of, and a constitutive justification for technological society'.

Also related to the sublime is the nature of reality and its presentation in art. In the case of technology, virtual reality (and its cultural links in the developed West) is a good paradigm to represent technology as a sublime sentiment. Virtual reality overturns 'rules' which define reality. The loss of referentiality corresponding to technology leads to a situation where reality disappears as the criterion of presence. It creates objects of reality as commercial products and presents the unpresentable not in terms of categories of experience but as pure experience itself. Technology as pure experience is the search for the sublime sentiment of this totalizing world of technology. This response is a cultural response specific to the particular culture of late capitalism which generated this technology.

The Indian response is characterized in terms of its awkward relationship with technology and includes the cultural habit of offering prayers to machines. This only points to the growing system of beliefs which accrue to technology. This is a project which may not fit within the modernist discourse, but that is for the better. There is no need for technology to stand as a culturally neutral adjudicator between competing cultures or get too intimately tied in with the preoccupations of scientific epistemology.

The Cultural Space of ITs

The just concluded discussion was an analysis of what constitutes culture of technology with specific reference to Indian society. With this background, we can now look specifically at Information and Communication Technologies (ICTs). In what follows, I shall focus only on ITs. We can begin by asking what is the specific cultural domain inhabited by information technologies. What is the product of the interplay between the cultural spaces of IT and the Indian society?

To understand these issues, we need to first clarify some special characteristics of this technology. Debate about technological impact on society, as tools for social change, and so on, occurs whenever a new technology enters our societies. Thus, even in the case of IT, much has been said about whether the claims on its behalf are tenable or not.[8] To understand the issues involved in the case of ITs, I will begin with a brief discussion on how we relate to some technologies.

Many technological products are produced and given in some completed form. These machines perform functions which they are designated for, with some leeway for an individual's choice. Consider two simple examples: fan and radio. The scope of a fan's use is very limited. There is little interface between a human subject and this product. As a technological product, the fan has its prescribed role and in the interaction with a culture, this prescribed role is not significantly negotiable.

So, also for the radio. An individual may decide which station she wants to listen to, but has little choice regarding the prescribed role of a radio. The interface between human use and technology, as far as its functionality is concerned, is once again minimal. (This does not mean that the radio does not get appropriated into a larger cultural space.)

When we compare these examples with a computer, we can see a fundamental difference. In the case of a computer, the interface between the user and the machine is quite open-ended. One can do a host of things with a computer which is not always restricted to a prescribed role. Not surprisingly, the idea of custom-made software actually reflects the custom-made nature of this technology itself. This specific nature is not like choosing the colours for a fan. It is something essential, something akin to customizing the design of a technological product. The degree of interface between the user and the machine is significantly much higher in comparison to almost any other traditional technology.

In fact, it is this strong aspect of interface that promotes the feeling that ITs are very people-friendly and can be a potent tool in governance and social development.

This strong sense of interface is also what allows a stronger cultural symbiosis in the use of this technology. In standard examples of technology, the user technology interface is limited at all levels. In most cases, the user can only partake, or at the most modify within generally rigid parameters, the already prescribed output of technology. On the other hand, ITs are much more malleable and allow the user to create the output she desires. This can be accomplished by a user who does not have to be a specialist in some restricted sense. The creative interaction between the 'ordinary' user and the machine, where the creative autonomy of the individual has not been lost in this interaction, marks an essential difference between conventional technologies and IT.

It is not surprising then that it is the computer (and the set of other technologies associated with it in various ways) which is closely aligned with the human in many ways. A common trait that is particularly potent as a symbol of humans is the idea of intelligence. While people may quibble about the nature of intelligence in computers, it is nevertheless true that most of the functions we do and see in some sense as an intelligent act can actually be duplicated by this technology. This exhibition or even simulation of intelligence by these machines generates a new level of ***kinship*** with them, a bond which psychologically manifests itself in various ways. For example, addiction to various facets of this technology, such as the Internet, is being recognized today as a serious psychological disorder. Because of this much more human character of these technologies, the cultural aspects become more significant.

This quasi-human aspect of these technologies is of seminal importance if we want to understand their use in governance and social development. One of the defining images of technology in the human imagination is the superiority of humans over machines.

This distinction is reinforced by privileging human characteristics such as intelligence and emotion, which are not possessed by machines. But with IT, which illustrates many common characteristics of intelligence, the relationship between humans and machines becomes more complex. A very simple example will suffices to illustrate this: playing games against the computer—be they chess or 'speed' games—generates a range of emotions in the human, which is similar to (and sometimes more intense than) the case when two humans play against each other.

Thus, the new, defining characteristic of IT lies is the creation of a domain of machines which are not human yet humanlike in many profound ways. This generates troubling issues when we consider the role of these technologies in social intervention.

How do we normally understand the role of technologies in society? A dominant position is that technologies are tools and agents of change, under control of humans. This instrumental view, which, in spite of many academic attacks against it, is still popular in the public imagination, sees technology as being the tool which does the work we want it to.

At this juncture, it is useful to consider an analogy from philosophy of language. One of the traditional and dominant views of language has been that language is a tool to represent and describe the reality around us. Language as a tool implies that language tries to faithfully, or at least as faithfully as it can, represent what we see and experience. Language does not, in this view, contribute to the creation of the reality which is being described. In other words, the use of language to describe—say a tree—does not in itself contribute to the reality and meaning of the tree. Thus, language is a sterile tool to be used by humans in ways we see fit.

Placed in opposition to this view is one which says that language constructs reality and that there is no 'objective' reality independent of the language we use to describe it. Suppose we describe something using some concepts. Now these concepts, which are human creations, exhibit a reality which is dictated by the structures of a particular language. The role of language in this view is that language is all that there is and the reality it describes is dictated by the strengths and limitations of a particular language. A common example is the use of over twenty words to distinguish the types of snow by Eskimos. Each one of these terms describes a more nuanced description of snow, which to most others in the world is well and truly captured by just one word—snow.

There is an intermediate position which would view language as being co-constitutive of reality. In this position, language does not 'create' reality, nor is our description of reality entirely independent of language. The ways in which we see the world—conceptual categories, descriptions of our sensory experiences, and so on—actually capture a part of objective reality, but the language which is used also contributes to that reality. This is a powerful idea that accepts the importance of language in a more fundamental sense than merely being a tool. It claims that language and reality are co-constitutive of each other.

Technology, in many traditional accounts, has been viewed in the same way that language had been earlier, namely, as a tool for humans to achieve certain ends. We have also had theories about technology which say that only technological development defines the creation of a society, a position similar to the constructivist view of reality. That is, technology is not merely a tool but actually fashions social reality.

I prefer to hold the co-constitutive view of technology and society. This would mean that technology is not the effective cause for the world to be as it is now, nor should it be seen merely as a tool of humans. Just as the world is shaped by technology, so is technology shaped by the world.

Consider the discourse of governance. We have various views on what good governance could be. We also believe today that ICTs are the new magic technology which enable efficient governance. In the ways in which we talk about this new technology, we can clearly discern an instrumentalist view of technology, which makes the claim that these technologies are tools for better governance. But, is it not the case that the very notion of governance undergoes a significant change depending on the technologies used in the task of better governance?

How exactly does IT, for example, enable better governance? We hear, in many projects of IT and governance, how one can send e-mails to politicians and bureaucrats, how information on prices of commodities is immediately made available to the villager, and so on. Why would we think all these contribute to better governance? *More importantly, is there a subtle shift in our very understanding of governance that is now being shaped and mediated by the specific character of these technologies?*

My answer to this question is a clear yes. These technologies are not merely subservient to some notion of governance that is already there. Rather, they begin to dictate what good governance should be and modify the discourses of governance and development to suit their strengths and weaknesses. What they dictate will very much depend upon the nature of these technologies. For example, one of the important characteristics of ITs is speed, speed by which information is transmitted, databases are checked, and so on. This element of speed now becomes an important element of good governance—not necessarily because that is the way we conceive of good governance, but because that is what these technologies are best suited to do.

Once we understand that the nature of these technologies actually begin to take over and then dictate the very notion of governance, we see how the co-constitutive nature of technology begins to work. Thus, some of the characteristics of these technologies should be first clearly understood before jumping in with a naïve view that these technologies are great for good governance. The essential connection of ICTs and speed, ability to process enormous volume of data very quickly, and so on, points to the usefulness of these technologies in having speedy governance and not necessarily for effective or humane governance, for example.

Also, these technologies do not arise from nor are responsive to concerns of governance, particularly governance in countries like India. They are more responsive to dictates of science and technology and the forces that shape them. That is, the origins of these technologies, the reasons for the development of these technologies and the motivations that drive new developments in these fields have very little to do with good governance, especially governance that is meant for the betterment of rural or urban poor. So here, we have a situation where a dominant technology, which is influencing the nature of governance, is not developed with governance in mind, nor is that changes to these technologies arise as a response to concerns of governance. Rather, it is driven by reasons of capitalist economy, research programmes relating to certain characteristics of computers, defence use, and so on. Governance is not on the list of priorities in the creation of these technologies. Thus, for governance to become excessively dependent on these technologies and moreover begin to be dictated by them is for governance to be hijacked by technologies which are totally insensitive to the concerns of good governance. In other words, while this technology is being touted as that which will enable accountability in societies, these technologies themselves are not accountable to the society they belong. We are again confronted with the classic cliché: who governs the governor?

The language of governance today emphasizes the importance of ideas such as ownership and stake of partners in any process. What is the ownership and stake of citizens in the development and creation of these new technologies? Absolutely zero. This is not just for the rural poor but also for the urban rich. The concerns of these technologies are articulated elsewhere with different motivations. We have seen this already in the case of computers. Machines bought two years back do not have parts available now. Chips that were in use till recently have been replaced by much faster chips. A faster chip also dictates what kind of things you want to do and can do. There is a simple question, which I am sure many of us have asked a million times. Why have all this speed in our PCs? What dictates the frenetic race for higher and higher speeds? What does this speed have to do with effective governance? And moreover, when we have no control over these technologies, we are stuck with running behind and incorporating any new development in these machines without any dialogue or feedback on the process.

These technologies are inherently not democratic. The large number of users have very little say in how these technologies develop or are used. Given the enormous resources needed to develop and maintain such technologies, it is clear that these technologies are driven by the idiosyncratic concerns of their developers and very few 'outsiders' have a say in how they should be developed with an eye towards social development. This trend is exhibited in the development of almost all technologies, including more traditional ones. But ICTs are being touted as the nirvana technology for governance, democracy and social development. It is ironic that technologies that are seen as harbingers of social development are developed with no concern for social development. It should be a matter of concern that these technologies which are supposed to empower democracy are inherently non-democratic in character. They are top-down; they are given rather than shared; they are imposed rather than negotiated.

This leads to a troubling question: what kind of democracy will we have when the tools of these democracies are inherently non-democratic? This is like political parties in India. If the parties are not democratic in their inner functioning, how seriously can we expect them to uphold the ideals of democracy in the country? If these technologies are inherently not responsive to the concerns of social development, democracy or good governance, how can we really expect them to contribute in a significant sense to social development of a poor country?

Taking these characteristics along with the points discussed earlier, we can see the complexity of the process of 'ITization' of our country and in particular, the enormous belief invested in them as technologies that are going to 'save' governance and democracy. I believe that the colonial experience with technology continues to influence the way we understand and appropriate technology. But our unique cultural responses to technology, as illustrated in the example of offering prayers to machines, generate differing responses to the way we see and use these technologies, and thus suggest that before we become too gung-ho about new technologies there is a need for serious introspection on the nature of technology.

Acknowledgement

I thank Dhanu Nayak for many useful discussions.

Notes

1. Sociology of science has developed many arguments for incorporating sociological and cultural factors in the development of science. As a consequence, there have also been arguments that one could have 'feminist science', 'Indian science', and so on. For material on this, see Barnes et al. (1996). Increasingly, there is also an emphasis on the nature of scientific discourse, especially its strategies of discourse formation, which points to the commonality between science, literature and arts. For references on this, see Halliday and Martin (1993), Locke (1992) and Sarukkai (2002).
2. The dominant belief about neutrality of technology pushes societies and cultures into adapting to technology rather than the other way around. This is the objectivization of technology, where a non-objective system is made to look objective by this kind of cultural adaptation. The Indian experience with technology illustrates this adaptation. An electrical machine from the US works at 110 volts whereas in India an adapter to convert 110V to 220V would be needed. Analogously, ***cultural adapters*** are routinely being used in our relationship with machines across cultures, many times without our conscious awareness.
3. I do not want to enter into a discussion of what constitutes ideology and the problem of making visible the 'invisible' ideology. Ideology as used here does not need more clarification than the following remarks by Zizek (1994): 'We are within

ideological space proper the moment this content—'true' or 'false' (if true, so much the better for the ideological effect)—is functional with regard to some relation of social domination ('power', 'exploitation') in an inherently non-transparent way: the very logic of legitimizing the relation of domination must remain concealed if it is to be effective. In other words, the starting point of the critique of ideology has to be full acknowledgement of the fact that *it is easily possible to lie in the guise of truth*'.

4. Rather than use the phrase 'catching-up', I prefer to see this process as 'pulling-up'. Catching-up implies that the lesser developed nations are of their own will developing to catch-up to the more developed nations. But in this race, the definitions of development arise from developed nations and the rules of the race are designed to keep the distance constant or at least make it near impossible to catch-up. So it is not a game of 'catch-up' that the developing countries are playing. Instead, the phenomenon is best described as 'pulling-up'. This acknowledges that unless they have their own conceptions of progress and development they are only playing games whose rules will not allow them to win.
5. Interestingly, there is another festival day, following this particular day, called *Saraswati Pooja*, when books are kept in front of the gods and prayers offered to them.
6. This is the usual way in which this process takes place: a priest comes with a plate, goes to each vehicle, smears pastes, places flower on them, breaks a coconut if offered by the worshipper and offers camphor flame in front of the vehicle. This offering of camphor flame mimics the offering of prayers to a god's idol. The rituals of worship to the vehicles follow, to a significant extent, the rituals of worship to the gods. There is also the phenomenon of pictures of gods adorning the insides of three and four-wheelers. In two wheelers, it is quite common to see stickers of gods or related symbols stuck on the vehicle. This habit is widespread and is followed by Hindus, Muslims, Christians and other religious denominations. I have seen this across the spread of the country.
7. For some useful essays on postmodernism, see Docherty (1993).
8. Castells (1998), Madon (2002), Saith (2002) and Sreekumar (2002) supply useful analysis of some of these issues.

References

Adas, M. 1989. *Machines as Measure of Man*. Ithaca: Cornell University Press.

Alvares, C. 1991. *Decolonizing History*. New York: The Apex Press.

Barnes, B., D. Bloor and J. Henry. 1996. *Scientific Knowledge: A Sociological Analysis*. London: Athlone Press.

Castells, M. 1998. *The End of Millennium. The Information Age: Economy, Society and Culture, Volume 3*. Oxford: Blackwell Publishing.

Dharampal. 2000. *Indian Science and Technology in the Eighteenth Century, Collected Writings, Volume I*. Goa: Other India Press.

Docherty, T. 1993. *Postmodernism: A Reader*. London: Harvester Wheatsheaf.

Halliday, M.A.K. and J.R. Martin. 1993. *Writing Science: Literacy and Discursive Power*. Pittsburgh: University of Pittsburgh Press.

Heidegger, M. 1977. 'The Question Concerning Technology', in D.F. Krell (ed.), *Basic Writings*. San Francisco: HarperCollins.
Kroker, A. 1992. *The Possessed Individual*. New York: St Martin's Press.
Locke, D. 1992. *Science as Writing*. New Haven: Yale University Press.
Lyotard, J. 1984. *The Postmodern Condition: A Report on Knowledge*. Minneapolis: University of Minnesota Press.
Madon, S. 2002. 'The Developmental Impact of E-Governance', paper presented at the IDPAD funded International Seminar on ICTs and Indian Development, held in December 2002, Bangalore, and organized by Institute of Social Studies, The Hague and Institute for Human Development, New Delhi.
Saith, A. 2002. 'ICTs and Poverty Alleviation: Hope or Hype?', paper presented at the IDPAD funded International Seminar on ICTs and Indian Development, held in December 2002, Bangalore, organized by Institute of Social Studies, The Hague and Institute for Human Development, New Delhi.
Sarukkai, S. 1995. 'Humanism and Technology', in B.V. Subbarayappa (ed.), *Facets of Humanism*. New Delhi: Affiliated East–West Press.
———. 2002. *Translating the World: Science and Language*. Lanham: University Press of America.
Sreekumar, T.T. 2002. 'Civil Society and ICT-Based Models of Rural Change: History, Rhetoric and Practice', paper presented at the International Seminar on ICTs and Indian Development, Bangalore.
Zizek, S. 1994. 'The Spectre of Ideology', in S. Zizek (ed.), *Mapping Ideology*. London: Verso.

3

Sustainable Internet Connectivity in Rural India

Ashok Jhunjhunwala
Anuradha Ramachandran

Indian Telecom Industry—Current Scenario

India has seen an unprecedented growth in the telecom sector over the last one-and-a-half decades, with the number of telephones increasing from a mere 5.07 million in 1991 to over 100 million as of April 2005 (see Figure 3.1). Over 50 per cent of the telephones in the country today use wireless, either in the form of fixed wireless or mobile wireless.

Though there are several reasons for the quick growth in the industry over the past five years, a significant contributing factor has been the falling tariffs, which made telephony in India affordable to the larger masses. The fall in tariffs has primarily been driven by a reduction in the capital expenditure required to provide telephony, from Rs 40,000 per line a decade ago down to about Rs 10,000 per line today.

Yet, for a country whose population is close to 1.1 billion, the teledensity is just over 9 per cent. Moreover, the distribution of telephones

FIGURE 3.1 **Growth in the Number of Telephones in India (1948–April 2005)**

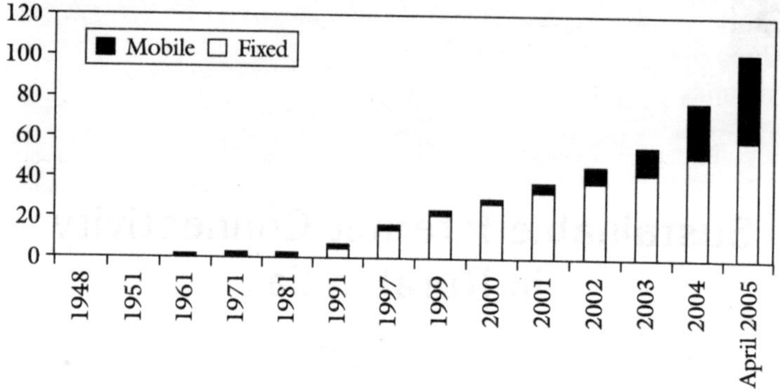

Source: *The Hindu*, 14 April 2005.

within India is highly inequitable, with rural teledensity being below 1.5 per hundred as against an urban teledensity of 20.7 as of 2004. This is especially alarming, given the fact that over 70 per cent of India's population continues to live in rural areas.

Will the Growth Sustain?

The telecom boom, which swept through the Indian telecom industry, is predominantly urban in nature and has not addressed the needs or the opportunities in rural India. Also, the more developed and industrialized states have received the lion's share of the attention of the operators, while the others have been relegated to lower priority. As a consequence, there is a huge discrepancy in the subscriber density of rural areas as compared to urban India, particularly in the major metros.

According to the Indian Telecommunication Statistics, 2004 released by the Department of Telecommunications (DoT) of India, almost half of India, with as many as 13 out of the 27 circles, have missed the telecom revolution, having a much lower teledensity than the pan-Indian average of 7.02 as on 31 March 2004. The telephone penetration per 100 inhabitants in these 13 circles ranged between 1.5 and 5. The teledensity growth in the less developed, predominantly agrarian states has been dismal.

Of the 27 circles in India, 11 circles, including Assam, Bihar, Jharkhand, Orissa, the North East (comprising six states and union territories), West Bengal, Rajasthan, Madhya Pradesh, Jammu & Kashmir and Chhattisgarh, account for a meagre 6 per cent of the total number of telephones (see Table 3.1).

TABLE 3.1 **States with Lower Teledensity (Number of Telephones per Thousand Population)**

States	*Teledensity as of January 2004*
Chhattisgarh	1.58
Bihar	1.59
Assam	1.97
Jharkhand	1.97
West Bengal	2.53
Northeast-II	2.57
Uttar Pradesh east	2.74
Orissa	2.81
Jammu & Kashmir	2.9
Northeast-I	3.16
Madhya Pradesh	3.76

Source: Indian Telecommunication Statistics, 2004 (Policy Framework, Status and Trends), Ministry of Communications, The Department of Telecommunications, Economics Research Unit, Government of India.

Rural India is Still Far Behind

India, the largest democracy and the second most populous country in the world, is predominantly rural in nature. Of the 1.027 billion-strong Indian populace, 741 million live in 638,365 villages scattered across the diverse terrain of India.

The rural GDP now stands at US$ 140 billion, contributing to 24 per cent of the national GDP. With 138.2 of the 191.9 million households living in rural areas, the GDP contribution is highly disproportionate. With an average of five people per household, about 85 per cent of households earn less than US$ 900 per annum, and with the per capita income at a meagre US$ 50 per day, the heart of India remains poor to a larger extent.

Seventy-four per cent of rural households in the low-income bracket earn less than US$ 700 per annum, 23.1 per cent in the middle-income bracket earn between US$ 1,800 and US$ 4,800 per year and

FIGURE 3.2(a) **Income Distribution of Rural Households in India**

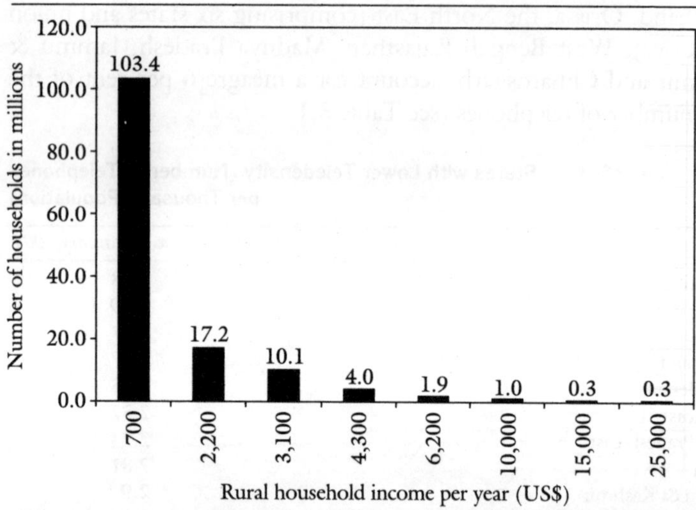

Source: India Market Demographics Report, 2002, Market Information Survey of Households, National Council of Applied Economic Research (NCAER), India.

a paltry 2.9 per cent in the higher-income bracket bring home more than US$4,800 in a year.

Obviously, the Average Revenue Per User (ARPU) supported is very low and rural teledensity is meagre, languishing at less than 1.5 telephones per 100 inhabitants. Further, the mobile coverage in rural India is limited to the periphery of towns and national highways. Interestingly, wherever coverage does exist, the number of mobiles have increased dramatically, often defying logic. But ARPU continues to be low and operators have little incentive to increase the coverage.

Figure 3.2(b) illustrates how the divide between urban and rural teledensity is widening, thereby creating an increased divide between the haves and the have-nots.

Overview

Rural India—A Brief Snapshot

India is one of the emerging powerhouses of the 21st century. Along with China, India garners the most attention for its sheer size, the

FIGURE 3.2(b) **Urban versus Rural Teledensity: The Widening Digital Divide**

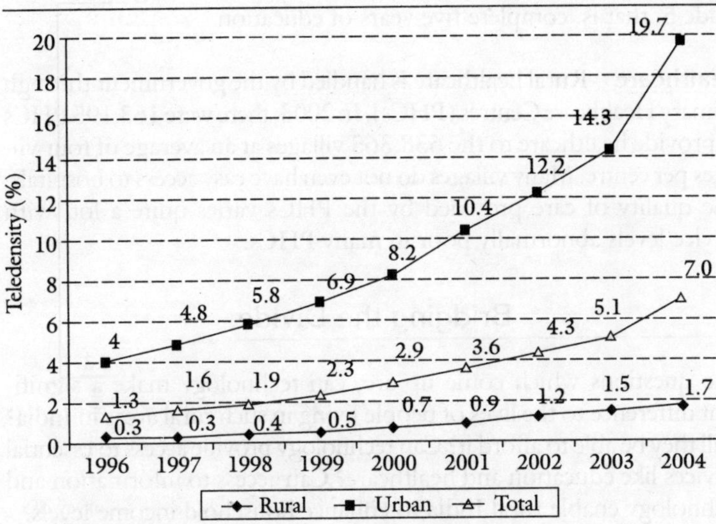

Source: Indian Telecommunication Statistics 2004 (Policy Framework, Status and Trends), Ministry of Communications and IT, Department of Telecommunications, Economics Research Unit, Government of India.

diversity of its huge population, the challenges that its complex economy poses and at the same time, for the potential of rich dividends that investors from the developed world look forward to. India is home to 1.027 billion people who speak 18 official languages apart from 1,600 minor languages and dialects.

Education and Literacy of Rural India Rural India is largely illiterate, with the literacy rate at 49.4 per cent compared to the urban populace, where the literacy rate is 70 per cent. The education system has an inherent inefficiency in the lopsided ratio of a huge number of students to a teacher at 43:1. In rural areas, more than 75 per cent of schools have one teacher taking simultaneous classes for students placed at different educational levels. Many villages do not have schools or lack the proper infrastructure for running one. The villages that do have schools often cater to the needs of students of several villages nearby. Yet, access to these schools is difficult, with no proper roads or transportation facilities. An alternative available to parents is to send their children to schools in bigger towns nearby, leading to overcrowding of these town-schools. Invariably, the quality of education imparted suffers as a result. The dropout ratio is very high, with only 40 per cent of primary school entrants managing to reach grade 5, that is, complete five years of education.

Healthcare Rural healthcare is handled by the government through Primary Healthcare Centres (PHCs). In 2003, there were 163,195 PHCs to provide healthcare to the 638,365 villages at an average of four villages per centre. Many villages do not even have easy access to hospitals. The quality of care provided by the PHCs varies quite a lot, with service levels abnormally poor in many PHCs.

Bridging the Divide

The questions which come up are, can technology make a significant difference to the lives of people living in such rural areas in India? Will they be able to afford it? Can technology provide access to essential services like education and healthcare? Can access to information and technology enable rural India to enhance household income levels?

n-Logue: Bridging the Divide

The Genesis of a Rural Service Provider

The Telecommunications and Computer Networks (TeNeT) group is an 18-member academic team from the fields of Electrical Engineering and Computer Science, Indian Institute of Technology, Madras (Chennai). The mission of the group is to create economical technologies for emerging markets. TeNeT has over the past few years incubated around 14 companies which have developed different technologies suitable for the needs of the developing countries.

TeNeT believed that access to communication and Internet technologies can make a significant difference to the lives of the rural population in India. Technology could provide rural India access to education, health, governance and such essential services. Information Technology (IT) could also enable rural livelihood through access to information, finance, markets and insurance. But the issue was, how to implement this?

India has over 700 million people living in over 600,000 villages. Therefore, any programme that is implemented in 100, 1,000 or 10,000 villages, is miniscule and makes very little difference to the entire rural economy of India. The challenge was to build a model with *the potential to scale to half-a-million plus villages*. Anything that is not scalable is simply an experiment, probably worth watching, but not of significant relevance.

Building a Scalable Organization

There are three factors that are mandatory to build a scalable and successful business in rural areas. These are:

- **Technology**, which is cost-effective, affordable, robust, scalable and capable of delivering the relevant applications.
- **A business model**, which enables each participant to earn from the effort.
- **An organization**, which is exclusively focussed on the rural market, which thinks and acts rural. This is very essential, as an organization which attempts to build infrastructure in both urban as well as rural markets, sooner or later tends to forget the rural areas and starts to focus only on the urban market.

Connectivity Over the last 15 years, the state-owned incumbent telecom operator, Bharat Sanchar Nigam Limited (BSNL), has taken fibre to almost every taluka (county town) in India. These fibre lines are infinite bandwidth pipes. Further, if a wireless coverage of 15 to 20 km is established from these county towns, one can cover almost 85 per cent of the villages in India. In India, there are typically 300–500 villages within a 30 km radius from such towns.

Further, wireless technologies have constantly evolved over the past decade—costs are coming down and bit rates are going up. The corDECT technology, jointly developed by TeNeT (Telecommunications and Computer Networks Group of IIT, Madras) and Midas Communication Technologies Pvt Ltd, has proved to be a major breakthrough for the cause. An exchange and a base station are installed at the taluka or county town where fibre is located. This exchange functions at a temperature of upto 55 degree centigrade, and does not require air-conditioning. The total power requirement for connecting around 1,000 subscribers is only 1 kW. Also, a small generator can cater to the vagaries of power supply in small towns. corDECT is capable of offering simultaneous voice and Internet access and can deliver a 35/70 kps connectivity to villages which are within a radius of 25 km from the fibre-connected taluka.

The start-up cost for corDECT installation is low. Last year, US$ 200 million worth of this technology was sold and deployed in India and other developing nations including Brazil, Argentina, Nigeria, Tunisia, South Africa and Singapore. An upgrade of this technology, providing 80/160 kbps dedicated connection to each subscriber, will be available soon.

Such wireless technologies work wonders for 85 per cent of the villages of India, which lie predominantly in the flat areas. In case of rough terrain, primarily mountainous regions and forests, fibre fails to go deep and the problem gets complicated. The TeNeT group is working on a solution that combines satellite and terrestrial wireless, to provide low-cost connectivity even in such villages. In other words, the technology to connect most of the villages in India is already available.

Business Model However, technology is only the first leg. The second leg is a business model, which allows this kind of set-up to scale up to 600,000 villages. The issue that needs to be addressed is, in villages where affordability is low, can a business scale? Most of the existing telecom operators have declared rural connectivity as an unviable business.

The clue to a successful business comes from what was done in the mid-1980s in India. During that time, obtaining a telephone connection even in urban areas was very difficult. People had to endure a 7–8-year waiting period to acquire a telephone connection.

At that time, an innovative idea was thought of—find a shop in every street in an urban area and convert it into an operator-assisted telephone booth or a Public Call Office (PCO). The PCOs were set up at street corners so that no one needed to walk more than 50 metres to reach the nearest one. The PCO was manned by an entrepreneur, who kept it open for 16 hours a day, 365 days a year. Such PCOs spread rapidly. The approach made connectivity viable and pervasive while it also created a stream of entrepreneurs. The success of the PCO revolution can be gauged by the fact that until recently, 25 per cent of India's telecom income came from these PCOs. Today, 300 million people, who do not have a telephone in their house, use these PCOs. The lessons learnt from the PCO revolution were several—aggregation of demand, presence of entrepreneur-driven business, proximity to a facility for greater access.

These three factors provide the basis of a viable business model for rural Information and Communication Technology (ICT). Demand aggregation would address issues of affordability in a village, while the entrepreneur and the easy access would ensure a steady stream of users. Aids and grants cannot create the scale to reach 600,000 villages, whereas successful enterprises can.

Organization This leads us to the third leg of a successful rural ICT service—an organization called n-Logue, a rural service provider (RSP) whose entire focus is rural India. It focuses on providing commercial telephone and Internet connection to every village. The company is prohibited from operating in urban areas by virtue of its charter.

n-Logue uses an approach similar to the 'operator-assisted PCO' model in rural areas to provide telecom and Internet services. Demand aggregation is through an Internet kiosk, which is owned and operated by a rural entrepreneur. n-Logue chooses an enterprising local person from the village itself to set up and run the kiosk. The kiosk operator (KO) is trained to use and to help other people use various applications which are made available by n-Logue. The KOs are typically youngsters, mostly women, who have passed their 10th standard. They are not required to have any prior knowledge of computers. But as entrepreneurs, they need to have good local language communication and networking skills, to be able to serve the village better.

Each kiosk is equipped with a multimedia computer with a web camera and speakers, an Internet connection, a printer, power back-up and software required, including local language software and video-conferencing software. The kiosk is the hub of rural connectivity providing communication services (e-mail, chat, browsing), as well as other much-needed applications like education and training, health-care, agriculture consultancy and e-governance.

The aforementioned equipment, software, training the kiosk operator and six-month Internet charges, all come for a package cost of about US$ 1,000. The kiosk operator is required to bring in around US$ 200 and the rest is funded through bank loans. The kiosk needs to earn about US$ 70 to US$ 80 per month from telecom, Internet and desk-top services to break even, which in a village of about 1,000 people roughly translates to just 7–8 cents a month. At this price level, the service becomes affordable to the rural populace and a sustainable business is built. The next issue that needs to be addressed is, who would service the kiosk operator? What happens when the computer gets a virus? Where are the technicians who are equipped to deal with such problems? Such bottlenecks would need to be addressed immediately. Also, kiosk operators would need regular support in terms of maintenance, connectivity and with applications/services provided. n-Logue, therefore, created a middle tier of Local Service Providers (LSPs) to service the needs of the kiosk operators in every county or taluka. The LSP is located in a town not more than 30 km from any village. This proximity enables the LSP to reach a kiosk in about 60 to 90 minutes in case of an emergency. The LSP covers a 30 km radius, 3,000 sq km/300–500 villages and is stationed in the place where the tower/access centre is located. The LSP makes a partial investment, but has total charge over local operations and is made a 50 per cent partner in the business by n-Logue. The LSP is also assigned the task of identifying an appropriate entrepreneur in the village and is required to help them, train them, and commission and maintain the equipment. The business model is such that the LSP earns money only if the kiosks are up and running.

n-Logue provides the connectivity backbone in the operations, co-ordinates with multiple technology providers for relevant applications and content, trains LSPs and kiosk operators and supports them through all operational/technical and other issues, sources critical supplies for kiosks including the available hardware and software, co-ordinates with regulators and policy maker to ensure service availability and markets the services to the community with the help of the LSPs and kiosk operators.

n-Logue, thus, adopts a three-tier model for rural connectivity. Tier I is the kiosk operator in the village who does the primary customer interface. Tier II is the LSP who services the kiosk operators and is

FIGURE 3.3 **The Structure of the n-Logue Model**

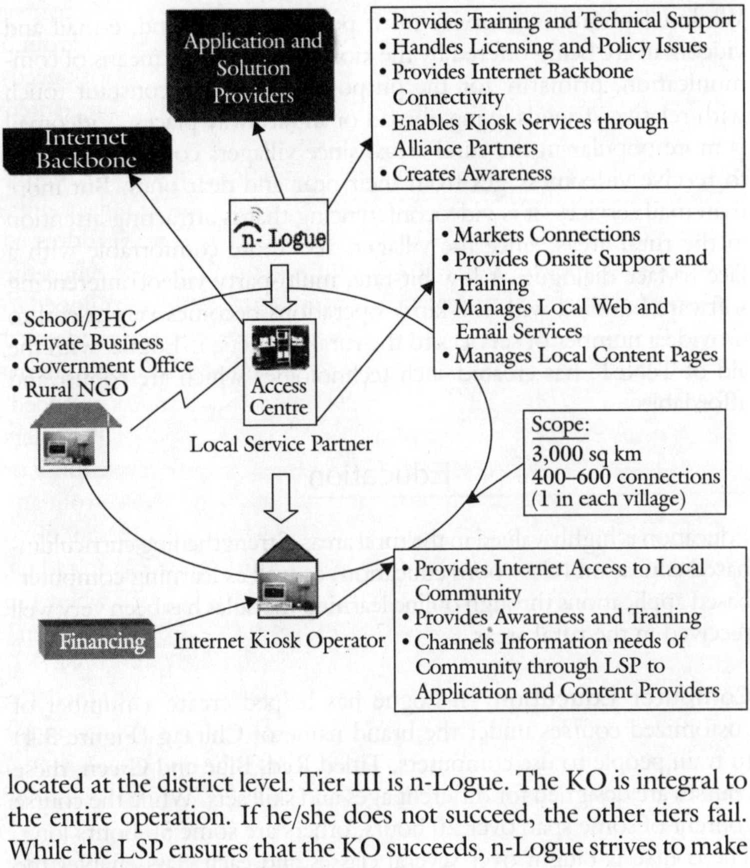

located at the district level. Tier III is n-Logue. The KO is integral to the entire operation. If he/she does not succeed, the other tiers fail. While the LSP ensures that the KO succeeds, n-Logue strives to make the LSP functional.

Description of the Services Provided

The important factors in creating a scalable, sustainable business model for a RSP was clearly established. The next imperative was to create a need among the rural people to use the connectivity. It was therefore necessary to provide a host of services through the kiosks to ensure regular usage by the villagers.

Communication

While telephony is a service most people comprehend, e-mail and videomail are being offered by the kiosks as additional means of communication, primarily for the purpose of being in constant touch with relatives/friends living abroad or in far away places. Videomail is more popular in the rural areas, since villagers consider it a treat to receive videomessages from their near and dear ones. But more than mail services, it is videoconferencing that is attracting attention in the rural areas, since the villagers feel more comfortable with a face-to-face dialogue. A low bit-rate, multi-party videoconferencing software bundled with the kiosk operations becomes very useful to provide a number of services to the rural markets; n-Logue, with the aid of TeNeT, has created such technologies which are robust and affordable.

Education

Education is highly valued in the rural areas. Strengthening curriculum-based education (classroom education) as well as learning computer-based applications through online learning modules has been very well received in the rural areas.

Computer Education n-Logue has helped create a number of customized courses under the brand name of Chiraag (Figure 3.4) to train people to use computers. Titled Red, Blue and Green, these courses are designed for different ages and skill sets. While the course content of some span over 20 hours, others are some 50 hours long. The course is taught over several classes and each class enables the rural populace to get familiar with computer basics. These packages have proved to be extremely popular in rural areas, as they are seen as means to employment.

Remote Tutorials The online tutorial is specifically focussed on assisting children to prepare for their school examinations. Unfortunately, there are not too many good teachers at the village level and the remote tutorial attempts to bridge this gap through three modes—learn, practice and test (Figure 3.5). The remote tutorial is based on a question and answer format. The lessons are supplemented with

FIGURE 3.4 **Computer Education**

FIGURE 3.5 **Remote Tutorials**

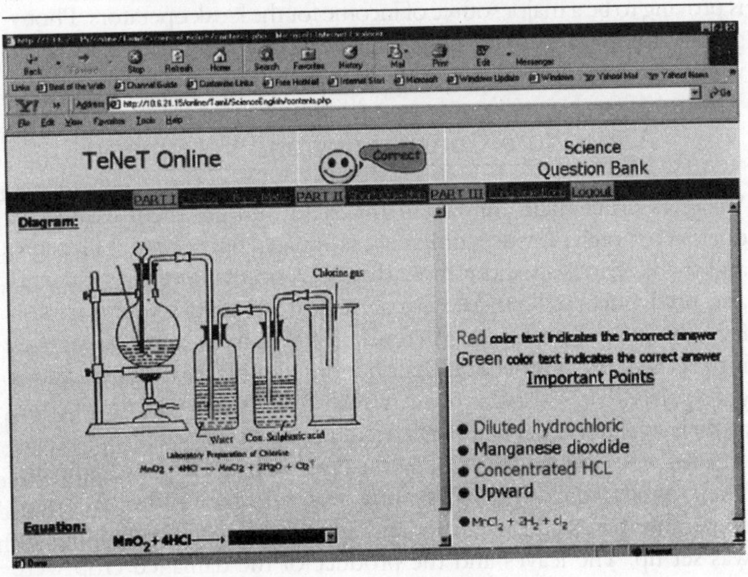

a voice-over that makes learning easy. The online tutorial has focussed on several subjects, including Mathematics, Social Science, English and Science.

Computer-Assisted Spoken English Course Villagers consider Spoken English as an important credential in acquiring a job. Having understood this significance, Spoken English module has been specially created to help children and adults improve their English speaking ability.

Typing Several children come to the kiosk to learn basic typing. They pay upto a US dollar and attend 8–10 classes a month for this purpose. Similarly, adults avail themselves of the kiosk's facilities to acquire typing skills, which they again find useful in obtaining employment. Apart from this, villagers also learn some simple but useful skills like making a resume, how to apply for a job, application for a bank loan, and so on.

Photography

Most kiosks double as photography studios, where photographs are shot using a digital camera and prints are made available. Photography is proving to be a major source of income for the kiosk operators. Photographs, for a variety of purposes, including for government forms and veterinary applications, are being shot at the kiosk.

Agriculture Consultancy and Veterinary

Villagers bring their animals to the kiosk and get them treated by doctors remotely. A videoconference is initiated between the villager(s) and the veterinarian and remote diagnosis occurs. Similarly, farmers also hold videoconferences with agricultural experts to get attention for crop-related problems. At times, farmers from different villages interact with each other and their queries, doubts and apprehensions get resolved. These services are extremely important to the rural people as their livelihood depends on agriculture and cattle/poultry.

A recent example is the case of the farmers of T Ulagapichampatti. Their okra (ladies finger) produce was turning yellow. A videoconference between the farmer and agriculture specialists in the city was set up. The leaves and the produce of the damaged crop were

FIGURE 3.6(a) **Picture of a Diseased Plant**

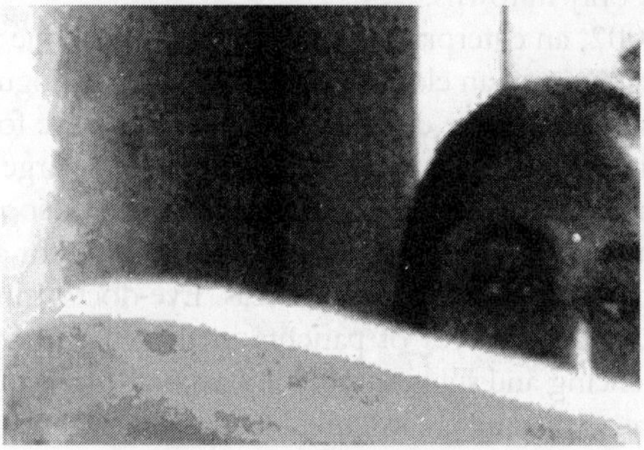

FIGURE 3.6(b) **After Treatment**

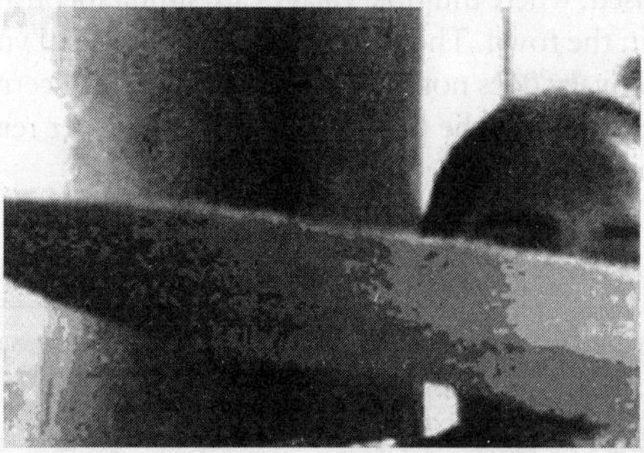

shown through the web-cam, the kind and amount of fertilizers added were also discussed. The experts diagnosed it to be yellow mosaic. Appropriate treatment was administered and the farmers were able to prevent a loss of Rs 140,000.

n-Logue is also planning to enhance the services provided to the farmers through videoconferencing by accessing a greater number of resources and through additional services like weather reports, crop price and other market-related information.

Healthcare

Even though Public Health Centres (PHCs) exist in every three to four villages, many of these centres function poorly. Healthcare is a major concern in rural areas. Qualified doctors in reasonable numbers are available only in towns.

In May 2002, an enterprising kiosk operator took the initiative to send some pictures of an elderly woman's eyes to n-Logue. The lady was suffering from severe eye pain. The pictures were forwarded to doctors in the Aravind Eye Hospital (Figure 3.7), a large facility in a neighbouring city. The doctors used Internet-based videoconferencing to examine the patient. This provided the beginning for an Internet-based eye-care programme for rural areas. Eye-doctors in town now regularly examine the eyes of patients in the villages by means of videoconferencing and suggest remedies as might be required.

This process was quickly extended to regular healthcare. Doctors in towns use videoconferencing to provide medical advice to remote patients located in villages. A multi-party videoconferencing product is commonly used, where multiple villages are simultaneously connected to a doctor in the town. The doctor connects to several villages at the same time. Privacy does not seem to be an issue of concern for the villagers. When asked on the privacy aspect, one villager remarked that such a system makes the doctors more accountable.

FIGURE 3.7 **Remote Eye Care Consultation with Aravind Eye Hospital**

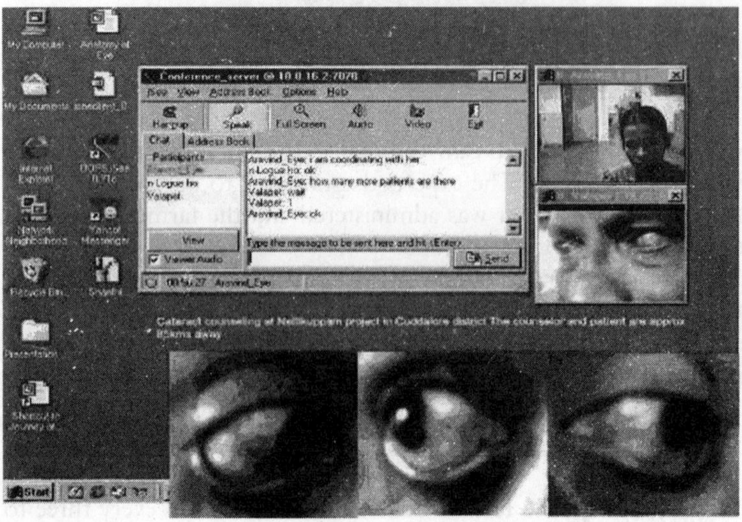

E-Governance

Similarly, the Internet is being used as an interface between the rural populace and the government. There is a well-known story concerning a handicapped person who sent a complaint to the Chief Minister's cell complaining of unfair termination of service. Having received the complaint, swift action was initiated wherein the person in question was handed back his job.

Other e-government services include online applications for services like widow's pensions, birth/death/caste certificates, ownership of land certificates, payment of bills like electricity bills and taxes.

While this service has worked commendably well in certain villages, it has failed to make a dent in several others. The performance of such a service depends on the district collector, who needs to drive the service.

For instance, in Thiruvallur, a small town on the outskirts of Chennai, the district collector started what is popularly known as a 'weekly web darbar' with the villagers. Through the darbar, the collector connects on videoconferencing to 50 villages simultaneously. Through these sessions, he answers queries and addresses grievances.

Technology Innovations which Enable Rural Services

One question that has been asked several times is—how has all this been done? What has made n-Logue capable of delivering such services? Today, n-Logue is present in around 40 districts and 2,000 villages. The TeNeT Group (at IIT, Madras) stands behind n-Logue and has incubated over 14 companies with over 1,400 engineers whose focus has been developing technology specifically for rural areas of developing countries. Such an approach has been purposely initiated because technology development for urban areas can totally absorb an organization, leaving the needs of rural areas unattended. Some of the major technology efforts are described next.

corDECT

corDECT, a Wireless-in-Local-Loop system, is one of the most cost-effective solutions that was developed by a research group at the IIT,

Madras in India, in collaboration with a Chennai-based company, Midas Communication Technologies Pvt. Ltd.

With a radio exchange and base station located in a taluka town, corDECT, using a fixed wireless system enables simultaneous telephone and Internet connectivity (at 35/70 kbps sustained rate) in villages within a 25 km radius (using Relay Base Stations). The system is extremely cost-effective with the total deployed cost including that of the subscriber unit being under US$ 250 per line. Interestingly, the system is extremely versatile in its ability to scale up and also to be deployed under even harsh conditions. A 1,000-line radio exchange and base station works even at 550C, requiring no air-conditioning and has a total power requirement of only about 1 kW which could be easily provided by a small diesel/petrol generator. The start-up cost for the system is only US$ 20,000, at which the first subscriber can be connected. Currently, infrastructure for over 2.5 million corDECT lines is being deployed by all major telecommunication service operators in India and as well as in more than 10 developing countries.

Soon, an upgraded version of corDECT which will allow connectivity at 100/200 kbps will be available and this would further enhance the kind of services which can be provided in the villages.

Language Applications

While connectivity was easily addressed, how were rural people to communicate if they had no knowledge of English, the *lingua franca* of the Internet? It was also found that existing language software was quite expensive for the village kiosk. TeNeT found an answer in CKShakti, an Office-suite package in the local language from a venture called Chennai Kavigal Private Ltd, which it helped create. CKShakti offers word processing, spreadsheets, database, presentation and e-mail client in eight Indian languages as well as in English and is priced at a fraction of the cost of the MS Office package. It also has a dual language option, which facilitates switching between the local language and English. This package played an important role in setting up the kiosks.

Videoconferencing Software

Initial studies at the Internet kiosks indicated that e-mail was considered a difficult communication medium in the rural areas and most users needed the help of the kiosk operator to type and also read out mails to them. To address this problem, TeNeT with a company, Objective Oriented Programming Systems Pvt Ltd (OOPS), created a videoconferencing software called iSee. This multi-party videoconferencing software can be enabled at very low bit-rates. This software has facilitated the use of health, education and consulting services between the rural people and experts located in towns.

Remote Diagnostics

As discussed earlier, videoconferencing is being used to connect villagers to doctors/hospitals. The doctors, who have used the Internet to deliver healthcare services to the rural poor, have often wondered whether they might be able to do a much better job if they could remotely monitor the patient's temperature, blood pressure and heart beats.

Recently, Neuro-synaptic Ltd, with help from the TeNeT group, has launched an interesting low-cost remote diagnostic kit. The kit is placed in the kiosk and enables a doctor to measure the temperature, blood pressure, pulse count and ECG of the patient remotely, and also to listen to heart-beats using a remote stethoscope. This combined with videoconferencing enables a doctor to significantly enhance remote medical treatment. The kit is being tested in several kiosks by n-Logue.

The Rural ATM—A New Technology

TeNeT believes that going beyond normal Internet applications and leveraging ICT to enhance livelihoods would be the primary means of sustaining rural kiosks. One of the innovative applications it developed to support livelihood is a rural Automated Teller Machine (ATM).

With the support of the TeNeT group at IIT, Madras, Vortex Engineering Ltd. has developed a low-cost ATM specifically for rural applications. The ATM will make use of an existing network of Internet kiosks that currently access hundreds of villages in India. Capable of dispensing both new and used notes, the ATMs will also offer non-cash transactions, micro-deposits, credit and a number of other services tailor-made for rural areas. The machine can be used with ATMs or SmartCards and also has the option of using fingerprint identification. The machine costs around US$ 1,000, just a fraction of the US$ 20,000 it takes to install an urban ATM.

Regulatory and Policy Issues

In its 10-point agenda released in May 2004, the Government of India has outlined the use of ICT to improve literacy through distance education and the quality of healthcare services through telemedicine and Internet in the remote rural areas. Regulations and policies to ensure execution of the vision are very critical and the government must play the role of a facilitator to ensure a sustainable model where competition and open market dynamics drive quality, relevance, affordability and acceptance of services. Policies that allow a free and fair play for the rural operators are essential. Policies that enable infrastructure sharing for the rural operators would help them reduce costs and offer customized services. Shared infrastructure is imperative at least until the rural markets attain sustainable levels of maturity and are able to provide economic returns in line with their urban counterparts. This chapter discusses the history of telecom regulation, suggests policies for improvement and also highlights some issues that need urgent redressal.

Neglect of Rural Telephony

Telecom operators in most countries historically concentrated on building telecom networks in urban areas and have generally considered providing rural connectivity as a liability. Telecom operators in India are no exception. Providing rural telephony and Internet connectivity has always been dismissed as a loss-making proposition. If at all, attempts to venture into rural areas have been done only under obligations imposed by regulatory or licence conditions. When the Indian telecom sector was opened up for competition through policy initiatives by the government as part of NTP 94 and NTP 99, private basic service providers were given licences to operate in the lucrative urban areas only on acceptance of the obligation to provide a certain level of connectivity to rural areas. Further, a significant part of the licence fee charged by the government consisted of a levy towards the Universal Service Obligation Fund (USOF) which was used to pro-vide connectivity in these rural areas.

Not withstanding all these, the new private operators barely ventured into rural areas, failing to carry out their obligations. Acceptance clearly did not translate into implementation of any of the conditions. The state-owned incumbent, BSNL, also expanded only reluctantly in some rural areas, albeit only while complaining loudly and bitterly about the burden they had to shoulder. Yet, for its part of the burden, BSNL was indeed compensated to a certain extent by government grants *in lieu* of licence fee and spectrum charges paid by them and also by Access Deficit Charges (ADC) levied on all calls.

In other words, all major telecom operators competed only for urban business and avoided getting into rural telephony as much as possible. Telecom companies would rather pay penalties than roll out services in the un-remunerative areas. Therefore, most of the existing rural connectivity is either because BSNL provided it under pressure from the government or because it exists simply as a spill-over from the infrastructure created to cover urban areas or highways. It is under these trying circumstances that the USOF had called for tenders from operators to provide rural connectivity. While private operators largely ignored the tender, even the incumbent was reluctant to bid for the tender on the pretext that the maximum bid amount stipulated by the USF was too low and they stood to lose money in connecting rural areas even with the maximum amount of subsidy provided.[1]

As mentioned earlier, NTP 99 required private operators to provide certain amount of rural connectivity as part of their licence conditions. Since the operators had failed to meet most of these conditions, in 2002, and again in 2003, they were threatened with the imposition of heavy fine or the cancellation of their licences. Acting under such pressure, some operators did take some minimal initial efforts (Box 3.1).

Strangely, in December 2003, while resolving the dispute on Limited Mobility and Full Mobility, the DoT acting on the advice of the Telecom Regulatory Authority of India (TRAI), had removed the rural obligations of the operators. Rural connectivity thus got relegated to the background in one go.

Internet in Villages

Rural areas at least have some modicum of telephony (rural teledensity is close to 1.3 per cent), thanks to pressure from the government over many years and also due of the efforts of BSNL. However, Internet connectivity is in a much worse shape. It is almost non-existent but for some of the recent efforts. These initiatives are also saddled with

BOX 3.1 **Tata Teleservices Serves Universal Service Obligations**

THE HINDU

Date: 22/3/2003 URL: http://ww.thehindu.com/2003/03/22/stories/2003032201581700.htmt

Business

Tata Tele ties up with n-Logue for rural telecom

By Our Staff Reporter

CHENHAI MARCH 21. Tata Teleservices Limited (TTSL), a private basic service operator, has tied up with n-Logue Communications to develop rural telecommunication in selected areas. As per the partnership, n-Logue Communications would set up the network infrastructure and provide last mile connectivity to TTSL for offering basic services in four Short Distance Charging Areas (SDCA) in Tamil Nadu.

Addressing a press conference here today, the Chief Operating Officer, TTSL, R. Balachandran, said this tie-up would improve the quality of rural connectivity in Tamil Nadu. With this partnership nearly 400 kiosks would be set up in four SDCA's by 2004. These kiosks would provide telecom services in some specific small towns and rural areas. The operations in the four SDCA's) Bhavani, Vaniyambadi, Gudiyatham and Mayavaram) would start in three months time. TTSL was partnering n-Logue, which would act as its franchisee in the areas where TTSL proposed to operate. n-Logue would in turn work with local entrepreneurs to set up the network infrastructure and provide connectivity, training, support and linkage to relevant Internet-based services. Further SDCAs might also be subsequently covered through a similar model.

According to P.G. Ponnapa, CEO, n-Logue, said at present a typical kiosk would offer both telephone and Internet services to the local community.

The kiosks would act as a local and STD/ISD telephone booth, an Internet centre, E-governance centre, E-education centre and also provide employment for educated people in the rural area. For its network, n-Logue would use CorDECT, a WLL-based technology. It would look at partnering institutions and organisations, largely on the local area, that could provide such services.

Tata Tele to Tap Gilat Satellite Network for AP Circle

Tata Teleservices has tied up with Gilat Satellite Networks to provide 1,000-site fixed rural satellinge telephony network to service basic customers in Andhra Pradesh. This new network, based on Gilat's DialAw@y IP(TM) VSAT

(*Box 3.1 continued*)

(*Box 3.1 continued*)

product, will serve public pay-phone and fax kiosks, supporting the needs of agricultural workers and other inhabitants of Andhra Pradesh. Ajay Pandey, chief operating officer of Tata Teleservices, said: 'VSAT networks is a superior platform for bringing affordable telecommunications services to rural populations. Gilat's technology is meeting our requirements and will enable us to serve our users.' DialAw@y IP is one of the only VSAT products capable of providing high-speed Internet connectivity and toll-quality telephony service at the same time. Each unit supports a PC/LAN connection and up to six telephone lines. According to the company, it is a low-cost solution for the fulfilment of universal service/access obligations, Public call office requirements and small office/home office requirements for bundled telephony and Internet access. This network will enable Tata Teleservices to meet its universal service obligations under its license agreement with the Department of Telecommunications for Andhra Pradesh. In the future, Tata Teleservices expects to use the network to provide high-speed Internet connectivity to many of those sites.

Sources: *The Hindu* dated *22 March 2003* and *Business Standard* dated 15 April 2002.

many problems. Technology is a major impediment. The land lines that exist in rural areas are capable of supporting 9.6 kbps or 14.4 kbps Internet connectivity on dial-up modem, severely limiting the number and type of applications that could be offered. CDMA (IS-95) is another option but, even this technology limits the data connectivity rates to 9.6 kbps. However, some strides in the path of improvement have been made using technologies like corDECT Wireless-in-Local-Loop which has the potential of providing 35 kbps or even 70 kbps dedicated (sustained) Internet connections.

The next biggest problem limiting the Internet connectivity in rural India is the affordability and availability of personal computers for rural families. Finally, there are very few applications that are truly relevant to rural areas rendering even connections where available, of little use. Therefore, it is hardly surprising that the Internet has barely made any impact in the rural milieu.

Regulatory History

It is not that this state of affairs of rural connectivity was not foreseen. Almost four years ago, some experts had suggested the introduction of a special category of operators called Rural Service Providers.

They had argued that for years to come, the focus of the larger private operators would be on getting a bigger share of the pie in the urban telecom market. It was therefore desirable to create special operators who would focus on rural areas and attempt to make rural telephony and Internet businesses viable (Box 3.2).

In fact, the efforts of ITC and n-Logue Communications have shown that there were companies who were undaunted by the challenging task ahead. Even though they were not provided a license to provide telephony, they have done a commendable job of bringing the benefits

BOX 3.2 New Policies can Unleash Entrepreneurship in Rural Telecoms

Currently, more than half of India's villages lack telephone connectivity, let alone Internet access; the arrival of the information revolution to India is in doubt. The 26 million phone lines (mostly business-owned) and one million Internet subscribers that do exist nationwide are highly concentrated in urban areas, leaving rural areas out of the loop and harming the interests of both groups. If over 55 million televisions were wired with cable in less than a decade, imagine how many homes and villages could be wired with the Internet, a system that offers a far wider range of services, generates more earnings, is more flexible in design, and is essential to social and economic development. Sadly, rural Internet and telephony is not taking off like cable television because the regulatory environment, rather than promote universal access, actually serves to discourage it.

We believe that a policy that creates and promotes community-based RSPs will fundamentally change the economies and realities of access for rural India. With the creation of a myriad of local, small, entrepreneurial, fast, and flexible RSPs, India can fundamentally alter the economics of rural access and in doing so, network the entire country. Because their operating costs are less than half BSNL's, small private RSPs can provide quality telephone and Internet service at affordable prices—but only if the policy environment allows them. Entrepreneurs will roll out reliable service quickly, help meet universal service obligations, add to BSNL's (and other BSOs and backbone providers) revenues, and reduce costs. If we have learned anything from India's cable television experience, now is the time to apply it.

Excerpts from: Best, Michael L., Ashok Jhunjhunwala and Colin M. Maclay, 2001. 'New Policies Can Unleash Entrepreneurship in Kural Telecoms', *The Economic Times*, 04 April.

of the Internet to the rural areas. Fortunately, they were enabled by the Internet policy of 1999, which allowed virtually anyone to become an Internet Service Provider (ISP) with a token sum of under US$ 2 as licence fee and a nominal amount for a bank guarantee. ITC and n-Logue have used this ISP licence to run their Internet operations in rural India. A well-planned and crafted regulation on the lines of RSPs would have enabled them to move faster and reach a wider section of people.

With the introduction of the NTP 99, the tottering mobile services industry (crushed under the burden of huge licence fees) was brought back on its rails. At the same time, the inexperienced policy makers, regulators and operators were constantly at tussle as several vexatious arose on the issue of 'Limited and Full mobility'. Factional disputes came to the fore between basic (or fixed line) and mobile operators. While the DoT and TRAI were completely focussed on resolving these complex and highly contentious issues, there was little time available to examine in depth the issues plaguing rural telephony. Intense lobbying efforts in the country's capital were also focussed on issues in urban areas only. Thus, the path-breaking suggestion of RSPs was largely ignored. The policy makers either hoped that the concern of rural connectivity would either be addressed on its own or would simply await its turn.

When the rural service obligations of the private operators were done away with, vehement protests were raised against ignoring rural interests (Box 3.3). But, these were silenced with arguments that either BSNL would take care of providing telephony or the USOF would. Even as counter-arguments flew that urban telephony bloomed (with both reduced tariffs and improved quality of service) only with competition among multiple operators, it was left to BSNL's devices to take care of rural telephony. This only meant that somehow competition would not be required in the rural areas and any benefits arising thereof are not important either.

Regulatory Changes Required

It is only in the recent few months that it has been acknowledged that BSNL is incapable or unable to make significant inroads into rural telephony and rural connectivity, and that the USOF would not be the magic panacea for the ailing rural telephony market. Attention is now again being given to the regulations and policy framework for

BOX 3.3 Letter to the Minister of Information Technology and Communications, Protesting Withdrawal of Rural Service Obligations

INDIAN INSTITUTE OF TECHNOLOGY MADRAS
CHENNAI - 600 036

Dr. ASHOK JHUNJHUNWALA
Professor

Telecom and Networks (TeNet) Group
Department of Electrical Engineering

REF: EE/AJW/962/03

01.12.2003

Shri Arun Shourie
Honourable Minister for Disinvestment & Communications
Ministry of Information Technology & Communications
1st Floor, Electronics Niketan, Lodhi Road
New Delhi - 110 001.

Dear Sir,

It was a surprise to us that along with the introduction of universal license in telecom (merging BSO and CMTS services), the obligations that the BSO had to expand their network in rural areas have been done away with. What is being stated is that the rural obligation will largely be driven in future by BSNL and USF will help in this regard. It is also being stated that the private BSO were any way not proceeding forward very much on the matter.

1.It is true that the private BSOs had moved very slowly on rural obligations. But under pressure from DOT, many of them had made a good beginning and started taking interest in rural connectivity. I understand that Reliance had already installed their POP in over 1000 SDCAs. Tatas had taken a massive effort to connect over 1000 hitherto unconnected remote villages in Andhra Pradesh using VSAT, and along with N-logue Communications were in the process of connecting over couple of thousand villages in Maharastra, which were so far not connected. Even in Gujarat, it was tying up with N-logue to work out a business model to connect every village in certain districts. Even the Govt. of Gujarat was participating in this milestone as Internet kiosk was to be put in each of these villages.

2. While much of these attempts were to meet obligations, the learning from these efforts would have enabled these BSOs to come up with innovative business models for rural areas. Unfortunately the new policy is killing these efforts in the bud.

3. Sir, you would agree that even in urban areas the telecom boom is taking place only after competition has been introduced. It is the competition that has helped reduce prices and enhance services. In a similar way the competition in rural areas was essential to make rural telecom a business proposition and help enhance services. Now by making the rural connectivity effort largely fall back on BSNL, one is taking a big step backward. Without competition in connecting rural areas, rural services are unlikely to significantly improve.

Sir, I do not think the objective of the merger of the licenses was removal of this rural obligation. This has happened perhaps due to oversight. I therefore feel that it is important that the removal of rural obligation due to the merger of licenses, be done away with immediate effect.

Thanking you,

Sincerely yours,

(Ashok Jhunjhunwala)

Email : ashok@tenet.res.in
Fax : +91-44-2257 0120

Telephone : +91-44-2257 0120 / 2257 8365 (O)
+91-44-2257 0202 / 2257 9365 (R)

rural connectivity. The RSP policy is getting a fresh look. The TRAI has also suggested the creation of a new category called 'Niche Operators' who will offer fixed-line telephone services in rural districts where the teledensity is less than 1 per cent. This is akin to the concept of RSPs mooted earlier.

Rural Service Provider Policy It is being increasingly recognized that smaller operators focussed entirely on providing services in rural areas are extremely critical. Regulations must therefore enable and encourage them to fully take advantage of the well-entrenched optical fibre connectivity existing in India. Interconnections to the national and international backbone network should be facilitated by carrying all rural traffic to the nearest fibre point (irrespective of who owns the fibre) and should have a single point of handover. The RSPs should be allowed to freely use any technology they deem fit to provide connectivity.

As an encouraging gesture, any levy of upfront licence fee should be waived; however, a percentage of total revenue earned (equal to that charged of other telecom service providers) could be charged. Of course, spectrum charged as paid by other service providers may be charged as well. RSPs should be entitled to a percentage of call charges to carry the calls and appropriate mechanisms put in place to share call charges as defined by the operators. Their operations should be restricted to providing any extent of telephone and Internet connectivity, although in the rural areas alone.

Such a broad-based regulation would truly enable the smaller operators focussed on rural areas to create a viable business operation.

Other Policies and Incentives

There are a number of policy initiatives which could help in enabling Rural India getting Internet connectivity at the earliest.

Duties, Taxation and Service Fees The primary reason for the reluctance of the telecom service providers in offering services apart from the low returns is the high cost of setting up the network and running it. Customs duties, sales and service taxes amount to a significant percentage of capital expenditure in telecommunications.

The government has reduced the taxes on computers to drive penetration of Internet. However, the taxes and duties on telecom equipments remain at a high level, resulting in a high infrastructure cost. In certain cases, it even amounts to 30 per cent of the cost of equipment. Such high rates severely hamper infrastructure investment and impact the expansion of rural telecom. In order to ensure the viability of rural telecommunication, considerable reduction of taxes for rural infrastructure may be considered.

The service tax on telecom and Internet service is high at 10 per cent. While it is supportable in an urban scenario, this cannot be sustained in a rural economy and may be waived as a special circumstance.

During the initial phases of the IT revolution, the government had identified Software and IT services as a thrust area and provided tax holidays for up to 10 to 15 years for companies operating in this domain. Such concessions, if offered to rural connectivity, would help drive penetration effectively.

Investment Offering telecom and Internet services in the rural areas needs three levels of investment.

Kiosks: Aggregation of demand is the viable option for offering sus-tainable services in the rural hinterland. Internet kiosks are a necessity to enable such aggregation. The set-up cost of a typical kiosk ranges between US$ 1,000 and US$ 2,000. To encourage rural entrepreneurs to invest in such kiosks, affordable loans should be made available through banks, which are flush with funds. Banking regulations do mandate the banks to invest up to 40 per cent of their funds in the priority sectors like agriculture, Small Scale Industries (SSI), and so on, to enable growth for the underdeveloped sections of the society. However, ICT for rural areas is not recognized as a priority sector and necessary actions need to be taken to enable the same. Also, concessional interest rates should be made available. The total investment needed for connecting all the villages (638,365 of them), at say US$ 1,500 or roughly US$ 1 billion, may seem high. However, a substantial part of this amount requested as individual loans and providing these loans could be easier for banks.

Telecom Infrastructure: Developing the infrastructure to provide connectivity to rural India will cost approximately US$ 100,000 for each rural area of approximately 2,000 sq km (with a radius of 25 km). In order to optimally cover rural India, 1,500 such areas need to be connected and thus the total capital investment needed would be US$ 150 million. Such funds should be made readily available to the RSPs.

Raising this money is not too difficult. The government has several schemes under Rural Infrastructure Credit and Investment. Cheap credit, tax holidays and, most importantly, hassle-free access to these investment devices would facilitate the capital inflow necessary for the RSPs.

Consumer Loans: The next level of investment would be in the form of consumer loans for individuals to acquire personal telephones, computers, Internet connections, and so on. The upper-income bracket among the rural population would easily earn enough to acquire these and also pass any creditworthiness checks by banks. Therefore, it is an easy task for banks to finance the purchase of equipment necessary at the level of ordinary consumers.

Problems and Bottlenecks

Even though the potential of connecting rural India appears bright, there are a few possible bottlenecks which could retard growth. Some of them are discussed in detail in the following sections.

Liberalization of Rural Markets Liberalization of the Indian economy began in 1991 and allowed Indian industry to be governed and led by market forces. Yet, there are many restrictions that throttle rural development. The rural market still remains regulated and controlled by the government to a great extent. There are government diktats that allow the sale of some varieties of farm produce only through government-owned mandis. For example, sugarcane can be sold only to the nearest co-operative sugar factories and cannot be processed or sold by farmers directly. There are crops that cannot leave the borders of the states they are grown in. And, there are some more that cannot be exported to the lucrative markets of other countries. These measures may have come into place to safeguard the interests of farmers. However, they are now passe and are not in synch with current realities. Some of these have since been done away with, but others merrily continue.

There is a ceiling on the size of land an individual can hold and this hampers the use of large-scale and intensive farming techniques. Farmers who subsist on the smaller plots are the ones who are hit the most as a result. Another quirk is that while farmers are now allowed to sell produce to buyers other than the *mandis*, no transparent price reference mechanism exists. This leads to dependence on unofficial sources, including middlemen, often leading to problems of a different kind. And, the farmers are unable to fully reap the economic returns that they deserve.

Thus, such archaic regulations and half-hearted, hastily planned measures stifle the growth of the rural economy and need to be done away with. The issues affecting rural economy need to be seriously looked into and addressed effectively to create a robust and vibrant rural economy.

Social Factors Rural society is highly fragmented with caste and religious differences existing even now. These differences raise important questions—who owns the kiosks? Where are they located? How do various sections of people use the kiosks? How do women, easily the most oppressed in a rural setting, gain access to the services in kiosks? These issues are complex and do not present easy solutions.

But, as some of the n-Logue kiosks have demonstrated, such differences are set aside when business is the prime driver for the kiosk operations. Kiosk operators who are keen on maximizing their revenues are capable of finding the common ground to enable all sections of society to utilize the services at their kiosks. This also necessitates the introduction of services tailored for various segments of people. While e-mail and browsing may be popular with the upper class and literate sections, applications such as veterinary care are more relevant to the low-income segments dependent on the farm animals for daily livelihood. E-governance services which benefit the poorer section of the village community is another area to be vigorously pursued.

Competition is another great leveller; areas with multiple kiosks see the operators striving harder to reach out to all prospective customers, enabling greater penetration of these services. The most critical factor driving success is the networking skills of kiosk operators with the local community.

Power Supply The Internet kiosks require power to operate. There are parts of India where rural areas have grid power for even 12 to 16 hours a day. Cost-effective options are available to offer power back-up for 4–8 hours to enable uninterrupted operation of kiosks.

But, in several other parts of India, power is available for barely 4–8 hours or often, even less. The power back-up would then not be adequate for the kiosks to be in operation for a significant portion of time. Solar power or diesel generators are the other available options. Even though the kiosks may not require more than 300 W of power, the cost of solar panels with back-up is in itself significant and could almost double the cost of the kiosks, rendering them economically unviable at least in the initial stages. The use of diesel generators could serve as a viable alternative as the initial investment is less, but the continuous operating cost is a worrisome factor. With extensive researches underway on bio-diesel, there is hope for availability of cheaper alternatives in the near future.

Solar panels, diesel generators and bio-fuel alternatives may address the vexing problem of power for Internet kiosks. But driving the rural economy through a large number of micro-enterprises would require 20 kW to 100 kW of reliable power. With grid-based power supply remaining inadequate, the only viable option would be to set up decentralized, low-capacity power plants in villages. The challenge lies in developing the technology to make them economically viable.

Conclusion

To conclude, there are compelling reasons to reach out to the rural market. The development of the rural market and improvement of the lives of rural people are themselves compelling reasons. For companies though, the fast saturation of urban markets presents the need to explore newer markets. Still, due to the high cost of telecom infrastructure, the diverse terrain that it needs to be deployed in, and particularly, the low affordability levels of the rural masses, these markets remain unexplored to a great extent. The answers lie in technologies that are cost-effective and easy to deploy, applications and services that are appropriate, business models that are scalable and a regulatory environment that embraces and applies itself to serve rural India.

Scalability

India has over 740 million people living in more than 638,000 villages. The number of people whose lives would be impacted by effective ICT intervention is therefore extremely large, almost 12 per cent of the world population. This also means that any work that is restricted to a few hundreds, few thousands or even a few tens of thousands of villages makes very little difference, given the sheer size of rural India in total. The key is to scale and to reach the most number of villages.

To provide all the applications and essential services discussed in various sections earlier, careful co-ordination with all constituents of the drive to bridge rural divide is necessary. Isolated efforts do not attain a critical size simply for this reason. A united front to link together various elements like educational institutions, hospitals, banks, government bodies, private enterprises, welfare organizations and the like would help reach a critical mass in a shorter frame of time. Then, a target of reaching 100,000 villages could be easily possible with concerted and carefully planned efforts. Else, any non-scalable intervention is only easily rendered irrelevant.

Technology

Fortunately, technologies to connect the villages of India do exist and can scale. A major asset is the fibre connectivity running deep into each taluka. Current wireless technologies can be used to connect about 85 per cent of villages. They are cost-effective and can easily bring down the barriers for entering rural markets. Future wireless technologies are also rapidly evolving and seem capable of meeting rural India's future needs.

In addition to wireless technologies, a whole host of other technologies would be required to make the desirable impact in rural India.

- Multimedia technologies are today at the stage to make a difference.
- Caching technologies at servers are required so that even remote villages are not deprived of useful information.
- Internet-based transaction technologies are continuously evolving and have a major role to play as well.
- Technologies for banking and finance using ICT are available, but need to be made cost-effective for widespread use in rural India. This is applicable for computer and Internet-based education and healthcare as well.
- Language technologies are crucial to reach out to the masses in India who speak a multitude of languages. While the last few years have witnessed some progress in this area, there is still a long way to go.

- Personal computers today are relatively inexpensive and rugged enough to be useful in rural India. But, the price points need to fall further, to facilitate movement from shared to individual usage.
- Probably, power supply is the one that requires most careful attention. ICT initiatives in rural areas cannot get far without solving the problems associated with power.

Services

It must always be remembered that ICT is merely the means and not an end in itself. Therefore, ICT must enable fulfilment of the real needs of rural communities and remain relevant to them. Education, health and livelihood are the primary requirements of rural areas.

- Today, rural people appreciate the value of education for their children better than ever before. While schools do exist in most parts, the quality of teachers and of the education imparted and the infrastructure is questionable. ICT can become an important tool in plugging these gaps.
- Despite the presence of PHCs in every three to four villages, healthcare is another major concern in rural India. Quality healthcare is still available only with doctors in the nearest towns who are often most sought after. Telemedicine through videoconferencing between urban doctors and rural patients provides this vital link. Using simple remote diagnostics, healthcare can be provided instantly and effectively.
- However, above all, ICT can definitely provide chances to rural people to enhance their livelihood. The demand for ICT and services can be sustained in the long run, only by generating rural wealth. The primary wealth creators in this context are agriculture and agri-processing, animal husbandry, small-scale handicrafts, food processing and, often, even IT-enabled services.

Regulations

The key role of the government and regulating bodies is to enable rural connectivity to happen at the fastest possible pace. At the same time, NGOs and other organizations focussed on rural areas must be enabled to provide the necessary services. While immense infrastructure (fibre and towers) have been built up to connect rural areas easily, these are sadly heavily underutilized. Existing telecom operators focussed on urban areas treat these resources as property encashable in the future—whenever they would be willing to consider operations in rural areas. However, the 740 million rural people cannot afford to wait.

Regulations should be framed to ensure that existing infrastructure is tapped to provide rural telephony; that open markets are allowed to flourish; that equitable compensation mechanisms are in place to protect the interests of all parties and that measures to develop rural livelihood and entrepreneurship are encouraged.

Is This a Possibility?

Rural India is at a critical stage. It is large in size; problems are aplenty; the challenges are daunting and thus can present a scary picture to an outsider. But if enabled, it has the potential to transform itself quickly. Technologies exist today to provide telephony and Internet in every village. The development of the rural economy and improvement of the lives of its citizens constitute a prime force for governments, NGOs and other organizations working in the rural areas. There are also proven business models to sustain and drive efforts in this direction. The ability of the Internet to positively impact the lives of rural people has been amply demonstrated. Thus, Internet and other communication technologies can empower the rural populace such that they can chart their own bright futures, very different from what anyone could have ever imagined or written about.

Note

1. Tender No. 30-130/2003, the Universal Service Fund.

4

Two Mutually Reinforcing Applications of ICT for Socio-economic Development of India

Rajesh Jain
Atanu Dey
Reuben Abraham
Vivek Padmanabhan

Introduction

The gains from any innovation or revolution in technology or process usually have little impact on the poor in any developing country. The benefits usually accrue to the rich—peoples as well as nations. The industrial revolution of the past is an example of this, where poorer nations are yet to fully benefit from the industrial revolution. So far, the benefits of globalization appear to have not had an appreciable impact on the poor. This should change and the revolution in information and communication technologies has the potential to help break out of this unfortunate scenario.

Poverty—income poverty as well as non-income poverty—is perhaps the most common characteristic that defines the populations living in the developing world today. Non-income poverty in terms of education, healthcare, access to markets, and so on, directly produce the income poverty that traps the average citizen of developing countries. The question of how to raise huge populations out of this poverty trap is a formidable challenge that governments, multilateral organizations and policy makers face.

Because income poverty is relatively easier to measure as compared to non-income poverty, it is more commonly reported and emphasized.[1] Income poverty and non-income poverty are closely related, of course. The problem appears almost intractable because the two kinds of poverty are mutually reinforcing. Any solution that does not address both kinds of poverty is unlikely to be successful in poverty alleviation.

In this chapter, we focus on two uses of Information and Communication Technology (ICT) which hold the promise of immense benefit to the rural poor, specifically in India, and more generally in other parts of the developing world. We focus on the rural population because the incidence of poverty is higher there than in the urban population.

The first application is the use of ICT in providing education. Specifically, primary and secondary education, increasing literacy and providing vocational education. The current system is unable to deliver due to a number of reasons, primary among which is its reliance on mostly individual content creation and delivery, essentially through the public sector. The economies of scope and scale attainable through the use of ICT tools would make education accessible and more affordable.

The second application relates to expanding market access for agricultural and non-agricultural products. This would increase rural incomes and thus alleviate income poverty. The Internet can efficiently provide access to a vast market for traditional handcrafted goods which can be sold worldwide, for instance. This would be an effective way of integrating the rural population with the globalized marketplace.

Poverty can be considered to be the result of two gaps: one, the ideas gap, and the other, the objects gap. Poor people have less material goods at their disposal as compared to rich people, hence, the objects gap. The ideas gap arises from the inability of poor people to most effectively and efficiently use the limited material resources they have. For any level of objects gap, an ideas gap amplifies the problem. Knowledge goods, efficiently produced and distributed by ICT, can bridge the ideas gap.

The two applications are complementary. Primary education has tremendous social welfare implications, but the return on investment is long term. Being a public good, the returns cannot be fully captured and consequently private investment is unlikely to provide primary education.

In contrast to that, creating market access increases the incomes even in the short run. Thus, provision of market access can be commercially sustainable through user fees and so the private sector can be relied upon to invest in that application.

Both activities require a common infrastructure in terms of power, telecommunications, human resources, and so on. Therefore, the common provision of these two can lead to lower average costs. Given the resource constraints, a mechanism for efficiently delivering the two would require some degree of concentration of investment in rural areas rather than thinly distributing the resources.

This chapter discusses the economics of the two applications with respect to the costs and the benefits. The need for joint provision of the two applications is shown to be commercially sustainable. We show that information technologies can potentially overcome the market failures that have so far not brought the benefits of globalization to the rural poor.

The chapter concludes with a brief description of a mechanism which efficiently delivers the applications in a commercially sustainable way.

We begin by first seeing what we can learn from the ICT initiatives that have been previously implemented.

Lessons from the Past

There have been various initiatives to take IT to the masses in India—Gyandoot, eSeva, Drishtee, Tarahaat, Bhoomi, eChoupals are some examples. At best, these have been success stories limited in size, scale or scope. The digital divide is far from being bridged. Where is the problem? There certainly does not seem to be a lack of vision, ideas or even resources. And yet, what is missing is a solution that has been rolled out on a mass scale to make a difference to millions—not between two generations (25 years), but between two elections (five years).

As we see it, the following are the problems:

- **Government as Financer:** This is perhaps the single biggest issue which limits scalability. The government can fund 100 or even 1,000 centres or kiosks costing Rs 100,000 (US$ 2,000 each.

But the need is for many more access points. That is where the government-funded model becomes impractical—there simply is not enough money to set up these across a state or a country. And so, without the scale, the costs of operation are high, the villagers have to walk many kilometres to get to the nearest centre and that is simply not going to happen.

- **Demo Mentality:** The thinking when the plans are drawn up is to create 'pilots'. The reasoning goes: let us do 10 or 50 or 100 such demonstration centres or showcases. Once the proof-of-concept is achieved, then we can look at scaling these up. This approach is one which is setting itself up only for a short-term success; it will not succeed in the long term. This is because it is much easier to put in all that it takes to make a few centres work because the aim is not to prove commercial viability but to showcase a local success to funding agencies or key decision makers. The approach is not geared to creating solutions that can be scaled up rapidly.
- **Silo Solutions:** Many approaches think of the problem too narrowly. We think of solving a 'telemedicine' problem or a 'land record' problem or an 'e-mail and Internet access' problem or a 'literacy problem' or the 'voting machine' problem. The computing infrastructure required for solving each of the problems is almost identical. And yet, we think of each in isolation, trying to create economic models which will work in the silos.
- **Internet-driven:** Many of the current solutions assume the existence of a Net connection, essentially functioning as Internet kiosks. This is a big limitation, because connectivity is one of the biggest bugbears in the rural areas. Without connectivity, the computer is crippled, seriously limiting its usage. While transaction services like bill payments and railway bookings which need real-time Internet connectivity can offer immense benefits to the villagers, these services can be hobbled by the lack of connectivity.
- **Incrementalist, not Disruptive:** The need of the hour is for disruptive solutions. Yet, the thinking that percolates is very incrementalist. That may be because there is an interest in keeping things nearly the same, or because we look at technology that exists today and not at what the future is bringing. The solutions tend to be driven more by what may have worked in the developed

world or in the urban areas, because they are the ones who are either funding the solutions or providing the technologies. The need is for a completely fresh and bottom-up analysis of the rural markets, keeping in mind the emergence of 'cold technologies' which reduce the total technology spend.

Rethinking ICT Solutions

Let us look at the requirements for the ICT solutions for the rural markets:

- **Mass market:** The solution needs to address the needs for tens of millions of people. In India, this has to be a solution which can, in a short period of time, impact each of the 600,000 villages to make a difference to hundreds of millions of people. It is a solution on a scale that has perhaps never been thought of before.
- **Scalable:** Being able to scale out the solution is very important, else we will have created yet another 'demo wonder'. Scalability will mean that there has to be a decentralizable element in the solution.
- **Emergent:** Going hand-in-hand with scalability and decentralization is the need for the solution to have emergent properties—where it is driven from the bottom-up and the whole is much greater than the sum of the parts. This can only happen if the solution is driven not by government, but by entrepreneurs who see a commercial motivation to own, deploy and grow the solution.
- **Low R&D Costs:** There is little time to go out and develop new solutions. The approach should be that of aggregation, not re-creation. This means looking around and pooling together existing ideas and technologies which may be just good enough, rather than spending years on creating what could be the perfect solution.
- **Extremely Affordable:** We are talking of the world's poorest markets. Affordability needs to be redefined keeping in mind these customers. These are segments of society we do not ordinarily think about. But they are the ones who are the world's next markets. Costs have to be a fraction of what we are otherwise used to considering or paying.

- **Technologically Forward-Looking:** The solution needs to look to the future rather than into the past. What is there under the hood is not as critical as giving the same kind of features and performance as the ones in the developed world are used to. In some ways, there is an advantage in terms of legacy—there simply is not an 'existing solution to upgrade', so there is no need for backward compatibility. This gives us an opportunity to 'leapfrog'.
- **Platform Orientation:** The solution must create an ecosystem in which multiple players can thrive. The approach must be that of creating a platform that others can build upon, without having to redo the groundwork from scratch.
- **Consider the Constraints:** We cannot forget the limitations and realities of the rural markets—intermittent and fluctuating power, connectivity which probably is not there, a market which does not necessarily speak or understand English and one which has been largely ignored and forgotten by the world (except by the politicians who need votes in a democracy). Since connectivity is not a guarantee, the initial focus should be on information and offline communications services, rather than real-time, database-driven transactional services.
- **Commercially Viable:** Above all, the solution needs to be economically sustainable, given the constraints of the rural markets. It must provide the rural entrepreneurs with a business model which enables them to not just make money but also grow the business with their own initiative and innovation.

As we think of the solution, we should keep these words by Stuart Hart and C.K. Prahalad (writing in Sloan Management Review) in mind: 'Disruptive Innovations compete against non-consumption—that is, they offer a product or service to people who would otherwise be left out entirely or poorly served by existing products and who are therefore quite happy to have a simpler, more modest version of what is available in the high-end markets'.

Building Blocks and Proposed Solutions

The problem then is how the ICT tools and knowledge goods can be delivered to the rural population and what mechanisms exist for facilitating access to them. We present two solutions. The first is related to primary and secondary education and adult education as well. This is through a model that is implemented at the village level, called a TeleInfoCentre (TIC). The second solution is related to improving market access and providing vocational training. The model is called Rural Infrastructure & Services Commons (RISC) and is implemented at a level of a cluster of villages. Like the problems, the two solutions are complementary as well.

There are four technology building blocks that we need to look at, as we understand the potential of ICT to transform rural areas. The four technology building blocks are thin clients, server-centric computing, open-source software and WiFi. Together, they make up what we have termed as the '5KPC' ecosystem, with 5KPC meaning a Rs 5,000 (US$ 100) Personal Computer. We will first discuss the building blocks, then show how they can be used to construct the TIC and RISC centres which can facilitate the delivery of both education and market access, along with other services.

5KPC Ecosystem

The computer is a multi-faceted, transformation device. However, so far, access to it has been limited in rural areas to setting up kiosks with 1–2 computers. Deployment has been restricted largely because of the cost of the computers and the high cost of servicing a highly distributed base, along with the lack of availability of reliable connectivity. The 5KPC ecosystem is a solution which enables the creation of an affordable computing and communications infrastructure. By being able to reduce the total cost of ownership and simplify management of the connected computers, it becomes possible to deploy this infrastructure cost-effectively across rural areas. The 5KPC ecosystem makes real the vision of providing 'a connected computer accessible to every family'.

Thin Clients

The first building block of the 5KPC ecosystem is the thin client. There is no local storage and only limited processing which happens on the thin client. It handles the user inputs via the keyboard and the mouse and provides the graphical display via the monitor. All keystrokes and mouse clicks are sent to the server for processing and the resulting screen is shown to the user. The thin client can also be thought of as a 'computer terminal'. Any computer produced over the past decade (a Pentium class system) can become a thin client.

There is a virtually unlimited supply of thin clients available. Millions of computers are being disposed off annually by users and organizations in the developed world as they upgrade their desktops and laptops. These disposed computers have turned into a recycling problem for the developed nations. Most of these computers are in good working condition—it is just that they are not fast enough to run the new generation of software for most users. The typical upgrade cycle for computers is 3–4 years and the annual consumption of computers in the developed world is more than 60 million units. This ensures not only a large supply, but also one which is continuous. A fraction of this supply is good enough to meet the needs of the developing countries.

Shipping these computers to the emerging rural markets solves two problems—the recycling problem in the developed world and the need for an affordable computing infrastructure in the developing world. It is possible to take the computers being disposed, invest a small amount of money in their refurbishing and ship them for use across the rural markets to serve as thin clients. The cost of this entire value chain will be no more than US$ 100 (Rs 5,000).

Smart software running on the thick server ensures cutting-edge performance from these thin clients. In addition, because the computers are so cheap, it is possible to keep a few spare units—in case one of the units stops working, it can be quickly replaced by another unit. There is no need for expensive maintenance engineers, which can be a problem in rural and remote areas.

Server-centric Computing

The second building block is server-centric computing. A 'thick server' handles the processing and storage. Moore's Law is creating very powerful computers at ever lower prices. A new desktop costing about US$ 500 (Rs 25,000) is more than good enough to become a 'thick server' and support upto four thin clients. Shifting the processing and storage to the server also simplifies the administration of the computing infrastructure—the server is the only system that needs to be managed.

Open-source Software

The third building block is open-source software. Over the past decade, Linux has emerged as an equally good alternative to Microsoft Windows, both on the desktop and the server side. An ever-increasing pool of Linux developers and applications is now available. What open-source software offers is not just free software, but also the freedom to make changes as applicable to the software (provided the changes are themselves released in the public domain). What open-source software does is to enable the leveraging of a vast applications base for a near-zero price.

WiFi

The fourth building block is WiFi. The 802.11 set of standards use open spectrum to enable wireless connectivity across a distance of upto 100 metres. While it is primarily seen as a local connectivity solution through hotspots, WiFi can become the fulcrum for connectivity in rural areas through the use of repeaters and antenna innovation. Recently, Media Lab Asia in India demonstrated how two points separated by a distance of over 50 km could be connected by WiFi. Thus, the connectivity layer for the 'last mile' for both voice and data applications could be WiFi.

Thus, taken together, the 5KPC ecosystem, comprising thin clients which are refurbished computers from the developed world, server-centric computing on new desktops, open-source software based on and around Linux to provide the applications base and WiFi to provide connectivity, serves as the basic building block for an affordable and pervasive computing infrastructure across rural areas.

TeleInfoCentre

A TIC is present at the village-level and can have between 3 and 10 computers. All the content and applications required are available at the TIC. It is not necessarily dependent on real-time connectivity for providing services. A TIC is entrepreneur-driven.

The TIC fulfils a multi-centric role: it is a computing and communications centre, has a digital library of documents, complements the teachers for school and adult education and serves as a small business office for entrepreneurs. Its additional value comes from applications that it can enable for citizen services and government interactions, making it an e-governance touch-point for the villagers.

As far as possible, the TIC should be able to work in the offline mode—that is, its dependence on Internet connectivity should be minimal. The server should mirror key applications and relevant data, making it possible for the clients to work without the need for an Internet connection. In fact, even the assumption that a TIC may have a few hours of Internet connectivity daily could be far-fetched. This makes the application development challenging, but it becomes an important pre-requisite, given the realities of rural India.

The offline mode entails updating through CD (or such an alternate device like a USB Memory Key). A CD will get written daily at the village TIC which has the day's e-mails and requests which cannot be served locally. This CD would then be sent by courier or through the postal system to the next level in the hierarchy, which is likely to have better Net connectivity. Similarly, a CD from there would bring updates to the village. Over time, WiFi will solve the network connectivity bottleneck.

Here are the costs (in Rs) for a 3-computer TIC (TIC-3) and a 10-computer TIC (TIC-10):

Cost items	TIC-3	TIC-10
Thin clients at Rs 5,000 per system	15,000	50,000
1 Thick server	25,000	30,000
Software and curriculum	5,000	10,000
Networking, modem, phone line	10,000	15,000
Printer, scanner, webcam, speakers	10,000	15,000
Power supply	15,000	30,000
Total capital costs	**80,000**	**150,000**
Monthly running costs (Telecom, Consumables, Staff)	4,000	8,000
Teacher costs	3,000	6,000
Capital cost EMI (12 per cent interest, five years)	1,780	3,337

AMC (10 per cent of capital costs, in monthly payments)	667	1,250
Total monthly cost	**9,447**	**18,587**

Rural Infrastructure and Services Commons

RISC is a much larger set-up, serving a cluster of about 100 villages in a 'bicycle commutable' radius of 10–15 km. In India, an RISC would thus service about 100,000 people. Each RISC serves as a local business centre where the downstream flow of information and material to the villages is complemented by an upstream flow of goods and services from the local village economy to markets that are global.

An RISC Centre

- Clusters economic activities in specific rural locations by facilitating firms' businesses.
- Provides a standardized reliable infrastructure platform in an economically efficient way.
- Co-locates a wide variety of services provided by market forces on the platform.
- Provides services on a for-profit basis.
- Serves as a focal point for the bi-directional flow of information and materials within rural areas.
- Coordinates the investments of the private sector, the public sector and multilateral organizations in rural India.
- Uses the tools provided by advances in ICTs.

Essentially, an RISC is a 'micro-city', an appropriately scaled-down version of a city. It acts as a focal point that provides a bi-directional flow of information and materials that are essential to the rural economy and which uses state-of-the-art tools and technologies to do so efficiently. It can be conceived of as a holographic projection of a city on a small scale at the rural location.

By using elements from the 5KPC ecosystem, it becomes possible to create an affordable technology infrastructure at the RISC, which can be leveraged by the service providers.

We will now look at how ICT tools and platforms like the TIC and the RISC can transform education and market access in rural areas.

Education

Education plays a paramount role in the process of economic development. Besides being instrumental in development, it is also an end in itself because it helps people lead better lives. For broad-based, sustainable economic development, primary education is critical. Neglect of primary education is endemic in developing nations.

Public support of education is often regressive. For instance, public spending on education for a set of selected developing countries by income quintile shows that the poorest income quintile receives around 14 per cent of total spending, while the highest receives around 28 per cent.[2] Systematic discrimination against the poor regarding public spending in education is also found in India. As shown in Table 4.1, public expenditure for elementary education is only 0.17 per cent of GDP for India.[3]

TABLE 4.1 **Public Expenditure for Elementary Education (2003–2004)**

	Figures in US$ billion[4]	*Percentage of GDP*
GDP	581	100.00
Aggregate govt. budget	93	16.08
All social services and poverty alleviation	14	2.49
All education	6	1.05
Elementary education	1	0.17

The public support of higher education primarily benefits the urban rich and the middle class. The policy choice of supporting higher education at the cost of the neglect of basic education is short-sighted. Policy makers must recognize the redressing of the imbalance as one of the most critical challenges facing them. This task is made more tractable by the wide availability of ICT tools. The leverage provided by these tools releases the severe resource constraints that bound the task of bringing primary education to the population.

Education can be categorized into primary, secondary, adult and vocational. We will focus on primary education, since the arguments can be easily extended to the other categories.

Primary Education

Primary education is a public good. Thus, the level of primary education provided by the market can be expected to be lower than the socially optimal level. Therefore, it is upto the government to step in and either provide primary education itself or subsidize its provision by the private sector.

The higher-income groups living in urban areas have the willingness and the ability to pay for primary education. The low-income groups in urban areas and most income groups in rural areas do not have the ability to pay for education.

One way of solving the problem would be for the government to provide credit to the poor so that they could pay for primary education. However, given the small size of the budget allocated for primary education and the immense size of the relevant population, it is a challenge that cannot be addressed without resort to technology-induced increase in productivity in the education sector.

To briefly review the broad scope of the problem of primary education, we note that literacy is only 80 per cent[5] in urban areas and 60 per cent in rural India.[6] About 36 per cent of all 7- to 14-year-old children are illiterate. That is, the total population in rural and urban areas that needs primary education is 340 million. The annual budget for primary education is only US$ 1 billion (Table 4.1). Therefore, per capita approximately US$ 3 per year is available for primary education. This sum is clearly inadequate even if utilized most efficiently under the current method of delivering primary education. Thus, if we consider that the budget constraint is hard, then the only way out is to innovate in the process of imparting primary education.

Just to provide primary education, India requires seven million teachers if one were to have a 1:50 teacher-to-student ratio. Not only is that number formidable, the problem is compounded by the fact that these teachers are mainly required in the rural areas where the current number of qualified teachers is extremely low.

Process Innovation: Distance Education

The availability of ICT tools has major implications for the provision of education.

The process of providing primary education involves the following steps at a minimum:

- Content creation
- Distribution of content
- Delivery of the content at the 'last mile'
- Testing and validation

Content Creation

A good primary education system has to be built on a solid foundation of content. Content creation involves a high fixed cost. Once created, if the marginal cost of duplicating the content is relatively small, the average cost can be made arbitrarily small depending upon how large the population is that is served by the content. Since the content for primary education is relatively stable, once created the content can be reused year after year. Therefore, the use of ICT tools for content duplication is a natural choice. For instance, once produced, the content can be digitized and then duplicated on digital media such as CD-ROMs and DVDs.

Distribution of Content

If content is produced centrally and then duplicated so as to reduce the average cost of content, then the next issue that arises is that of content distribution. CD-ROMs and DVDs can be mailed relatively inexpensively through the regular postal system. Or the content could be delivered to the point of use cable or wirelessly. This represents high fixed cost but a very low variable cost of distribution of content.

Delivery of the Content: The Last Mile

Ultimately, primary education has to be delivered to the hundreds of thousands of schools throughout the land by an impressive number of teachers. Training of these teachers itself is a formidable task. Again, ICT tools can come to the rescue, both for teacher training and for assisting them in delivering the content to the millions of students.

This component of the educational process involves high fixed costs and high variable costs. The high fixed costs can be reduced by facilitating the 'last-mile delivery' through ICT tools. ICT tools can reduce the total training that the teachers need by shifting the burden of content creation from them to creation of the content centrally and having the teachers facilitate the delivery of the content. For instance, the actual teaching could be done by a virtual teacher on a CD-ROM connected to a TV monitor, while the physical teacher is someone who mediates the delivery and maintains discipline and the schedule.

As we noted earlier, about seven million teachers are required for the primary education of those who are currently illiterate. Training those teachers alone is itself a formidable task. This task can be made tractable through the use of ICT in three distinct ways. First, the training of the teachers themselves can be mediated by ICT tools. And second, the teaching of students by these teachers can be more effectively done by the use of tools such as audio-visual material to supplement books that are currently in use. This not only reduces the load on the teachers, but in fact teaches the teachers at the same time as the students. Finally, it reduces the variation in the quality of the teaching delivered. This happens because the audio-visual material is professionally produced and the quality of the teaching imparted is not entirely dependent on the skills of individual teachers.

How TIC and RISC Facilitate Education

The purpose of education is to empower children with the requisite tools to go through life with the appropriate knowledge to make the right decisions. Not only is education important, computer education too is critical considering the role that technology plays in our lives and is likely to play in the coming years. While no computer can replace a good teacher, it is not always possible to get good teachers in schools in developing nations, especially in the interiors. This is where computer-enabled education can complement the teacher in the classroom. Besides, a 'digital library' and the Internet can help enhance and widen the learning process. It would be nice to see some of the scientific concepts brought alive through animations, making for a richer and more interactive learning.

A school is an ideal location for a TIC because it is already seen as a bastion of knowledge and is respected by most people. The TIC can be located at every primary and secondary school. During school hours, the computers are used to complement the teacher in providing IT and IT-enabled education to the students. After school hours, the centre can provide community services, some of which can be priced. This approach has multiple benefits:

- Computers will attract students to schools. As has been said: 'You bring computers into schools, you bring children to schools'.
- During school hours, the multiple computers in the TIC become educational terminals for the children, complementing the teacher. The computers can also assist in personalizing the education based on the student, with a single teacher catering to multiple levels of students. Thus, for example, the computer could remember the 'hardspots' of students and test them on those to ensure that they learn better.
- After school-hours, the computers could be used for various community services, thus serving the needs of the village as a whole. This ensures that no additional infrastructure is needed.
- By locating the TIC in the school, the image of the school and interest of the community in the school are both increased. This heightened participation will also result in better education for the children.
- The same TIC platform can be used to provide literacy for the village residents.
- The entrepreneur managing the TIC could be the school teacher, who could be provided the appropriate training. This will also result in greater respect for the school teacher in the village.
- In addition, various youths can be provided with possible jobs in data entry and other related areas, providing the prospect of using the village TICs as resources for IT-enabled services at the village-level. This employment opportunity will further create a positive environment for education.

We saw earlier that the monthly operating cost of a TIC-3 is Rs 9,447, while that of a TIC-10 is Rs 18,587. If a TIC-3 supports 100 students and a TIC-10 supports 1,000 students, on the assumption that each student can spend Rs 20 per month on computer and computer-enabled education, the TIC-3 will thus generate Rs 2,000 per month from education, while the TIC-10 will generate

Rs 20,000 per month from education. Thus, by itself, the TIC-3 will need to generate an additional income of about Rs 7,500 per month for break-even, while the TIC-10 can completely break-even on earnings from education itself (generating a surplus of about Rs 1,500 per month).

A TIC-3 is most likely to be at a primary school, while a TIC-10 is most likely to be at a secondary school. Primary schools are present in almost every village. Thus, the TIC-3 may need to be partially subsidized by the government if the earnings from the other services are not enough. This is to be seen as an investment in primary education.

There are multiple ways by which the TIC can generate additional revenue which can bring down the cost of student education:

- The TIC can charge a fixed subscription fee for a base set of services (for example, adult education, e-mail, and so on.)
- The TIC can take up data entry jobs or other such work to better leverage the computers that it has.
- The state/district can pay for some of the services that it uses. For example, on account of the TICs, the information collection and dissemination costs can be reduced. Part of those savings could be channelized through to the TIC.
- Some funds could be allocated from the village for the operation of the TIC, since the village administration will also be a significant user and beneficiary.
- Additional services can be offered for the villagers beyond the base set, which can result in more revenue.
- Advertisements can be shown on the screens to create a revenue stream from companies interested in reaching the rural markets.

By making computers available in schools at the point of delivery of education, TICs thus play a critical role in the facilitation of primary and secondary education. In addition, the same platform can be used for delivery of adult and vocational education.

The RISC centre, which would be within a distance of 10–15 km of the TICs, would function as a local support centre. The RISC is where teacher training can be conducted on a regular basis. Given the current state of the infrastructure in villages, the creation of RISC precedes or at least goes hand in hand with that of TIC in villages.

Increasing Market Access

Expanding market access for agricultural and non-agricultural output of rural areas is important for releasing the income poverty constraint. ICT tools such as the Internet and the world wide web are ideally suited to achieve this goal.

Much of rural India's production is handcrafted goods, the market for which is normally restricted to the surrounding areas alone. The demand is consequently very limited and therefore the price obtained is low as compared to the potential that could be realized if these goods were taken to a global marketplace. There are many instructive examples of small rural production fetching extremely high prices when exposed to international markets, using the Internet.[7]

The rural Indian economy, like that of all developing economies, is largely agriculture-based. Increasing the efficiency of markets involved in the inputs for agriculture, such as seed, fertilizer, pesticide and fuel would have serious implications for the incomes of farmers. The use of ICT can remove information imperfections and barriers that account for the inefficiencies in the market. So also, ICT can remove information problems in the output market. Both the input and the output markets can gain from the use of ICT tools. A partial list of the information requirement of the agricultural sector is:

- Prices of inputs and outputs
- Information on appropriate pesticide for specific pests
- Fertilizer use and application
- Weather information

How TIC and RISC Facilitate Market Access

An RISC provides those services that are usually available in urban areas to a rural population by concentrating resources in a centralized location in a rural region. In effect, the model concentrates economic activity at a location which is accessible to a large rural population to achieve agglomeration economies. It also achieves economies of scope by co-locating a wide range of user services on a standardized platform that provides shared access to basic infrastructural services such as power, telecommunications, the physical plant (building), water, sanitation and security. Finally, economies of scale are obtained at two distinct levels: at the level of the single RISC, by aggregating a large number of services, scale economies are obtained in the production and delivery of infrastructure; at the level of a large collection of RISC centres, scale economies are obtained by lower transaction costs and the sharing of large fixed design and development costs.

The RISC is the primary information and transaction access point. It facilitates market access by providing the following services:

- **Market making, market access and market information**: Efficient markets depend on information. These essential services have the potential to remove current inefficiencies and to expand markets so as to increase incomes.
- **Banking, finance and insurance**: Financial intermediation is critical to economic growth.
- **Agricultural extension**: Agriculture is the backbone of the rural economy. Immense opportunities for income generation exist.
- **Telecommunications, Internet and web access**: This is to bring the benefits of the revolution in information and tele-communications to rural India.
- **Marketing of consumer goods**: For the supply of consumer goods to rural areas.

Over time, as WiFi technology is able to provide linkages between the TICs and the RISC centre, the TIC will also become part of the network. This will ensure that technology comes as close to the people as possible, thus making itself a utility in their lives.

Conclusion

In this chapter we focussed on the use of ICTs for providing a two-pronged approach to the problem of poverty: education and access to markets. The former addresses an important component of non-income poverty, while the latter is concerned with income poverty. Primary education is widely recognized as one of the most significant factors that promote economic growth and social well-being. Lack of basic education is an impediment for economic growth. Access to markets directly affects the realized money incomes of populations. Both education and access to markets are necessary for economic growth and consequently for the elimination of poverty. Both depend on the use of knowledge goods and services to a large extent. While the use of ICT tools for addressing them is natural and obvious, there are important issues that need to be considered in this context.

To provide the ICT solution, we presented two models—TIC and RISC. The TIC depends on four building blocks: thin clients, server-centric computing, open-source software and WiFi. Using these blocks, we can build low-cost computing solution around the '5KPC'. A TIC is then a cluster of 3 to 10 such 5KPCs located in schools in villages to provide education and other computing-related services.

We also presented the RISC which serves the needs of a large population—say that of a 100 villages, or 100,000 people. Market access is one of the main functions (but not the sole function) of an RISC. Vocational education can also be delivered through an RISC.[8]

ICT deployed through RISC and TIC can help transform rural India by providing education and access to markets. The two applications should be considered as a major part of a complete programme for poverty alleviation. Our discussion was centred on India, but the applications are broadly relevant to any developing country.

Notes

1. For instance, about half the world's population, or about 3 billion people, have an average income of less than US$ 2 a day, and of that about 1.3 billion have a daily average income of US$ 1 a day. For India, the figures are even more stark: about 60 per cent of Indians, or 600 million people, live on less than US$ 1 a day.
2. World Development Report 2000/2001: Attacking Poverty, Oxford University Press.
3. Data from Government of India Expenditure Budget 2003–2004.
4. In 2003 the exchange rate was Rs 47 = US$ 1.
5. For urban areas, the male literacy level is 86 per cent and for females it is 73 per cent; the corresponding numbers for rural areas are 71 per cent and 47 per cent.
6. Data from Census of India 2001 and from the Azim Premji Foundation at http://www.azimpremjifoundation.org/html/statistics.htm
7. For instance, a small producer of 'mirror-work' neckties in Gujarat was able to advertise her product on the world wide web. A buyer from a department store in London came across the product and placed a large (relative to what the producer normally sold) order. The producer realized a handsome profit, and it was all based on a small investment in cash for the advertisement and a large investment in imagination.
8. There is need for demonstration of the concepts outlined here. It is our intention to create prototype RISC centers and TIC centers in multiple Indian states over the next 12–15 months. This will hopefully then work as the trigger for various Indian infrastructure and service organizations to participate in the rapid development and consequent transformation of rural India.

5

ICTs and Poverty Alleviation: Hope or Hype?[1]

Ashwani Saith

Introduction

Well over a century has elapsed since the advent of the Victorian Internet, in the form of the invention of the morse and the telephone; but still, about one-third of the world's population lives in poverty and till recently, about one-half has never made a telephone call. Could leapfrogging Information and Communication Technology (ICT) provide the solution, simultaneously linking the poor to the rest of the world, and providing avenues out of poverty?[2] Recently, Grameen Bank (GB), the originator of the other global panacea in the form of micro-credit for the poor, launched a project bringing ICT into rural Bangladesh through the agency of poor rural women, using its finance and institutional facilitation. Does this prototype crossing of micro-finance and the cellular phone signal a quick end to the scourge of rural poverty across the globe?[3]

Every microcosmic example of an Information Technology (IT) initiative in an unlikely deep rural location is picked up by its generators and protagonists and plastered across the web. Every online newspaper and related site in turn picks up the signal and re-transmits it to everyone tuned in once again. The awesome informational and dissemination power of the Net (to those on it) is immediately confirmed.[4] There is a need to deflate this virtual multiplier and arrive at some notion of real effects; in other words, to separate the hope from the hype in the potential of ICTs in overcoming rural poverty.

The unfettered optimism of technologists and corporate-related ICT sectors is not difficult to justify, given the phenomenal pace of change and the unprecedented possibilities it opens up. 'The truly revolutionary aspect of modern IT is believed to be the possibility it offers to unbundle information from its physical carrier' (Pohjola, 2001: 2). The information superhighway is 'about the global movement of weightless bits at the speed of light' (Negroponte, 1995, cited in Pohjola, 2001: 2–3). In turn, this aspect of weightlessness forms the key concept in the economics of the knowledge economy, where the ICT revolution introduces new possibilities of dematerialized production, bringing the consumer closer to driving the production process from the demand side. This partial escape from the constraints and limits of the material supply side is one of the hallmarks of the emerging weightless economy, according to its proponents.[5]

Development practitioners find other grounds for their optimism with regard to the potential released by ICT. United Nations Development Programme (UNDP) provides an optimistic but more balanced list of expectations in referring to ICTs as a significant factor in the growth and performance of economies as well as a 'novel and effective tool to help advance sustainable human development'.

> ICTs allow faster delivery and a more adapted content of technical assist-ance in a variety of sectors—ranging from long-distance education, to telemedicine and the creation of new livelihoods. ICTs can involve more people, hitherto unreached or underserviced, and accomplish a deeper geographic penetration, especially to rural areas, than is the case with traditional means and modalities. ICTs allow access to information sources worldwide, promote networking transcending borders, languages and cultures, foster empowerment of communities, women, youth and socially disadvantaged groups, and help spread knowledge about 'best practices' and experience. ICTs are indispensable to realize the global information society and the global knowledge society.[6]

Neither are non-governmental organizations (NGOs) far behind in identifying the transformative possibilities implicit in ICT. The M.S. Swaminathan Research Foundation, for example, a leading organization in this field in India, has launched the National Virtual

Academy (NVA) for Food and Rural Prosperity in August 2003, which is part of the ambitious initiative 'National Alliance for Mission 2007: Every Village a Knowledge Centre'. This 'virtual academy' aims to link farmers' organizations and village women to scientific and technological information, while communicating local insights and needs to the scientists.[7]

Following the meteoric projection and diffusion of micro-finance as a solution to poverty, a new global panacea for poverty alleviation seems to be taking shape. It needs to be noted though that unlike micro-credit, ICT is not a poverty-alleviation instrument in itself, whether in its motivation, or in its current usage. Thus, micro-finance made its entry into the development economy at the base of the pyramid, whereas ICT's injection is via the apex. This raises the issue of the channels and mechanisms through which the impact of ICT filters down to the lowest levels of the social pyramid of wealth and capabilities. What are the specific features of ICT that are poor-friendly? Will they come to the fore through open market competitive processes, or will they need non- or extra-market mediation? If so, by whom and in what manner? In this, what is the role of the poor themselves in actively acquiring the technology, or placing themselves in a position to benefit from it indirectly?

In pursuing these issues, it is useful to avoid grand generalizations in favour of specifying the particular linkages and mechanisms between ICT and poverty alleviation strategies. This chapter is not concerned with the potential role played by ICTs in the political sphere of civil society initiatives and interventions by or for the poor.[8]

Poverty is a global phenomenon, but it does not have one face or a single form. First: poverty is not just about food, material-related basic needs and income gaps. This is undoubtedly the core material expression of most forms of poverty, but it is not the only one. There is a need to incorporate other dimensions of deprivation, social (and spatial) isolation, marginalization and exclusion as well. In the final analysis, poverty is about powerlessness. Thus, poverty can be expressed in a common lexicon across societies, but the specific real-life counterparts need to be understood against the social context from which they are drawn. This emphasizes the relative nature of poverty, a dimension critical in any global analysis. Here, a caution is necessary: discussions of relative poverty do not travel well and conclusions drawn from one country context might be very misleading in a different setting. This is especially true in the case of ICT applications, since these solutions are globally used in 'an extraordinarily, diverse technological and cultural matrix that changes almost continuously', due to its innovative character often into ways unimagined at the moment of invention (Weber, 2003: 3–4). Thus, an identical ICT package is likely to work differently for differently specified groups of 'the poor' operating in different economic, structural and institutional environments. It follows that ICT could offer rather different prospects for poverty alleviation in the rich, as against the poor countries.

This chapter analyzes the complex interface between IT and poverty alleviation, which embraces several mechanisms and forms of transmission. Analytically, it is useful to distinguish between the indirect effects which are conveyed through the impact of ICT on the economy in general and the impact which arises from direct ICT usage by the poor.[9] Issues pertaining to the indirect impact of ICTs on the poor are reviewed in the section on 'Indirect Impact on Poverty'. The next three sections deal with the arguments concerning the central focus of this chapter, namely, the benefits that the poor are supposed to derive as direct users of ICTs. This theme is explored through the instrumental and illustrative use of some high profile, recent field initiatives that endeavour to bring ICT directly to the rural poor in developing economies. The final part of the chapter moves away from a narrow distinction between the poor and the non-poor to a broader focus on inequalities between and within nations. There is a marked tendency to ignore or overlook inequalities and to focus attention almost exclusively on the incidence of absolute poverty. However, citing the potential of ICTs to assist the absolute poor in the developing economies should not form the basis for sidelining the issue of the massive digital divide between the rich and the poor nations and within nations.

Indirect Impact on Poverty

The indirect impact of ICTs on poverty is briefly discussed at three levels: first, through its contributions to the overall economy; second, at a sector level through some general remarks on the possibilities this new and growing sector of the economy provides in terms of jobs and business opportunities for the poor; and third, by scrutinizing the 'facilitating' role of ICTs resulting in a more efficient service delivery, particularly for the poor.[10]

Via Aggregate Economic Performance

What is the general impact of ICTs on aggregate economic performance and growth, and through this, subsequently on the poor?

The extensive studies focussing on this issue—mostly in the developed economies—do not throw up a clear consensus. If anything, the central organizing hypothesis, following Solow, has been that of the 'productivity paradox', whereby the highly visible expansion of ICT remains invisible in economic accounting (David, 1990). Several alternative explanations of this paradox have been propounded: the diffusion-lag hypothesis, the mis-measurement hypothesis, the management-mismatch and the small-capital share hypotheses, and finally, one which argues that there really is no paradox in the first place (David, 1990; Pohjola, 2001: 20–24). Different results are obtained depending on the methodology and levels of aggregation of the data sets that are used, with positive outcomes reported in inter-temporal industry-specific studies.

Recent decades, in which ICTs have become increasingly important, are also marked by structural changes in the production process, in general terms, with a shift into knowledge, de-materialized, weightless, or intangible forms of value generation. These involve a sharply increased emphasis on learning and corresponding reliance on skilled labour, thereby increasing wage inequalities. This tendency is accentuated by the emergence of a low-skill, low-wage easy-entry, easy-exit service sector due to a parallel process of liberalization and deregulation of labour markets.

While many advanced market economies have displayed widening contours of poverty—understood in its relative sense—it is difficult to link this outcome causally to this pattern of structural change, since several other factors, including changes in government policies, are likely to have contributed to this result. On balance, such ICT expansion as has occurred and diffused to the poorer sections of society in the advanced economies has not been sufficient to overcome the observed persistence of levels of poverty. Such an aggregate judgement inevitably remains silent over the *net* impact of ICT on the incidence of poverty.

There is very little evidence concerning developing economies in this regard, and the relative contribution to GDP of spending on IT was nominal even in the more advanced developing economies.[11] However, in some cases, such as India, the impact of ICT software exports on total exports has been dramatic. In general, it is difficult to argue at present that ICT could have had any noticeable influence on poverty through its impact on aggregate macro-economic growth. This also implies that there are currently few fiscal dividends in the form of raised revenues (deriving from faster growth) that the government might be able to spend—if it was so inclined—on poverty alleviation projects.[12] More research using appropriate macro-economic methods is necessary on this aspect.

Via Possibilities of E-commerce

Does e-commerce provide new jobs and business opportunities to the poor?

This burgeoning market comprises a 'dematerialized', 'weightless' sector, where the 'product' is, say, a software programme or a piece of music or the latest chapter of a serialized thriller; more generally, it refers to the commercial exchange of goods where the location and interaction between buyers and sellers, as well as the final contract or transaction, is conducted on the Internet. When goods and services cannot be 'consumed' directly on the Internet, they involve the physical despatch and receipt of articles through conventional forms of transportation and delivery. The 'e' component is then limited to the informational dimensions of the transaction, that is, to market search, product information, and to buyer or seller identification.

Several aspects invite comment. First, this medium, especially in its dematerialized form, has allowed the utilization or exploitation of cheap brain power in developing economies for generating value in the developed. But even while inequalities within and between countries could well have been exacerbated, these 'wages of exploitation' have been very significant in several developing economies, including notably, India and Costa Rica. Three types of activities and labour are involved: cheap semi-skilled labour in Least Developed Countries (LDCs) providing data entry, accounting and book-keeping services relocated from developed economies; skilled labour producing software and other higher levels of knowledge products from LDC sites for the ICT industry located in the rich countries; and a third category, where the nature of skills demanded requires direct involvement in the production sites, thereby requiring a migration of highly skilled knowledge workers from such LDCs to the various silicon valleys in the development economies (Varma and Rogers, 2004). All three types have made a significant impact on foreign exchange earnings, through service income earnings, direct exports, or through remittances, further supplemented by inflows of foreign direct investment into the ICT enclaves and pockets that have emerged in these countries. The impact on local labour markets has also been non-negligible.[13] Therefore, at the aggregate level, some economies have developed strong linkages, though these form the exceptions; most developing economies are out of this loop.

From the point of view of the impact on poverty, there are two questions to consider. What is the profile of those *directly involved* in these three activities—are they generally poor? In general, the short answer to this must be negative. Even in the sweated component, for example, data entry/processing, the labour involved has to be fairly well educated, and while the households from which these (often predominantly female) workers are drawn are likely to have below-average incomes, these are unlikely to be below the poverty line. The other issue pertains to the *indirect impact* on the poor: such diffusion of benefits could occur through local linkages and spill-over effects in and around the ICT enclaves and pockets within the country, or through a beneficial impact on the general macro-economic performance of the economy, which in turn would need to trickle down to the poor. There is little empirical evidence available. Speculatively, it seems plausible that the local effects could well be significant via their expenditure multiplier effects stimulating the service, construction and related sectors. At the macro-level there are clearly powerful effects through the balance of payments; the weak link, perhaps, is whether there are effective trickle-down mechanisms to the poor.[14]

A second dimension of e-commerce is the exchange of (conventional) goods through the medium of the Internet as the virtual marketplace which facilitates the search, location and transactions between buyers and sellers—whether from business to business involving very large transactions, or to final consumers, including ordering your lunchtime sandwiches from the local deli on the Internet. This involves an internal structural change in a wide variety of industries and sectors, for example, advertizing, real estate, banking, travel, merchandizing. The net impact of this on employment is unclear, and the broad falsification of the earlier concerns that computerization in the services sector would reduce labour absorption cautions against quick generalizations. The particular question, whether any such internal restructuring involves a distributional shift away from wages and especially from poorer unskilled sections of the workforce, also remains an open question. It is unlikely, though, that the possibility of ordering lunch by the Internet will mean that people eat more, or that Amazon.com will imply that people buy more books. The net impact on expenditure and consumption behaviour in affected sectors needs focussed research.

To what extent does such e-commerce open new opportunities for poverty alleviation? While there could clearly be advantages for small-scale businesses and for geographically isolated (buyers and) sellers, these do not translate automatically into benefits for the poor. It is always possible for an international development agency or an NGO to facilitate the sale of the produce of the poor via the Net. But the issues are, whether this merely provides an advantage to some of the poor (so assisted) over the rest; whether such efforts are economically viable; and further, whether these can be replicated to scale without individual mediation in every case. The answers to these questions are not immediately encouraging and lead the debate back into questions of inequalities in effective access to ICT.

When viewed at the level of countries, another dimension has to be addressed. What are the pre-conditions for a successful e-transaction? Are these adequately met in the poorer developing economies? The seller has to be assured of payment in full and in time. The buyer needs a guarantee of the product meeting the attributes advertized. The former requirement is met, barring problems of Internet security, by the use of credit cards or other similar guaranteed payment instruments. The seller also ships out the products only after receiving the payment. For the buyer, there are greater hazards, in that, an Internet image might well not match product expectations for a variety of reasons; there could also be damage or loss in transit. The difficulties here are not entirely dissimilar to those connected with purchase via mail order catalogues. There is clearly a need for effective and reliable insurance mechanisms to cover such eventualities and also systems of penalties, enforcement and loss recovery. These services and guarantees are non-viable for anonymous distant buyers and sellers for one-off transactions. This has induced the emergence of e-markets, which then develop and act as enforcement agents of a set of rules that imitate the institutional requirements of conventional markets. But these e-markets have been legally resident in systems which have the requisite institutional capacities and the required market size to make the exercise feasible and profitable. While the elite in the poor countries, with access to international credit cards, might feel confident enough to place their trust, say, in a reputable e-market based legally in the USA, the opposite is a highly unlikely proposition. Further, there is the independent issue of the reliability and trustworthiness of the postal customs services, especially in the receiving country. This territorial and institutional dimension is another element of the international digital divide. In 1999, 90 per cent of all e-commerce purchases of the USA were made within the country; the corresponding figures for Europe, Asia-Pacific and Latin America were 59 per cent, 35 per cent and 26 per cent, respectively (Petrazzini, 1999). In 1998, the USA accounted for 93 per cent of Web revenue, and 70 per cent of Websites. Latin American, Asia-Pacific and all other developing and ex-socialist bloc countries together accounted for 2.4 per cent of the revenue, and 5 per cent of the sites (UNCTAD, 1999: 5).[15]

Thus, with regard to the link to poverty alleviation through the labour market impact of ICTs, the evidence suggests a substantial impact, but only on a specific and circumscribed section of the labour market—for example, the dramatic visible manifestation in the form of the heavy outflow of skilled ICT brain workers from India to the advanced economies. While there are likely to be strong linkage effects of this migration, these will be largely sectorally and spatially contained and even their wider effects are unlikely to affect the overall incidence of poverty in the country.[16]

E-government and E-services

The argument here is that the induction of ICTs has made processes of government and administration more accessible, transparent, responsive, time- and cost-efficient and accountable; applications to the functions of taxes, passports, records, bills, information, regulations, and so on, are the ones most often cited. An analogous argument is extended to the case of the services sector, including both economic (banking, travel, and so on) as well as social sectors such as education, health, welfare, and so on. When demonstrating the potential power of e-government and e-services, the cases cited are usually of well-performing governments; in India, the cases are usually those of Karnataka, Andhra Pradesh and Kerala, all of which boast several such initiatives. E-government is likely to reflect the biases exhibited by non-government in any political formation insofar as the role of bureaucracy, politicians and citizens is concerned. It is hardly surprising that in a highly politically aware and mobilized population such as Kerala, these initiatives work relatively well. To demonstrate the intrinsic transformative power of ICTs, it would be more convincing if their capabilities could be established in the context of, say, Bihar, or Uttar Pradesh.

There is an extreme paucity of studies which reliably establish the incremental impact of the induction of ICTs into these sectors, but anecdotes abound and tend to substitute for the former. A second difficulty in arriving at clarity in this area is the tendency to confuse e-government with e-governance. A third problem is the ubiquitous assumption that if some governmental functions can be processed indirectly by ICT use, the role of rent-seeking by government functionaries is *ipso facto* curtailed. And fourth, it is incorrectly assumed, implicitly or by default, that these services are likely to be relevant equally, and generate similar rewards for every citizen, that is, there is no difference between the benefits for the rich urban elite, and the poor excluded rural groups of the population—forgetting somehow that the poor do not pay house or income taxes, do not apply for passports, and do not look up detailed regulations for applying for government permissions, or make banking transactions, or download health or educational information from the Internet or from government sites.

No doubt, many public administration and private commercial services become much more efficient; the point here is that it is very difficult to assume from here that this efficiency premium will indirectly trickle into the pockets of the poor in any predictable or demonstrable manner.

The Poor as Direct Users and Beneficiaries of ICTs

What about the poor as direct users of ICT, whether as consumers or as producers? Could this be a route to poverty alleviation?

There are three separate questions that have to be addressed here. While the three questions are general, in attempting answers to them, it is necessary to differentiate and relativize the positions of the poor in a developed market economy as against a developing economy.

Access Constraints

First: What is the extent of access to ICT that the poor enjoy?

The answer to this question differs greatly across the world, reflecting the different circumstances of 'the poor' in the rich and the poor countries and is likely to be further complicated by geographical context (for example, urban–rural) and various social–cultural factors like gender, generation, caste, ethnicity, and so on. This raises the full range of access and threshold issues resulting in a digital divide both between and within nations that will be further discussed in the final part of this chapter. The following paragraph moves to the level of intervention, assuming that 'the poor' have gained access.

From Access to Content

Post access, what are the substantive content-based informational benefits that the poor can derive from using this access?

This raises issues of relevance with respect to the Internet content and usage—whose interests and demands shape the Internet content?

Judging by the current pattern of use of the Internet, the interests of the rural poor of the developing economies do not appear to dominate use. E-mailing, chat lines and pornography sites are the most common uses globally. But actual usage at this level is no indicator of untapped potential, since usage is demand-driven and is silent on the profile of supply-side availabilities. The seemingly endless possibilities of job-search, market information, telemedicine, distance learning, problem-solving, networking, quality-of-life enhancing information provision, and governmental access have been frequently invoked by the protagonists of this new anti-poverty instrument.

However, bearing in mind the substantial search costs, the monumental volume of information on the Net needs to be sifted and oriented towards the interests and needs of the poor. This orientation could require two steps: interpretation and augmentation.

Interpretation The first involves *interpretation*, since such information is either context-free, or other context-specific. Converting generic information into local-context relevant material requires an intermediate intervention.

Thus, searches on *Google* immediately produced the following numbers of sites for the keywords entered:

TABLE 5.1 **Information Overload on the Net?**

Search descriptor used	*No. of sites in 2002*	*No. of sites in May 2006*
Telemedicine	185,000	10,400,000
Distance learning	593,000	162,000,000
Wheat price	275,000	14,800,000
Rice price	450,000	38,900,000

(*Table 5.1 continued*)

(*Table 5.1 continued*)

Search descriptor used	*No. of sites in 2002*	*No. of sites in May 2006*
Urea price	18,700	1,940,000
Panchayat	34,200	1,740,000
Measles	134,000	13,300,000
Nosebleed	41,900	2,520,000
Snakebite	29,300	991,000
Snakebite (India)	1,040	81,300
Eye wash	358,000	22,100,000
Hog wash	28,400	1,650,000

The top emergency snakebite site in the 2002 search provided 16 detailed instructions of what to do in the event of a poisonous snakebite (though many of these involved branded medical applications and accessories); 11 instructions for what not to do; six points on what reporting should be done at the hospital.[17] This leaves one with a final question: when bitten by a snake, should one race to the Internet, or to a trained village medical worker? This assumes, of course, that there is choice in the first place.

Augmentation and Value-addition The second is *augmentation* or *value addition*, whereby additional types of information that are particularly relevant in the local context or for specific target user-groups, namely, the poor, are made accessible on the Net. This is a critical step, since it leads to purposive additions and sites which directly cater to target-group needs and demands. It influences who uses the facility; but it also reflects who 'owns' the facility in the first place, putting the most marginalized in a very unfavourable position.

It needs to be emphasized that the conveyance and use of information occurs within a live social space; it involves a transaction between parties who themselves occupy embedded, if not entrenched, positions often characterized by conflicts and opposed interests in local society. The generation, transmission and use of 'information', therefore, can hardly be a neutral, objective, scientific exercise. There are several links in this chain. First, these needs and demands need to be articulated, aggregated, refined and prioritized. This involves social processes and calls for participatory approaches; the weaknesses of these methods in engaging the poorest are increasingly well-documented. Second, this information needs to be interpreted and conveyed to the potential user. This could be simple for an educated youth, but could take the form of a tutorial if the illiterate and disempowered poor are addressed by the facility managers. It could also be quite time-consuming, especially if the enquiry is customized and not general in nature; and there may be language, gender, cultural, and class and caste biases to overcome in this social transaction.

The answers with respect to content suggest that from the vantage point of the poor in developing economies, the pattern of availability on the Net is likely to reflect distant priorities. There is also likely to be an information overload. This calls for a need for filtration and sifting on the one hand, and for the purposive generation of dedicated websites on the other. Such dedicated sites in turn are likely to generate content reflecting the relative political weight and social priorities of local power structures, with the interests of the poor being peripheral to the exercise. There are likely to be difficulties upstream, since the poor are likely to suffer acute obstacles in access. To the extent that these are somehow mitigated and the poor do manage to gain access to relevant content, the focus shifts to the third dimension: the effective utilization of such informational content.

From Information Access to Transformation of Livelihoods

Having accessed the Net and located useful information in the form of de-materialized weightless bits, how can this information be used to transform the real material conditions of existence of the poor? At some point, the weightless intangible idea has to transform itself into a material object or tangible effect in order to impact upon poverty, unless of course poverty is defined altogether as a gap in ideas-space rather than in terms of objects, commodities and services.[18] What is the power of information as information?

This is where perhaps the greatest challenges arise. There are several problems to contend with. Does the knowledge of superior dietary practices provide a sufficient strategy against hunger? Are the causes of poverty and deprivation and avenues of improvement, unknown to the rural community and to the poor—what is the role of external knowledge here? Does information about safe child-birth practices, or the need for immunizations act as a substitute for village-level medical facilities and the availability of medicines? The number of questions can be multiplied at will. They all query the implicit fundamental assumption embedded in much of this discourse that information is the key missing gap[19] and the provision of adequate information through appropriate institutional environments and mechanisms constitutes a sufficient strategy for overcoming social and economic constraints at the grassroot level. This implies a somewhat disempowering if not patronizing view of the knowledge of the poor with regard to their own condition and location in the fabric of society. Moreover, it sheds critical light on the 'simple and cheap road' of development through non-competitive, welfare-enhancing uses of ICT relying on older-generation hardware.

In the following sections, some rather distinct types of interventions will be considered as illustrations of the potential that ICTs hold for poverty alleviation and empowering the poor. While each reflects a very different approach, they are not alternatives and could well be combined as elements of a more integrated strategization of the role of ICTs *vis-à-vis* poverty. In discussing the interface between each approach and rural poverty, the three sets of issues listed earlier need to be borne in mind: the first concerning access to the technology; the second, relating to nature of knowledge-augmentation that is enabled by such access; and the third pertaining to the translation of such additional knowledge into empowerment and poverty alleviation at the individual, household or group levels.

The arrival of pioneering prototypes is usually heralded by strong claims of success, and in an arena where successes are at a premium, these usually also attract international attention, support and finance. The first intervention discussed is a programme of Village Knowledge Centres (VKC) initiated by the M.S. Swaminathan Research Foundation (MSSRF) in rural India. This intends to bring the knowledge banks of the Net to the heart of the Indian village through setting up of ICT capacity within selected villages. The second ICT intervention, the Grameen Bank's Village Pay Phone programme in rural Bangladesh, focuses not on access to the Net-knowledge, but on creating communication links with the external world through which the villagers stand to improve their economic situation or their well-being. The key instrument here is not the computer, but the mobile telephone.

The VKC Initiative of the M.S. Swaminathan Research Foundation (MSSRF) in India

This IDRC[20]-funded project of the MSSRF embedded in a 'pro-poor, pro-women, pro-nature orientation to development', aims at delivering electronic knowledge to the poor,[21] which is locally relevant and inclusive of the needs of both poor men and women

(Arunachalam, 2004: 961). Since 1998, information shops, now re-labelled as VKC, have been set up in 10 villages (as of 2005) in the Pondicherry region.[22] Plans are drawn, under 'Mission 2007—National Alliance Every Village a Knowledge Centre', to extend this experiment, as a joint collaboration of several initiatives and organizations, to more than 600,000 villages in India by the year 2007, which marks the 60th anniversary of India's independence (Chapman, 2004: 7).

In the context of deep rural locations in developing economies, some insight can be gained into the question of access for the poor by considering the experience of the MSSRF Village Knowledge Centres initiative in the Pondicherry region. A UNESCO report[23] states that as of June 2000, a total of 15,651 users have visited the VKCs, of which 2,832 (18 per cent) are women; 4,571 (29 per cent) landless; 392 (2.5 per cent) illiterate and 3,674 (23 per cent) below the age of 14. Out of this grand total, nearly a quarter of the visitors (3,674) were the so-called 'one-time users', a relevant question would be whether these stem predominantly from the poor or the non-poor group. Furthermore, since women and illiterates are clearly under-represented, this suggests that the poor are unlikely to be the dominant users.

However, these statistics do not provide satisfactory answers to a range of relevant questions: what was the frequency of use of the poor *vis-à-vis* the non-poor? What is the comparison in terms of time usage and repeat users? Who are the individuals from the poor families that use the facility: the school-going young adult male? And are these drawn from the poorest families or, as in the case of most other projects, disproportionately from the least-poor of the poor? What are the trends: are there drop-outs, and from which sections? And what are the patterns of use in terms of content, that is, expressed need, for the poor and the non-poor? What is the rate of participation of the Dalits, the untouchables? Do users have more years of schooling? Are they drawn disproportionately from some age groups? In view of the special efforts at placing women operators for the facility, what percentage of the users, the poor and the non-poor, are women?

With regard to the issues of content, value-addition and use, what initial insights can be elicited from the early experience of pioneering prototypical interventions in this area? A good example is provided by the MSSRF Village Knowledge Centres. Inhabitants of the villages covered by the project can receive a variety of information gathered or collated and put on the net by the Value Addition Centre located in the hub village. What is this information? How is its provision likely to influence the lives of the poor inhabitants of these villages?

The following 18 items or categories of information have apparently been made available through the dedicated website.[24] These have been grouped here, according to affinity, into six groups.

- Advice on ecological ways of growing local crops and protecting them from diseases.
- Pest management plans for rice and sugar cane.
- Daily market prices for these crops in different markets in the project area.
- Costs and availability of agricultural inputs, for example, seeds, pesticides, fertilizers.
- Directory of insurance plans for both crops and families.
- Directory of local veterinarians, cattle and animal husbandry programmes.[25]

All these items of information are of interest and relevance mainly to landowners, not to the rural poor who are overwhelmingly landless, whether as agricultural labourers, or impoverished service and artisanal groups.[26]

- **Local weather forecasts:** These are picked up via the Internet from the US Navy and transmitted to the villages which have large fishermen communities. This helps them to avoid dangerous seas and to plan their outings. This is clearly a useful service, especially when it is broadcast over an audio system across the fishermen's village.[27] However, there does not appear to be exclusivity with regard to the use of the Internet, per se.[28] A community radio-station could suffice, and indeed this could warn fishermen at sea as well—all between favoured programmes of music, news, and so on. This is especially relevant for poor, traditional fishermen who could encounter changing weather conditions from day to day at sea. While the benefits of accurate and timely weather warnings on the radio (or the Internet) are obvious, it should not be overlooked that many poor fishermen go to sea even in the presence of considerable risk for lack of choice—they have families to feed.

- **Regional timetable for buses and trains:** This is passive and a relatively neutral information. Usually, the rich are more concerned about precise timings than the poor; they also use these facilities rather more for business in neighbouring towns. A notice board anywhere in the village could be more accessible.

In any event, the reliability of such services being such as it is in India, the intrinsic value of such information is a little in doubt.

- Women's healthcare.
- Child bearing and rearing.
- Targeted dietary information, based on an analysis of local nutritional deficiencies.
- A directory of local hospitals, medical practitioners and their specialities.

There might well be very useful items of information here for women in general, from which poor women could potentially benefit as well, subject to other pre-conditions being met.

- Information on about 130 government welfare programmes to aid the poor.
- Government lists and details of families eligible for benefits and living below the poverty line.

The latter item is in English, and therefore not really accessible to most of the population, especially the poor. Even if it was accessible, it is difficult to see what precise use the poor could make of it. The former, effectively a compendium on government welfare programmes, again seems like very useful general information if its provision enables some poor people to register with some of these programmes. On the whole, it is unlikely that 130 programmes would be running in any particular village! As such, from the point of view of any poor family in a village, most of the information might be irrelevant, and the Net content for them could well be limited to a few schemes which operate in the village and about which the poor have direct knowledge of modes of access and operation.

- **Copies of documents and statements from local official records, for example, land registration:** Such needs and uses are again likely to be strongly biased in favour of the non-poor. As far as print-outs of land records are concerned, the problem in rural India is to do with the control, manipulation and corruption surrounding the use and misuse of official records. The poor usually are helpless onlookers in this process anyway.
- **How to start up new family enterprises, such as manufacturing incense sticks:** No further details are provided. Capital and entrepreneurship for such manufacturing enterprises are almost

invariably drawn from the rural or peri-urban non-poor households. It appears almost fanciful to argue that the Internet would provide the missing link and form the basis for the poor to enter such economic arenas.[29]

- **Cricket information is much sought after through well-known websites:** This seems closer to Indian reality and, along with film, pop music and dance, perhaps forms the common fabric of an ever-widening swathe of Indian society, including the rich and the poor. That said, and the obvious arguments about the importance of catering to information concerning culture, the quality of life, and so on, notwithstanding, such information—sought usually when there are matches on—is more directly available through the virtually non-stop television coverage of cricket in India. Further, even if cricket is argued to be a bond (age) of Indian nationhood, it is still the preserve of the middle and upper classes. Information on how much of the total usage was for cricket is not known.
- **Results and marksheets of 10th and 12th standard school examinations for 931 students:** This saved a week of anxiety for these students resident in or near the project villages. Clearly, information which improved the quality of life rather superficially in the short run, it did not manage to change the numbers on the score sheets, or the numbers of boys and girls from poor households who got to school, let alone to Class 10 or Class 12.

What about the 'transformative capacities' of the use of such Net-knowledge? In terms of content, then, local value-addition can be seen to have a potential for improving the quality of life in various ways for various categories of villagers. However, its impact on the lives of the poor, and on poverty alleviation, is more difficult to demonstrate at present.

Jayalakshmi, the milch cow, has virtually become a household name after her life was saved in calf-birth, by a vet located on the Net; and Minakshi, an old woman, was saved from blindness. These successes have served the needs of advocacy and media projection, but unfortunately, there seems to be a shortage of any systematic independent evaluations of the initiative, especially in terms of its projected economic viability, and its impact on poverty alleviation.[30] And laudable as such an intervention is, the objectives, orientation and operation of the project also raise some unsettling general issues concerning the social location of, and control over the access, interpretation, dissemination and use of such knowledge. This touches on the ongoing debates over the domination of indigenous knowledge by external forms congealed or concealed in projects of modernization that obliterate the indigenous, implying also a loss of control over their lives and paths of social and cultural reproduction.

The Village Pay Phones (VPP) Initiative of the GB in Bangladesh

The VPP programme is an initiative of the GB, and it attempts to cross the demonstrated power of its micro-credit approach with the development potential of ICTs. It rejects the notion that the rural poor cannot exploit the advantages inherent in the new ICTs, and also that it is too expensive to hook them into the information loop.[31]

Since 1997, the GB, through its VPP initiative, has been providing mobile phones to selected rural poor households—usually female members of the GB—for the sale of telephone services to the village population in general.[32] The profits of these are expected to directly relieve poverty for these households; and the benefits of communication are expected to indirectly contribute to poverty alleviation for the other rural poor residents of the village.

However, delivering telephony to rural Bangladesh is by no means done for philanthropic motivations solely; it is above all a sound business scheme (Tambo, 2004: 8). The GB has created two telecom companies. Grameen Telecom, a not-for-profit company which administers the VPP, and Grameen Phone, a for-profit company which owns the network and holds a nationwide licence to operate and maintain mobile telephony in Bangladesh (Tambo, 2004: 10).[33] Alongside the various socio-economic benefits that technology may bring, there are considerable economic motivations for this Business–NGO partnership to tap the market for rural telephony, which constitutes 70 per cent of the population and at least 45 per cent of the economy hitherto ignored by established telephone operators and tapping into the lucrative long-distance market created by the Bangladeshi diaspora (Sobhan et al., 2002: 16).

The discussion that follows is limited to the VPP project of Grameen Telecom, largely based on information provided by three substantial evaluations of its progress.[34] Richardson, Ramirez and Haq, in their 2000 study (RRH00), list the many rural development 'firsts' of the VPP programme. It constitutes the first intervention for micro-enterprises based on ICTs; the first use of micro-credit programmes for diffusing digital wireless telephony; the first private sector rural telecom programme that targets poor village women as micro-enterprise entrepreneurs; and the first private sector telecom initiative 'with an explicit purpose of rural poverty reduction'. Clearly, the VPP project provides an innovative model. Bayes, Von Braun and Akhter, in their 1999 study (BBA99), conclude that 'the possibility that relatively poor people can also own and access modern IT may go a long way towards reducing poverty and restoring a more equitable distribution' (BBA99: 44). This evaluation offers 'two basic conclusions: first, pursuance of pragmatic policies can turn telephones into production goods, especially through lowering transaction costs; and second, the services originating from telephones in villages are likely to deliver significant benefits to the poor. VPPs also seem to have perceptible and positive effects on the empowerment and social status of phone-leasing women and their households. For villagers in general, phones offer additional non-economic benefits such as improved law enforcement, more rapid and effective communications during disasters, stronger kinship bonding, and so on. GB's style of managing communications can help significantly to expand access to this vital information input for all segments of the population, reduce inequality and thus enhance the broad-based, pro-poor orientation of rural development activities' (BBA99: 1). With minor variations, the second evaluation report (RRH00) underscores these conclusions (though often citing the findings of the previous report in support).

> The marriage of GB's revolving loan system and network of branches with cellular phone loan scheme is unique and working very well; Grameen Telecom staff are committed and providing excellent service to users; there are significant positive social and economic impacts, including relatively large consumer surplus and immeasurable quality of life benefits. (RRH00: 48)

An empirical validation of the professed orientation and claims made in terms of the impact on the rural poor would be significant and reassuring. Does the VPP bridge the digital divide at the lowest level?[35] How potent a weapon is it in the struggle for overcoming rural poverty?

With regard to the 'pattern of access', the VPP programme has to be divided into three clear categories: the 'owners of the village mobile phones' who operate the commercial service within any village; the service buyers, namely, the 'phone users in the village' who pay for using this mobile phone service as and when they want to; and the residual population, that is, the 'non-users'. Of course, the phone owners are also phone users. BBA99[36] and Sobhan, Kaleque and Rahman, in their 2002 study (SKR02),[37] provide empirical evidence from their respective field-survey sites in rural Bangladesh on the socio-economic profiles of the owners, users and non-users (though the last category is not included in the design of the BBA99 enquiry). The findings are summarized in Table 5.2.

Although absolute amounts and measures used differ between the two studies, a clear pattern emerges, with a stark contrast between users and non-users with regard to income, land, fixed assets and literacy rate, while the operators float somewhere in-between in income terms, but rank top of the bill in terms of literacy and sanitary health status. In relation to the wider context, using the BBA99 data, it shows that the operators' average per capita income of US$ 273 was just slightly below the average for all Bangladesh of US$ 280, while the group of 'users' with average US$ 451 per capita per annum, was about 60 per cent better-off than the average Bangladeshi. As such it seems unlikely that VPP operators belong to the group of 'poor', while the users could very well come from the better-off, if not rich, households, a conclusion which finds empirical support in the SKR02 study.

TABLE 5.2 **Household Characteristics of Phone Operators, Users and Non-users**

	BBA99 (p. 16)		*SKR02 (pp. 26–27)*		
Characteristics	*Operators (n = 50)*	*Users (n = 356)*	*Operators (n = 24)*	*Users (n = 72)*	*Non-users (n = 24)*
Average size of household	6.08	6.18	4.80	6.50	6.26
Annual per capita income $	273 (average)	451 (average)	462 (mean)	506 (mean)	320 (mean)
Average per capita land	0.07 ha	0.16 ha	0.69 acre	1.02 acre	0.57 acre
Percentage who are					
Extreme/very poor	0%	6.1%	0%	0%	0%
Moderately poor	14%	9.3%	4%	16%	37%
Non-poor	86%	84.6%	67%	43%	46%
Rich (only in SKR02)	Not specified	Not specified	29%	41%	17%
Literacy rate	94.0	89.3	78	64	49
Mean age (only in SKR02)	–	–	32	41	47
Average amount of fixed assets (US$) (only in SKR02)	–	–	680	1,200	590

Sources: BBA99: 16 and SKR02: 26–27.

Could it be argued that the financially superior position of operators, compared to the group of non-users, or to overall poverty incidence in Bangladesh,[38] is the direct result of the VPP? Aminuzzaman reports a huge increase in income due to the ownership of the VPP, such that a few months of operation could well see a doubling or trebling of average income for the poorer end of VPP-owning households. However, the data for incomes before the ownership of the payphone also show that most owners were clear of the poverty line (Aminuzzaman, 2002: 169). A closer look at slower moving indicators gives a better idea of the situation prior to VPP, and hence to what extent VPP can claim to have lifted 'the poor' out of poverty. The literacy rates are superior among operators and users, compared with non-users and with an overall adult literacy rate of 30 and 49 per cent for female and male, respectively, in 2000 (UNICEF, 2005), with the figure for the rural poor women probably considerably below this level. In addition, BBA99 shows that all operators' households had electricity and the use of tube-well water for drinking purposes; 80 per cent had sanitary latrines, and had had their children immunized; these rates being generally superior to those characterizing the richer user households (BBA99: Table 4.1, p. 16). It is further worth noting that the VPP-operating households on an average own 1.2 acres of land and also that at the time of the survey, no one of the heads of these households engaged in agriculture or non-agriculture labour as an occupation (BBA99: Table 4.2, p. 17), thereby setting them in dramatic contrast to the occupational profile of rural poor households in Bangladesh.

VPP owners are selected from the GB membership, and since this is 95 per cent female, most owners are women. But those selected have to meet a set of special criteria. First, they must have had a very good record of repayment of earlier GB loans; second, 'they should have a good business, preferably a grocery store, and have the spare time to function as the VP operator'; third, they should be literate, or at least have children who can read and write; and fourth, their residence should be near the centre of the village (RRH00: 9–10). In addition, they should obviously be in the good books of the local bank manager, since he is asked by the Grameen Telecom manager to recommend appropriate candidates based on their prior credit rating and their ability to successfully run a small business venture. It would indeed be remarkable if any rural poor woman at all met these criteria, and it is thus not surprising that below-poverty line households are virtually absent in the group of phone owners. And should there have actually been the odd such household, it is very likely that it was in temporary income poverty, or was only marginally in poverty. Of course, once an owner, the profits, associated status and opportunities, propel such households on to a steep upward path of socio-economic mobility well beyond the threshold of the poverty line. For the few who own it, the mobile phone becomes the instrument of acquiring relative affluence more than a lifeline to escape from absolute poverty.

Paradoxically, a prominent example held out—as usual in the form of an individual case history substituting for systematic data-based analysis—for proving the potential of the VPP for poor rural women is that of Anwara Khatun, who has done brilliantly well from the business opportunity provided by the pay phone scheme (PovertyFighters.com, 2003). But at the time when she was selected for the VPP scheme, was Anwara the median poor rural woman, even if not drawn from the poorest of the rural poor? It would appear not: she had already received 13 general, five seasonal, as well as one housing loan. Her eldest son was working in Singapore and sending in very substantial remittances, the middle son had completed a B.A. (Honours) degree, and the youngest son had his Secondary School Certificate. Her family was probably light years away from the destitution experienced by the poorest of the poor. On top of her considerable income, Anwara went on, from her participation in the VPP scheme, to make a lot more bucks. This is all excellent, but the only difficulty is that it is very unconvincing as proof of the poverty-alleviating capacity of the VPP scheme. Regrettably, somewhat analogous problems arise with the proof of the empowerment theorem. Aminuzzaman remarks in this respect that despite the fact that 91.8 per cent of the VPP owners are female, in his study only 4 per cent of the VPPs are operated by these female owners, the remaining 96 per cent by male family members. Furthermore, only 22 per cent of the VPP users are female (Aminuzzaman, 2002: 171); the figure is 30 per cent in BBA99 (Bayes et al., 1999: 20).[39]

What are the uses, and the consequent gains, for the many who only use the phone by paying for its services? There are varying pictures sketched here.

BBA99 lists a wide range of economic and quality-of-life benefits (Table 5.3).[40] Each of these is backed-up by some, often scant, anecdotal, evidence or story. While each of the channels listed could potentially have the declared effect for some households occasionally, the question is what all these add up to and what their social incidence is. All these need to be viewed especially from the point of view of the poor in the village, who, as it happens, are very poorly represented in the category of phone owners, or as phone users. This suggests a strong pattern of the limitation of the specific-use benefits of the telephones to

TABLE 5.3 **Assessment of Selected Benefits**

Variable	*Target village (n = 50; averages)*	*Control village (n = 10; averages)*
Prices		
Paddy (percentage of final consumer prices)	70–75	65–70
Eggs	Tk 13/hali	Tk 12/hali
Exchange rate	Tk 12.50/ryal	Tk 11.50/ryal
Cost of information	Tk 17	Tk 72
Chicken/Ducks	Higher	Lower
Chick feed	Lower	Higher
Supply of inputs		
Diesel	Stable	Fluctuating at imes
Fertilizer	Regular	Occasional problems
Others		
Poultry mortality rate	Lower	Higher
Law and order situation	Improved	Same
Communication during disasters	Quick, effective	Slow, less effective
Communication with relatives at home and abroad	Anywhere, any time, any day	Anywhere, but fixed time, fixed day
Transmission of new ideas	Improved	Same
Mobility of people	Higher	Lower
Spoilage of perishable products	Less	More
Access to healthcare services	Faster/effective	Slower/less effective

Source: BBA 99: 31.

the haves of the village, with access to communications acting usually as a multiplier effect on existing endowments and capacities.

Virtually all available studies on rural telephones mirror the findings for Bangladesh: ownership and access, usage and claimed benefits, all accrue disproportionately to the richer, more educated, externally connected, land-owning households, whereas the poor households are weakly registered as phone users. Being disadvantaged in several respects gives particularly the poor very little reason to call. 'Having no one to call' was the main reason (78 per cent) for being a non-user in RRH00 (p. 24), a trend also observed in SKR02 (Table 5.4).

TABLE 5.4 **Reasons Why the Non-users do not Use the Phone**

Phone not available	*Phone too expensive*	*No access to the phone*	*Queue too long*	***Nobody to call***	*Do not know how to use*	*Others*
0	16.7	20.8	0	**45.8**	0	33.3

Source: SKR02: 39.

Intermediaries as Access Bridges and Benefit Conduits

The two programmatic approaches discussed earlier can be contrasted with a third type of process through which ICTs interface with rural poverty and the rural poor. This third approach is not really a unitary project intervention, but more an organically emerging process in which a range of intermediate development organizations use the potential of ICTs to enhance the range, quality, responsiveness and reach of their interventions in favour of the poor. Here, the poor are indirectly benefited through the strengthening and empowerment of the link agencies and organizations working for them.

In this alternative perspective, such information provision can act as a stimulus only within the framework of social mobilization of the poor, and then alongside a process of overcoming the structural constraints which impede their life chances. In this context, the excessive optimism over distance learning and telemedicine needs to be moderated. ICT devices can easily be visualized as powerful tools in augmenting capacities of functioning educational and health infrastructures, but it is unrealistic to regard these as substitutes.

These link institutions, groups, or organizations could be of various kinds. For example, ICTs have the possibility of enhancing the mobilization, organizational networking and operational capabilities of a powerful member-based body such as SEWA in India. It could enhance internal communications capabilities; shorten the time-frame for problem recognition and problem solving significantly; all facilitate its linkages with the external world for purposes of networking, advocacy, finance-raising, and so on. In turn, there could be a significant general gain in the form of external effects whereby other organizations can gain information and interact meaningfully with SEWA and exploit its cumulative expertise.

Differently, within well-functioning village councils, ICTs could provide the basis for local institutional capacity building through networking possibilities with other councils. Such networks could be put to various uses, again ranging from programmatic interventions in poverty alleviation, to good local governance initiatives, to problem solving and experience sharing with regard to specific development interventions at village level, to the maintenance of a local but linked database which allows local bodies to monitor the progress of various development processes and programmes. It needs to be stressed that such potentialities are difficult to exploit where the local institutional and governmental structure is dominated by the rural rich. In addition, reliance on proprietary instead of open-source software also works as a potentially excluding factor. A certain degree of prior democratization and effective mobilization and organization of the rural poor at the local level could be a pre-condition for such pro-poor uses of ICTs within a village.

However, when such conditions are only partially met, the networks of external (to the village) development organizations and agencies, whether government or non-government, could use ICTs effectively in providing support to distant development workers operating in isolated environments. Ready access to information and to knowledge resources could enhance their capacity to subsequently deliver improved services to the poor.

The VKC and the VPP initiatives, constrained as they might appear to be at present in this regard, nevertheless can be seen to be foundations for the dissemination of ICT capacity to the rural sector. They could well be a significant first step which opens the path for more significant and wider uses of ICTs further down the development road. Unfortunately, both initiatives make remarkable claims with respect to their impact on the rural poor. Such exaggerated declarations test credulity, especially since they rely almost exclusively on evaluations that are either internal to the organizations, or sponsored by donors who underwrite the projects. While the general welfare-enhancing effects of such initiatives cannot be doubted, their magnitude and limits are often overstated. Most of all, the linkage to the lives of the poor is made more by assumption rather than by clear demonstration. Here, all too often, the general case is argued on the basis of a single positive intervention narrated in the form of an anecdote. There is a danger in such cavalier use of the method in assessing impact, while the outcome might be more beneficial for the projects than the really poor, from whose name these projects draw their motivation, their legitimacy, as well as a good deal of their finances.

In general, the third approach rests on the premise that Net-knowledge and mobile phone communication are not sufficient in themselves as substitutes for the wherewithal of development, namely, jobs, incomes, clean water, healthcare facilities, well-functioning rural schools, and so on. Knowledge and communication tools could significantly enhance the capacity of the ongoing mainstream development processes addressing these real issues; but the virtual cannot in the end substitute for the real.[41]

The Digital Divide: A Bridge Too Far?

After having looked in detail into claims and achievements of two ICT programmes, this final part of the chapter addresses the wider context within which these programmes operate. As argued earlier, empowerment comes from effective usage, not from simple access to ICT infrastructure. But without access, there is no use—access is thus the first of a two-stage process. In what terms can the digital divide be profiled? What are the underlying factors which determine the observed cross-sectional patterns at various levels? And on the basis of which factors can one prognosticate over these gaps in the medium term? A very stylized picture is sketched here, with the accent on identifying the underlying explanatory factors and processes, rather than on an empirical measurement of the various gaps that together constitute the digital divide.

The Digital Divide: Some Stylized Findings

The gap in computer and Internet use can be attributed in the first place to differences in per capita income, closely followed by 'disparity in telecommunications infrastructure' (Chinn and Fairlie, 2004: 22), resulting in differences in telephone connections; ownership and/or access to computers; density of mobile telephone and higher generation information-communications instrumentation; density of the Internet subscription and use; and volume indicators of e-commerce trans-actions. The central underlying concept is technological diffusion, with the logistic curve the single most comprehensive summarizing statistical device for profiling and comparing diffusion across units and over time. Each particular technological innovation has its own curve for a specific population and ICT innovations generate the same S-shaped patterns of diffusion—the differences between them are captured in the parameters of the logistic curves applicable to the respective populations.

It is necessary to observe some simple rules for comparison in order to avoid misleading deductions about widening or narrowing of gaps and about future expectations. After an initial period of delay during which the gap with the leading group widens, the late-comer group could enter a period of rapid diffusion. This phase could well overlap with the tendency in the leading group for further diffusion rates to slow down as near-saturation levels are reached. This would imply a narrowing of the absolute gap. But just as the earlier widening could not be read as a indication of continual stretching of the divide, the later catching-up cannot be interpreted as proof of convergence, since the different populations could have varying long-term 'equilibrium' levels of adoption. Also, it is helpful to note the position of the specific innovation being used in the comparison within the sequence of the adoption of a chain of related innovations. This is important, since the early adopters might well have moved ahead on to the next stage even as laggards catch up with them on the previous one. In such a case, data on the late phase of the former cycle could be read as evidence of a convergence, that is, of the original laggards having successfully caught up, while in reality, the pioneers have already moved a few rungs up the next ladder. This also introduces the possibility, much debated, of late-starters accelerating the catching-up process by leapfrogging intermediate stages and jumping on to more advanced technological platforms. In this context, the length of time taken in narrowing the gap, catching up, or leapfrogging is of great importance, since the leading group is able to gain cumulative competitive advantage over the laggards over this period; so, when eventually the late-starters do 'catch up' in terms of diffusion indicators, they might find that inequalities had widened considerably, and that they are still, and perhaps even more disadvantageously placed with respect to the next innovation diffusion cycle. Much depends on the locus, pace and nature of the process of generating technological innovations and breakthroughs; in turn, this process has to be viewed within a historical and structural context.

The empirical evidence, revised as it is continuously in order to keep track of a fast-moving target, all confirm the existence of a chiasmic divide: this applies to the different elements of ICTs; and then for comparisons between continental regions; within advanced and poor economies; within each country to the enormous gaps between rural and urban populations; within urban regions to the wide divide between the megapolitan centres and large cities on the one hand and the small towns on the other; within cities to the different categories of suburbs that house different social groups. The divide can also be specified in terms of ICT access and use gaps between various social categories of the population within any given system: thus, between households with different levels of income and assets; different levels of education; differences in occupations; by language, as English continues to dominate usage and software development (Hariharan, 1997); by racial/ethnic group; by gender; and by age. On the whole, the pattern of inequalities with respect to ICTs mirrors the wider cumulative patterns of social privilege and exclusion.[42] Technology-optimists then have the onus to demonstrate how ICTs could both bypass this inherited and oft-repeated pattern, and then, for the extreme optimists, become an instrument for reversing such entrenched socio-economic and power inequalities. Evaluating the likelihood of such possibilities calls for a closer scrutiny of access constraints at different levels.

Unpackaging Access: Factors Determining Threshold Requirements

The concept of access needs to be unpackaged if it is to have a meaningful and operational interface with the design of appropriate policies addressing the digital divide. The starting point of such an exercise must be recognition of some attributes of technological change, especially as embodied in physical capital and infrastructure. Physical capital is not malleable, not putty-putty, in real life; it is usually lumpy in terms of scale and is characterized by significant indivisibilities; it is not universally usable, in that it could require considerable prior orientation and training; it is not usable everywhere, since it could require various infrastructural pre-conditions for effective use; and it is usually relatively expensive, that is, has to be capitalized over a lengthy life span.

Put together, these imply that technological packages could have strong pre-conditions for entry and use, and these determine the pattern of access to different potential user groups. These pre-conditions apply generally at the level of the system, that is, society or economy, and further down the line, separately, at the level of the individual (or small-group) user.

Factors at the System Level

- Infrastructural pre-conditions for adoption: telephony, connectivity.
- Institutional requirements: banking, credit cards, postal services.
- Skills pool: of users and trained workforce.
- Markets: trust, verification, redress; real and legal possibility; transaction costs.
- Policies: customs rules and regulations; nature of market regulatory frameworks; and so on.

Table 5.5 shows the marked differences with regard to telephone and computer density, which are both heavily dependent on infrastructural pre-conditions such as electricity, fixed lines and/or wireless networks.

TABLE 5.5 **Telephone and Computer Density in Selected Countries**

	US	*UK*	*India*	*Bangladesh*
Telephone main lines per 1,000 people in 2001	667	592	38	4
Mobile phones per 1,000 people in 2001	451	770	6	4
Personal computer per 1,000 people in 2003*	658.9	405.7	7.2	7.8

Source: World Bank, 2001.
Note: * Data from Telecommunication Union's (ITU) 'World Telecommunication Development Report' taken from World Bank '2005 World Development Indicators'.

Factors at the User Level

- Cost of machine
- Cost of using, maintenance
- Length of life of machine, rate of obsolescence
- Training, re-training needs
- Language
- Physical/Cultural/Educational Capabilities
- Willingness/Interest

Direct cost of machines, software and user costs can be significant barriers, and are the main rationale for community forms of access such as the earlier-discussed VPPs and VKCs. The poor in rich countries face these individual costs as well; however, due to a higher level of infrastructural development (as indicated earlier), their relative ICT deprivation is qualitatively different from that endured by the poor in poor countries. The poor in rich countries can fall back on a variety of other avenues of group access to ICT facilities. These could take several forms: the local library, community centres, post/telegraph offices, cyber cafés, or specialized booths and centres created by local authorities and government or non-governmental agencies. Increasingly, paid access is also available at a range of commercial establishments, such as supermarkets, airports, and so on. Moreover, since there are fairly well-developed second-hand markets in such countries, it might even be possible to set up individual access, even if not fully up to speed.

A small, unrepresentative survey[43] on cheapest market prices of PCs with Pentium 4 processors across the globe indicated considerable differences in pricing and brought various informal strategies to the surface which lower the price considerably, however dependent on large markets. Prices for PCs ranged from under US$ 400 in Dhaka, Bangladesh, and even cheaper if a Celeron processor is used, to US$ 1,300 for a similar type of PC in Lusaka, Zambia. A trend could be observed with low prices in South Asia, based on a system of local assembly,[44] and highest prices in the African countries surveyed with no apparent system of local assembly. The Latin American countries surveyed hold middle way with prices around the US$ 600, generally relying on local assembly. The two 'transitional economies' stood out in availability of components and whole PCs which were non-branded and of brands unknown in 'the West', nevertheless PC prices were still not compatible with those in the South Asian countries surveyed.

Costs of software and particularly Internet use have generally decreased across the globe; however, especially if taken as share of local incomes, a great gap between the rich and the poor nations can still be observed. For example, US$ 560, the price of a single licence fee in 2003, according to amazon.com, for Windows XP together with Office XP, works out to be only 0.19 'GDP months' in the US, while 14.53 'GDP months' in India, 7.37 in China, 18.99 in Zambia and 2.31 in Brazil (Ghosh, 2003).[45] In addition, the total monthly Internet price as a percentage of GNI per capita, using 2003 figures, turns out to be 0.5 per cent of GNI per capita per month in the US, while it is 21.9 per cent in India and a staggering 66.8 per cent in Bangladesh.[46]

The combination of the micro and macro factors determines the contours of effective access in different systems and can be used to explain digital divides between and within countries. Much is made of the rapidity of the expansion of ICTs in the two Asian giant economies. However, even a cursory glance at the data presented in Figure 5.1 confirms the yawning chasm between the rates of ICT usage between China and India compared with, say, the UK—the range would be considerably wider if the least and the most advanced economies were compared.

It is worth emphasizing that the internal digital divide between the privileged and the under-privileged social groups within countries is as dramatic as that between countries. Often, the inter-country catching-up reflects simply the reduction in the distance between the elites of the developed and the less-developing economies. This picture also emerges from the trends in teledensity in India, presented in Figure 5.2. Increase in teledensity stems almost entirely from an urban increase. Furthermore, looking at state-specific figures, even these general patterns constitute great variations. The national teledensity for urban areas is 25.9 per 100 people, as of December 2004; however, rates vary from 64 in Himachal Pradesh to a poor six in Chhattisgarh. The rural national average is 1.69, with variations ranging from 9.47 in

FIGURE 5.1 Average Internet Usage in Five Selected Countries

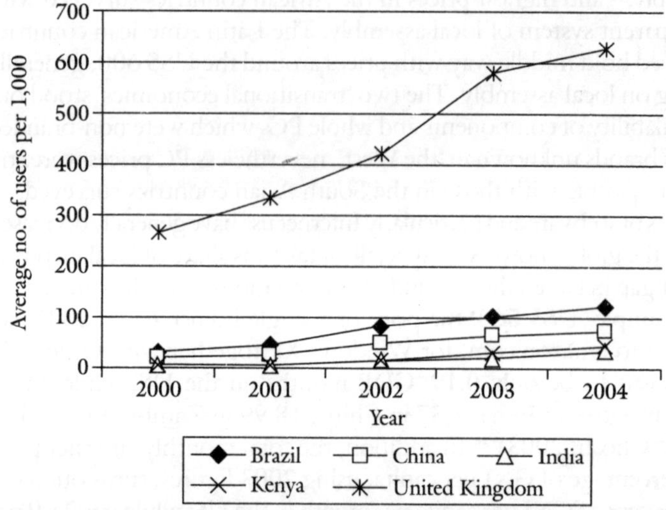

Source: The World Bank 'World Development Indicators 2005'.

FIGURE 5.2 Teledensity in India (1995–2004)[47]

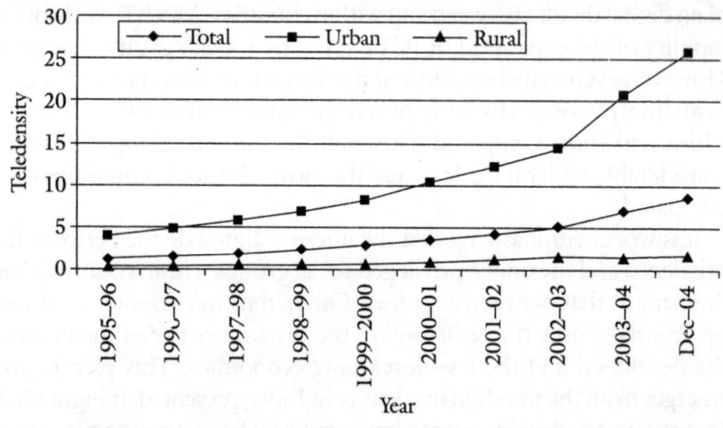

Source: Adapted from Annual Report 2004–2005, Ministry of Communications and Information Technology, Department of Telecommunications (2005), Government of India, New Delhi (p. 98).[48]

Kerala to 0.47 in Jharkhand. This teledensity picture gives a good indication of the extent of inequalities in Internet connectivity across India, since teledensity strongly correlates with Internet use. In this respect, prospects look rather bleak for rural India; particularly bleak in some states, despite much attention having been paid to a handful of innovative rural projects.

Mechanisms for Lowering Access Thresholds

There are very many processes through which the constraints to access are eased, whether at the level of the system, or especially for the individual user. Some of these apply to technological factors, some to economic policies, and others to institutional factors. These are listed as follows:

- Technological innovations allow mobile connectivity, reducing hurdle set by limited landline infrastructure.
- Economies of scale in production of hardware and software lines leads to decline in unit prices.
- Open access through the web allows access to free or low-cost software.
- Increased competition within industry along with deregulatory policies lead to falling unit prices for a range of products and services.
- Reduction in import tariffs for ICT inputs and final products.
- Cumulative acquisition of skills and independent capacity to service own systems.
- With widening of markets, larger language pools are increasingly catered for, though software and service systems lag.
- Telephone costs declining steadily on account of e-com and advertisement revenues and increased competition between operators and licensees.
- Shared access:
  - via shared facilities at home, within the family;
  - via the employer at the work place;
  - via educational establishments, for example, schools, colleges;
  - via training centres, unemployment exchanges;
  - via social/community organization, local community centre, local library;
  - via commercial channels, local cyber café, local telecom shops.
- For non-cutting edge, non-competitive usage, access to recycled machines and hand-me-downs.
- Second-hand market, though with high risks and transaction costs.

- Through public–private financial partnership arrangements.
- Through various NGO, CBO, CSO initiatives.

It is worth noting that many of these mechanisms are triggered by the process of diffusion itself, for example, fall unit prices with accelerated adoption; and that other threshold-lowering devices are induced by the persistence of the gap, for example, governmental or non-governmental initiatives to address the divide. The question of what all these collectively add up to, and how far they go in bridging the chasm, is eventually an empirical one. While there is evidence of an impressive and expanding array of such initiatives, they are as yet mostly indicative of potential rather than accomplished possibilities. It is also paradoxical that such extra-market processes are far more pronounced in the developed economies than in the less-developed ones. For example, per 1,000 people, the US has, on an average, 57.1 PCs installed in schools, the UK 30.7, South Africa 8.3, Brazil 5.1, China 1.6 and India 0.2.[49] Yet, with generally a greater share of the population in school-going age, developing countries tend to perform even worse than these indicators suggest.

Is Access Getting Easier?

What about the dynamics of access thresholds? Are there discernible trends in the above variables which suggest that access to ICT infrastructures and hardware could be getting easier for those on the underside of the digital divide? As ever, there are contending positions and pointers.

There certainly are grounds for optimism with respect to several factors.

- Absolute cost of hardware falling rapidly.
- Cheaper, quicker telephone connections.
- Telephone use charges falling.
- Some national, or systemic infrastructural requirements being reduced or bypassed by wireless satellite technology; cheaper connectivity to remote areas, especially with the use of solar energy technologies.
- The availability, albeit selectively, of cheap or free software from the net, not least through the efforts of movements promoting open, free access to software on the net.
- The drive towards deregulation, along with new patterns and sources of technological innovation—often imply an increased

role for competition, which in turn accentuates the factors leading to a decline in threshold levels.
- With widening institutional and commercial points where ICT facilities can be accessed, increasing possibilities of shared access.

But, there are strong forces which operate to limit the possibilities of easier access, or the benefits that might accrue from falling thresholds:

- Fast pace of obsolescence means computer system and infrastructural life is dramatically reduced.
- Specialized software costs higher with wider 'indispensable' needs written into systems; these universal trends need to be kept in view to balance the optimism generated by open software access initiatives.
- Increasing presence of pay sites on net.
- Cost of connectivity declining, but there are fresh heavy costs with frequent system changes.
- Start-up costs of hardware, software systems—market research, selection, search, install, learn to run and maintain.
- The mobile is not equivalent to the PC plus wired telephone, since the former is individualized; has inferior sharing possibilities.
- Hidden costs, for example, built-in obsolescence of other related electronic systems at home.
- Stamping out of piracy raises user costs.

At both the level of the individual as well as the system, the access constraints are daunting for the less-developed economies. All evidence suggests a differentiated profile, with some groups of developing economies faring relatively well in the ICT stakes. However, with respect to the vast majority, the digital chasm has been continually widening over the past decade and could be expected to widen further in the foreseeable future.

That said, in recent years, some creative new initiatives have emerged, which challenge some of these structural barriers. The corDECT system, invented by the Indian Institute of Technology in Chennai, offers a considerably cheaper alternative to one of the main challenges in developing countries: lack of telecommunication infrastructure. Instead of conventional fibre and copper network technologies that require an investment of approximately US$ 1,000 per subscriber, the corDECT system makes use of fixed Wireless-in-Local-Loop technology which brings down the investment cost to US$ 300 per subscriber.

Importantly, corDECT system enables both voice and data transmission, thus serving phone and Internet applications (Best, 2003: 110, 113). The corDECT system is currently implemented by n-Logue Communications Ltd, in collaboration with various private and (non) governmental partners in several districts of seven Indian states (Paul, 2004). The corDECT system has also aroused much interest in Africa and Latin America, where similar infrastructural conditions prevail. Another Indian invention, the Simputer,[50] addresses universal access at the level of Internet usage by coming up with a 'stripped-down', but fully functional PC, which, instead of a hard disk, relies on a network for storage, and can be obtained at just over US$ 200. This brings down the price of a PC considerably, particularly since it is designed for shared usage. An Open Source Linux operating system, a purpose-built user interface with rural-relevant context and text-to-speech capabilities in major Indian languages (Best, 2003: 110) make the Simputer not just a cheaper but also a more relevant and sustainable ICT application for the rural population. The Simputer, after a delay of nearly three years, was launched in March 2004, branded as Amida Simputer (BBC, 2004). It is too early to assess how far these two systems might go in reaching the poor effectively and, if successful in this respect, how such projects could be scaled-up.

Conclusion: The Real and the Virtual

The answers to the three sequential questions, namely, concerning access, content, and utilization, vary considerably for the poor in rich, as against in poor countries. It is important to note such differences in order to avoid superficial and untenable generalizations with regard to the potential or real impact of ICT on the lives of the poor.

Nevertheless, one outcome that can be predicted with some certainty is that the early phases of IT adoption and diffusion are likely to be accompanied by an increase in inequality, and thereby relative poverty. Initial adopters are overwhelmingly from non-poor countries and non-poor sections of society, and early adoption gives them a head start in this high-returns game. Depending on which prognosis one subscribes to, with respect to the future pattern of development of the technology, as well as in socio-economic and infrastructural factors that govern its diffusion, it is possible to hypothesize that this initial widening of the gap will be reversed as the poorer countries and poor people leapfrog technological stages and catch up. But this must remain an optimistic speculation. Would technologies, and the non-poor, be stationary in the interim, thus allowing this catching-up to occur? It might be more plausible to argue that the pace and sustained nature of technological advance in ICT are more likely to extend the gaps for the foreseeable future and thus have a deleterious impact insofar as relative poverty is concerned. Looking across countries and socio-economic strata, the probability of such an outcome is likely to be reflecting the width of the initial digital divide: where the initial gaps are low, it is likely that corrective interventions could be effective in the short run, thereby inducing a rapid catching-up process. However, where the gaps are massive, bridging them is likely to call for the prior satisfaction of a variety of social, economic and infrastructural pre-conditions; in turn, these could require substantial investments, forms of institutional development and other enabling measures to permit the poor to access ICTs on a broad, sustained and effective basis—all time-consuming processes.

The findings of a major project of research on ICT and development comes to the conclusion that:

> To be able to reap the benefits from investment in IT, developing countries have to build up a mature stock of physical infrastructure, which enhances and amplifies the effects of IT. As advances in technology tend to be embodied in the latest vintages of capital, investments in physical capital and in IT are complements also for the reason that increasing amounts of microelectronics are embodied in physical capital. Consequently, national IT strategies should include policies promoting investment in physical infrastructure.[51] (Pohjola, 2001: 26)

Apart from these, investment in education, especially linked to IT, 'is the key to economic development. This is, of course, an old policy prescription in the economics of development. What is new in the analysis…is the view that the IT content of these investments should be high' (ibid.: 29).[52]

Specifically at the level of the poor and devising strategies of poverty alleviation, the argument here has been to query the power of information per se in overcoming poverty in the social and material spheres. The potentiality of ICT has to be exploited within the framework of ongoing strategies of poverty alleviation, which confront the deep-rooted but mutable dimensions of poverty. In this, the poor might occupy a location too distant in terms of the digital divide in order to exploit these possibilities. Thus, strategies of widening effective access are critical. Here, it is argued that the non-competitive, welfare-enhancing potentialities of ICT are best transmitted to the poor through augmenting the IT and institutional capacities of intermediate agencies, whether governmental, non-governmental, community-based, or organized by the poor themselves. What is critical here is that in these efforts at translating virtual potentialities into real break-throughs for the poor, the voices and priorities of the poor should provide the motive force for emancipatory change, rather than a legitimizing device for perpetuation of the social divide in the brave new digital era.

Notes

1. Earlier versions of this chapter have been presented in seminars at the ILO, Geneva, at the Centre for Development Studies, Thiruvananthapuram, at the Institute of Social Studies in The Hague, at the International Conference on ICTs and Indian Development in Bangalore, and at IFAD, Rome. Comments from the participants in these presentations are gratefully acknowledged, in particular from Rashid Amjad, Arjun Bedi, K.P. Kannan, Atiqur Rahman, Eric Ross, Alakh Sharma and M. Vijayabaskar; in particular, I would like to thank Roy Huijsmans both for comments and excellent research support in the revision of the chapter.
2. See for a good coverage of issues related to ICTs and development, Mansell and Wehn (1998).
3. Muhammad Yunus, the man behind Grameen Bank's micro-credit scheme, remarked at a recent (2003) conference on ICT and Poverty at Harvard University that 'the quickest way to get out of poverty right now is to have one mobile telephone' (International Development Research Centre website).
4. Illustrative in this respect is a simple Google search on two initiatives discussed in this chapter. A search on 'viliage payphones grameen bank' results in 12,100 hits and 'ICT MS Swaminathan Research Foundation' produces 11,200 hits. However, a more thorough search displays that all these different pages tend to refer to not much more than a handful of serious studies conducted on these interventions.
5. Ricardian constraints in the form of diminishing returns in agriculture were deemed to set a dampener on growth in the 19th century. These natural constraints were arguably substantially overcome by industrialization with its inbuilt economies of scale, though the resource-intensive and environmental dimensions of modern manufacturing have emerged as key limiting factors. The new economy could be viewed as a third generation of technological and production reorientation, partially extending the frontiers of growth.
6. Taken from UNDP at <http://www.undp.org/info21/index5.htm> accessed on 29 August 2005.
7. See Balaji (2000) for more on knowledge systems and food security.
8. This role can be quite powerful in specific situations, as was demonstrated by the movement of the Chiapas; the various anti-globalization organizations; in

experiments of democratization of government and governance; and so on. This is a large and rapidly developing theme, both in terms of substantive activity, as well as research.

9. A horizontal break between the poor and the ultra-poor has been observed. The latter category displaying features discontinuously inferior, or different, from those in the upper strata; and further that these attributes reflect weaknesses in the ability of households on the lower side of the kink in the distribution(s) to engage productively with the growth process. The question emerges if ICTs can be meaningful for this group of ultra-poor.
10. See for a discussion on the indirect and direct impact of ICTs on poverty, Keniston (2001).
11. In the Republic of Korea, it was 4.7 per cent; in India, 1.5 per cent; in Hong Kong 2.1 per cent; in Brazil 1.6 per cent. Only a handful of IT pioneer economies posted impressive figures: Finland, 36.7 per cent; Sweden, 24.8 per cent, UK, 12.1 per cent and USA, 10.0 per cent. The contribution refers to GDP change over 1991–96 (these data are from Pohjola, 2001: Table 1.3, p. 12).
12. Such a notion of 'fiscal trickle-down' has sometimes been invoked by neo-liberal growth—first protagonists as a second line of defence in the face of evidence that the growth process is distributionally biased against the poor.
13. This impact would vary from country to country. Where the supply of local, high-skilled ICT knowledge workers is limited, their emigration could be damaging for the further development of the sector. This could apply to small economies. In contrast, India seems to offer an unlimited supply of high-skill ICT workers at low wages—with no serious likelihood, at least thus far, that this emigration will set up a supply bottleneck in the economy. If anything, the large stocks and massive annual inflows into the pool of young, educated job-seekers have been especially attracted to this sector with its low level of entry requirements.
14. In this context, a comparison with the Gulf migration episode is useful. While the profile of the migrants was wide, the dominant outflow from South Asia was of unskilled and semi-skilled workers. Their withdrawal was on a scale which influenced local labour markets, pushing up wages for several categories of workers. The initial income levels of these workers also placed a lot of them close to, or below the poverty line; and even if they were some way above it, they were effectively pushed below it through the debts they had to incur for buying their passage. The remittances were therefore to families in a lower-income stratum. To what extent there are benefits to the non-migrant local population depends on the ability of the local economy to meet the demands set up by the new expenditures. Paradoxically, in Kerala, a major labour-sending pocket, many of these effects leaked away to neighbouring states or to the wider economy on account of the inability of the local economy to provide an adequate supply response.
15. Up-to-date and comparative data on these issues are hard to find; yet, reports from various key-organizations working on 'Electronic Commerce' suggest that the patterns presented here have not changed significantly over recent years (see, for example, OECD, 2002; UNCTAD, 2004).
16. In the case of the outflow of skilled migrants to the Gulf after the oil-boom, there is evidence that the labour-market gaps created by the withdrawal of skilled and semi-skilled workers, such as carpenters, masons, and so on, opened-up possibilities for lesser skilled and poorer workers who graduated up the domestic labour-market ladder to fill their positions. But those were mechanical and artisanal skills, readily

picked up at an inferior level; it is unlikely that the withdrawal of ICT migrant workers to the US and elsewhere will induce any similar impetus in favour of those below the poverty line.

17. The original site is no longer available on the Internet. This in itself highlights the ephemeral nature of a significant part of the knowledge resources provided by the Internet at any particular point.
18. The idea that poverty and deprivation might be defined in a direct relation to an ideas-gap is perhaps less implausible than might at first appear. That thought should precede deed is hardly odd. It could appear odd, however, to suggest that the poor need to surf the net to discover the blueprint for their emancipation. See also Romer (1993).
19. The World Bank has recently declared 'social capital' to constitute the key missing link in development; micro-credit *a la* Grameen Bank, has been treated in much the same way.
20. The International Development Research Centre.
21. Website MS Swaminathan Research Foundation (http://www.mssrf.org/special_programmes/ivrp/ivrpmain.htm 9-9-05).
22. For a brief introduction of the Village Knowledge Centres, see Balaji et al. (2002).
23. Available from <http://www.unesco.or.id/apgest/pdf/india/india-bp-it.pdf> accessed on 8 August 2005.
24. This list has been culled from seven Internet reports (between 1999 and 2000) on this project, including some emanating from the project itself. See also other MSSRF (a, b, c) reports.
25. While draft cattle are owned by landowners, some milch animals might indeed be owned by families below the poverty line.
26. It is stated that 'the information [on market prices] was also of interest to female agricultural workers, who receive part of their wages in grain' (Shore, 1999: 2), but this appears to be rather speculative and untenable.
27. This system was effectively used at the time of the Asian tsunami in the village of Nallavadu (Yoon, 2004).
28. It has been argued that the fishermen benefit from ICTs indirectly, since the improved weather forecasts incorporate the benefits of the use of such technologies by the weather forecasting systems which serve as the source. This argument applies analogously also to many other activities and processes, for example, the postal system, television broadcasting, banking, and so on. The key issue is not whether the use of ICTs improves weather forecasts, but whether this can be claimed as a new benefit which is transforming poor people's lives through the use of village-based ICT facilities which access the Internet.
29. This industry is usually located in the unorganized sector, is not subject to labour laws, and provides some very flagrant cases of child labour. It also tends to pay wages to its workers, which translate into below-poverty line incomes even when reckoned on a full-employment basis. The only way to combine decent work with market imperatives is to work within a co-operative framework where the workers also obtain income through claims on the profit component of value added. It can only be presumed that the advice provided to potential rural entrepreneurs would cover such rights-oriented issues.
30. Two independent studies, specifically focussing on the impact of micro-level ICT projects on poverty alleviation, both arrived at sceptical conclusions. On the

SARI project in Madurai district in Tamil Nadu, see Kumar (2003)—Social and Economic Impact of Technology Intervention in Rural Communities: Case of Sustainable Access in Rural India (SARI). (Department of Urban Studies and Planning, Cambridge (US), Massachusetts Institute of Technology)—and on the Gyandoot Project in Madhya Pradesh, see Kinoshita (2004) —Does ICT Really Give Power to the Poor People? A Case of Gyandoot Project in India (Institute of Social Studies, The Hague).

31. See also: Grameen Phone Website on the Village Phone Project, Lawson and Meyenn (2000) for a brief overview and early evaluation, as well as Bayes (2000).
32. An OECD report, published in 2004, states that since 1997, the VPP project has provided 45,000 telephones to 39,000 villages in Bangladesh, thus bringing VPPs within physical reach of approximately 70 million people—on the basis of UNFPA 2004 figures, almost half of the Bangladeshi population (Tambo, 2004: 7).
33. Grameen Phone is a joint-venture company in which Grameen Telecom, the not-for-profit component, has a 35 per cent share. The other partners are Telenor Invest AS Norway (51 per cent), Marubeni company Japan (9.5 per cent) and Gonofone Development Corporation USA (4.5 per cent) (Bayes et al., 1999: 12). Grameen Phone is, as of 2003, Bangladesh's largest provider of telecommunication services with 1.3 million subscribers, thus ahead of the 700,000 fixed lines provided by the Bangladesh Telegraph and Telephone Board (Tambo, 2004: 11).
34. The three studies are: (1) Bayes et al. (1999), 'Village Pay Phones and Poverty Reduction: Insights from a Grameen Bank experience in Bangladesh'. This pilot study is conducted by researchers at Jahingirnagar University, Bangladesh and the Center for Development Research (ZEF) of the University of Bonn, Germany. It evaluates, using the VPPs in Bangladesh as a case, the role of telecommunication within the contexts of rural development and poverty reduction. (2) Richardson et al. (2000), 'Grameen Telecom's Village Phone Programme in Rural Bangladesh: A multi-media case study'. This study was commissioned by the Canadian International Development Agency and set out to 'investigate the impact of the Grameen Phone and Grameen Telecom provision of micro-credit cellular phone service on poverty reduction and the socio-economic situation of women Village Phone operators and users'. (3) Sobhan et al. (2002), 'Factors Shaping Successful Public–Private Partnerships in the ICT Sector in Bangladesh'. This study was commissioned by the Commonwealth Policy Studies Unit of the University of London and undertaken by the Bangladesh Enterprise Institute. This study has evaluated the VPPs with the objective of identifying the 'factors shaping successful pro-poor information society partnerships'.
35. The brief evaluation here is second-hand, in that it relies mainly on three major available systematic evaluations of the VPP project, BBA99, RRH00 and SKR02. However, instead of merely reporting on the findings of these, the discussion here goes on to identify issues and questions on which there could be grounds for disagreeing with some of the key interpretations of these evaluations, using the same database as reported in these documents.
36. This study surveyed phone owners and users. Through random sampling 50 phone owners in 50 different villages at distances of 40 to 50 km from metropolitan Dhaka were selected, constituting 60 per cent of all phone owners in the research area at the time of study. The 400 phone users surveyed were selected through random sampling based on lists provided by phone owners of persons who'd used the VPPs

in the two weeks preceding the study. The sample of VPP users constituted 27 per cent of all VPP users in the study villages (BBA99: 5).

37. This study is based on a survey in the Northern Mymensingh region, a relatively underdeveloped area which was not included in the earlier BBA99 study. The VPP placements that were to participate in the SKR02 research were selected by GB field managers. The total sample size is 120:24 VPP operators, 72 VPP users and 24 non-users. The report does not clarify which sampling method was used (SKR02: 3).
38. Correspondingly, extreme poverty in 1995–96 was 38.5 per cent vs 13.7 per cent; and in 2000, 37.4 per cent vs 19.1 per cent in rural vs urban areas, respectively—(2002). Poverty in Bangladesh: Building on Progress. Dhaka, World Bank, Asian Development Bank.
39. With regard to the gender impact of VPPs, Aminuzzaman (2002: 171) comes to the following sobering conclusion: 'Such exclusion from the VPP practicalities may lead to women's exclusion from decision-making involved in VPP operations and from the use of income from VPP businesses. When owners are not involved in VPP transactions, they are probably also less exposed to social interactions. Moreover, only 22 per cent of VPP users are female. It seems, therefore, that in rural Bangladesh, in both the use and operation of the VPP, the traditional seclusion of women from the wider society has been maintained'.
40. There is much that is questionable in the methodologies used for identifying and estimating the hypothetical benefits of the use of VPPs (see for example, the methodologies used in Torero et al., 2002; Bayes et al., 1999: 29; Bertolini, 2002; Muller–Falcke, 2002; Song, 2003). For instance, on the side of costs, the price of a phone call is known. But what 'control' or alternative method of obtaining information should it be compared with? Would the villager have to take a day off, incur travel expenses for a trip to the market/s? Do the prices obtained by the two methods not vary with respect to other transaction costs and risks? And, is not information obtained by other means shared? Are there no other information carriers who travel through? Can a deal be struck on the basis of price information exchanged over the phone without either party wishing to physically verify the quality of the merchandize? Thus, it is not easy to provide meaningful interpretations to the operational use of the concepts of opportunity cost and consumers surplus. The use of the 'willingness to pay' notion is equally full of hazards.
41. For a comprehensive empirical profile of the socio-economic divide underlying the digital divide, see Shariff (1999).
42. This has been demonstrated in the US in a report by the US Department of Commerce looking into Internet access by various characteristics of certain population groups (US Government, 2000).
43. Based on data from India, Ethiopia, Tajikistan, the UK, Bangladesh, Zambia, Colombia, Rwanda, Argentina and Albania.
44. The PC is assembled locally using elements (for example, CD-drive, modem, keyboard, and so on) from various branded or non-branded makes, in order to arrive at the lowest price.
45. These 'GDP months' are arrived at by calculating the number of months it would take to reach US$ 560 using local GDP per month.
46. Data from Telecommunication Union's (ITU) 'World Telecommunication Development Report', used in World Bank '2005 Development Indicators, Table 5.11',

available from http://www.worldbank.org/data/wdi2005/wditext/Tables5.htm, accessed on 21 August 2005.

47. Teledensity in this figure is presented as the average number of telephones per 100 people; this includes Public DELs, PVT.DELs, CMPs and WLL (fixed and limited).
48. Available from <http://www.dotindia.com/annualreport/english.pdf>, accessed on 10 August 2005.
49. Based on 2001 data calculated from World Bank 'ICT at a glance' and UNFPA population figures for 2001. Available from <http://www.worldbank.org/data/countrydata/ictglance.htm> and <http://www.unfpa.org/swp/2001/english/indicators/indicators2.html>, accessed on 10 August 2005.
50. Simputer stands for simple inexpensive multi-lingual computer. For more information, see <http://www.simputer.org>. The Simputer provides an inspirational story of value-driven, socially oriented technological innovation. Pioneered many years ago by a team of idealistic scientists committed to bringing the potentialities of ICTs within reach of an otherwise digitally excluded sections of the Indian population, the case also demonstrates the uphill task that such pioneering efforts have when it comes to the transformation of scientific breakthroughs into successful mass-produced product lines for wide dissemination. A loss of time in conversion of ideas into products allows other rival products to capture markets and quasi-rents to innovation. Thus, the Simputer, a product of scientists at the Indian Institute of Science, Bangalore, seems to be in such jeopardy in relation to a rival being launched by scientists at MIT; this emphasizes the crucial significance of overall systemic attributes in influencing the successful outcome of potentially economically viable technological innovations and prototypes.
51. See also Schware (1992).
52. The findings also suggest that there might be low returns and high risk in pursuing national leapfrogging strategies with regard to IT production, which is dominated by a few core countries.

References

Aminuzzaman, S.M. 2002. 'Cellular Phones in Rural Bangladesh: A Study of the Village Pay Phone of Grameen Bank', in Andrea Goldstein and David O'Connor (eds), *Electronic Commerce for Development* (pp. 161–78). Paris: OECD.

Arunachalam, S. 2004. 'Information and Communication Technologies and Poverty Alleviation', *Current Science*, 87(7): 960–66.

Balaji, V. 2000. 'Knowledge System for Sustainable Food Security'. Available at http://www.idrc.ca.telecentre/evaluation/nn/08_Kno.html.

Balaji, V., S. Kumaran and R. Rajasekarapandy. 2002. 'MSSRF: Village Knowledge Centers', *Gender, Technology and Development*, 6(1): 147–52.

Bayes, A. 2000. 'The Phone and the Future: An Evaluation of Village Pay Phones in Bangladesh', paper presented at Global Dialogue: The Role of the Village in the 21st Century, Hannover, 15–17 August 2000.

Bayes, A., J. von Braun and R. Akhter. 1999. 'Village Pay Phones and Poverty Reduction: Insights from a Grameen Bank Initiative in Bangladesh', ZEF Discussion Papers, Number 8, University of Bonn, June.

BBC. 2004. 'Simputer for Poor goes on Sale', London: BBC News World. Available at http://news.bbc.co.uk/2/hi/technology/3578309.stm accessed on 1-09-05, (Accessed on 1 September 2005).

Bertolini, R. 2002. *Telecommunication Services in Sub-Saharan Africa*. Frankfurt am Main: Peter Lang.

Best, M.L. 2003. 'The Wireless Revolution and Universal Access', in J. Alden (ed.), *Trends in Telecommunication Reform 2003: Promoting Universal Access to ICTs, Practical Tools for Regulators* (pp. 107–22). Geneva: International Telecommunication Union (ITU). Available at http://cbdd.wsu.edu/edev/edp/ITU-2003-Telecom-Reform-trends-UA-US.pdf (Accessed on 1 September 2005).

Chapman, R. 2004. *ICT Enabled Knowledge Centres and Learning in the Global Village*. A report based on the 'South-South Exchange Travelling Workshop', sponsored by Global Knowledge Partnership, 15–22 October 2005. Available at http://www.odi.org.uk/rpeg/research/Pol_Gov_&_Institutions/CKP/Documents/ICT_enabled_knowledge_centresmsrev.doc (accessed on 16 August 2005).

Chinn, M.D. and R.W. Fairlie. 2004. 'The Determinants of the Global Digital Divide: A Cross-Country Analysis of Computer and Internet Penetration', IZA Discussion Paper Number 1305, Bonn, IZA. Available at ftp://repec.iza.org/RePEc/Discussionpaper/dp1305.pdf (accessed on 11 August 2005).

David, P.A. 1990. 'The Dynamo and the Computer: An Historical Perspective on the Modern Productivity Paradox', *American Economic Review* 80(2): 355–361. Papers and Proceedings of the Hundred and Second Annual Meeting of the American Economic Association.

Ghosh, R.A. 2003. 'License Fees and GDP per Capita: The Case for Open Source in Developing Countries', *First Monday* 8(12). Available at <http://www.firstmonday.org/issues/issue8_12/ghosh/> (accessed on 29 August 2005).

Government of India. 2005. *Annual Report 2004–2005*. New Delhi: Ministry of Communication and Information Technology, Department of Telecommunications. Available at http://www.dotindia.com/annualreport/english.pdf (accessed on 10 August 2005).

Hariharan, V. 1997. 'The Other 99 Percent of the Digital Age', *Net.Columns: A Wednesday Feature,* 29 October 1997. Available at http://iw.sify.com/home/indialine/net.columns/column30.html (accessed on 17 May 2006).

Keniston, K. 2001. 'IT for the Common Man: Lessons from India', Second M.N. Srinivas Memorial Lecture, National Institute of Advanced Studies, Bangalore, 3 December 2001. Available at http://www.i4donline.net/issue/may03/kenneth_full.htm (accessed on 17 May 2006).

Kinoshita, M. 2004. 'Does ICT Really Give Power to the Poor People? A case of Gyandoot project in India', unpublished MA Research Paper, Institute of Social Studies, The Hague.

Kumar, R. 2003. 'Social and Economic Impact of Technology Intervention in Rural Communities: Case of Sustainable Access in Rural India (SARI) Project'. Report on Summer 2003 Internship under MIT program on Human Rights and Justice. Available at <http://mit.edu/phrj/internpapers/KumarReport1.doc> (accessed on 8 August 2005).

Lawson, C. and N. Meyenn. 2000. 'Bringing Cellular Phone Service to Rural Areas: Grameen Telecom and Village Pay Phones in Bangladesh'. Note 205, Washington: Public Policy for the Private Sector, The World Bank, March 2000. Available at

<http://rru.worldbank.org/Documents/PublicPolicyJournal/205lawson.pdf> (accessed on 17 May 2006).

Mansell, R and U. Wehn. 1998. *Knowledge Societies: Information Technology for Sustainable Development*. Oxford: Oxford University Press.

MSSRF (M.S. Swaminathan Research Foundation) n.d. a. July 2000–March 2002. Information Village Research Project—Second Phase—First Year Report.

———. b. March 2002–March 2003. Information Village Research Project—Second Phase—Technical Report.

———. c. 2002. ICT Enabled Development—South-South Exchange Workshop, 21–28 October 2002.

Muller-Falcke, D. 2002. *Use and Impact of Information and Communication Technologies in Developing Countries' Small Businesses: Evidence from Indian Small Scale Industry*. Frankfurt am Main: Peter Lang.

OECD. 2002. *Measuring the Information Economy*. Paris: Organisation for Economic Co-operation and Development. Available at <http://www.oecd.org/dataoecd/16/14/1835738.pdf> (accessed on 21 May 2006).

Paul, J. 2004. 'What Works: n-Logue's Rural Connectivity Model'. World Resources Institute Digital Dividend. Available at http://www.digitaldividend.org/pdf/nlogue.pdf> (accessed on 1 September 2005).

Petrazzini, B.A. 1999. 'Policy and Regulatory Issues'. Powerpoint presentation delivered at Electronic Commerce Seminar, 17 May 1999, organized by the ITU. Available at <http://www.itu.int/newsarchive/wtd/1999/ppt/bp/sld020.htm> (accessed on 18 May 2006).

Pohjola, M. 2001. 'Information Technology and Economic Growth: Introduction and Conclusions', in Pohjola, M. (ed.), *Information Technology, Productivity, and Economic Growth: International Evidence and Implications for Economic Development* (pp. 1–30). Oxford: Oxford University Press.

PovertyFighters. 2003. 'Grameen Phones and Anwara Khatun'. Available at <http://povertyfighters.com/successes/successStories.cfm?storyId=151> (accessed on 10 September 2003).

Romer, P. 1993. 'Ideas Gaps and Object Gaps in Economic Development', *Journal of Monetary Economics*, 32(3): 543–73.

Richardson, D., R. Ramirez and M. Haq. 2000. *Grameen Telecom's Village Phone Programme in Rural Bangladesh: a Multi-Media Case Study*. Ontario: TeleCommons Development Group (TDG)–Canadian International Development Agency (CIDA).

Schware, R. 1992. 'Software Industry Entry Strategies for Developing Countries: A "Walking on Two Legs" Proposition', *World Development*, 20(2): 143–64.

Shariff, A. 1999. *India: Human Development Report*. NCAER, Delhi: Oxford University Press.

Shore, K. 1999. 'The Internet Comes to Rural India'. Available at <http://www.hub.col.org/agricol99/0040.html>

Sobhan, F., M.H. Khaleque and S. Rahman. 2002. 'Factors Shaping Successful Public Private Partnerships in the ICT Sector in Bangladesh'. Bangladesh Enterprise Institute, 15 March 2002. Available at <http://www.cpsu.org.uk/downloads/Annex2.pdf> (accessed on 8 August 2005).

Song, G.S. 2003. *The Impact of Information and Communication Technologies (ICTs) on Rural Households*. Frankfurt am Main: Peter Lang.

Tambo, I. 2004. 'Grameen Phone Revisited: Investors Reaching Out to the Poor'. OECD, DCD/DAC/POVNET(2004)8/REV1, 5 October 2004. Available at <http://www.oecd.org/dataoecd/36/6/33962074.pdf> (accessed on 16 August 2005).

Torero, M., S.K. Chowdhury and V. Galdo. 2002. 'Willingness to Pay for the Rural Telephone Service in Bangladesh and Peru', ZEF Discussion Papers on Development Policy, Number 54, Center for Development Research (ZEF), Bonn, October 2002.

UNCTAD. 1999. 'Can Electronic Commerce Be an Engine for Global Growth? Electronic Commerce and the Integration of Developing Countries and Countries with Economies in Transition in International Trade'. Note by the UNCTAD Secretariat, TD/B/COM.3/23, 1 June 1999. Available at <http://r0.unctad.org/ecommerce/docs/c3d23.pdf> (accessed on 18 May 2006).

———. 2004. *E-Commerce and Development Report 2004*. New York and Geneva: United Nations Conference on Trade and Development. Available at <http://www.unctad.org/en/docs/ecdr2004_en.pdf> (accessed on 21 May 2006).

US Government. 2000. 'Falling Through the Net: Toward Digital Inclusion'. US Department of Commerce. Available at <http://search.ntia.doc.gov/pdf/fttn00.pdf> (accessed on 31 August 2005).

Varma, R. and E.M. Rogers. 2004. 'Indian Cyber Workers in the US', *Economic and Political Weekly*, XXXIX (52): 5645–52.

Weber, S. 2003. *Open Source Software in Developing Economies*. Berkeley: University of California. Available at <http://www.ssrc.org/programs/itic/publications/ITST_materials/webernote2.pdf> (accessed on 29 August 2005).

World Bank & Asian Development Bank. 2002. *Poverty in Bangladesh: Building on Progress*. Dhaka: World Bank, Asian Development Bank. Available at <http://lnweb18.worldbank.org/SAR/sa.nsf/Attachments/PovRpt/$File/BD+2002+Poverty+Assessment+-+Final.pdf> (accessed on 31 August 2005).

Yoon, C.S. 2004. 'Phone Call Saved Scores of Indian Villagers from Tsunami'. In *Digital Review of Asia Pacific*, Penang, Malaysia. Available at <http://web.idrc.ca/en/ev-68766-201-1-DO_TOPIC.html> (accessed on 8 August 2005).

Websites

Grameen Phone Website. 'Village Phone Project'. Available at http://www.grameenphone.com/modules.php?name=Content&pa=showpage&pid=3%3A11%3A1&submit.x=10&submit.y=12 (accessed on 17 May 2006).

International Development Research Centre website. 2003. 'A Dialogue on ICTs and Poverty. The Harvard Forum'. Available at <http://web.idrc.ca/uploads/user-S/10787612051Harvard_Forum_2003_-_Summary.pdf> (accessed on 11 August 2005).

M.S. Swaminathan Research Foundation Website. Available at <http://www.mssrf.org/>.

Simputer Website. Available at <http://www.simputer.org>.

UNDP Information and Communications Technologies for Development Website. Available at <http://www.undp.org/info21/index5.htm> (accessed on 29 August 2005).

UNESCO Website. Available at <http://www.unesco.or.id/apgest/pdf/india/india-bp-it.pdf> (accessed on 8 August 2005).

UNFPA Website. 'The State of the World Population 2001: Demographic, Social and Economic Indicators'. Available at <http://www.unfpa.org/swp/2001/english/indicators/indicators2.html> (accessed on 10 August 2005).

UNICEF. 2005. 'Statistics, Bangladesh'. Available at <http://www.unicef.org/infobycountry/bangladesh_bangladesh_statistics.html> (accessed on 30 August 2005).

Worldbank Website. 'ICT at a Glance Tables' and 'ICT at a Glance'. Available at <http://www.worldbank.org/data/countrydata/countrydata.html> (accessed on 8 August 2005).

———. '2005 World Development Indicators'. Available at <http://www.worldbank.org/data/wdi2005/> (accessed on 11 August 2005).

6

Decrypting E-governance: Narratives, Power Play and Participation in the Gyandoot Intranet

T.T. Sreekumar

Introduction

Construction of the notion of 'e-governance' along with such other 'epithetized phenomena' as e-learning, e-banking, e-marketing and so on, played a major role in shaping the futuristic e-topias[1] of the global information society discourse. Woolgar (2002: 3) points out that 'epithetizing' various existing activities and social institutions with notions such as 'virtual', 'digital', 'electronic' (or simply 'e'), 'cyber', 'tele', and so on, 'conjures up a future consequent upon the effects of electronic technologies'. Given the backdrop of the increased involvement of new media technologies in delivering e-governance, it is important to understand the social and historical specificities of the emerging technological systems that facilitate the construction of the notion. It is also, perhaps, pertinent, to take a closer look at the relativity of technical design and absorption to the culture and strategies of actors (Feenberg, 1999) in order to discuss e-governance from a non-essentialist perspective. Social constructivism provides some provisional but meaningful theoretical foundations to look at e-governance in non-essentialist terms. One of the important conceptual endeavours from the constructivist perspective to understand e-governance initiatives would be to disaggregate the question of technology from the differential perspective of the dominant and subordinate subject positions of the actors involved.[2] Arguably, e-governance projects are rationally planned and implemented by technocrats in an effort to exercise a far more effective control over resources and social organization. Nevertheless, common people encounter these technologies of systematization as part of their life world and appropriate, reject or force revisions in the designs as well as systems. As Feenberg (1999: xiii) argues:

> (T)he invariant elements of the constitution of the technical subject and object are modified by socially specific contextualizing variables in the course of the realization of concrete technical actors, devices and systems. Thus technologies are not merely efficient devices or efficiency oriented practices, but include their contexts as these are embodied in design and social insertion.

In the case of ICT-based networked governance, the possibility of the formation of a rural network society is imminent and its proto forms with deep crevices and conflicting layers of incorporation of different actors have appeared in the contexts of many e-governance projects where the person-to-State interface and to a limited extent, person-to-person relations are reaffirmed by the technology of the Intranet. In this chapter, an attempt is made to understand the social dynamics that underlie the practices of e-governance in India on the basis of an analysis of the Gyandoot Intranet, a massive e-governance project launched in India in the state of Madhya Pradesh. The relatively weak, but undeniable structuration of the rural network society is manifested in the narratives of networking facilitated by the design of the project with scattered nodes connected to a centralized service monitoring centre. Nonetheless, the rural network society is a complex social domain of opposing interest groups and a space where some of the political and ideological conflicts in the larger society manifest in newer forms. The fact that the technology itself gets enmeshed in the relatively autonomous logic of the network society partly explains the inertia that stops the inevitable progress envisaged in the visions moulded by concepts informing paradigms of rapid social change consequent on the implementation of e-governance projects.

The notion of e-governance as it is practised and promoted in third world contexts has close resemblances with scenarios of neo-liberal discourses of new media technologies that Armitage (1999) powerfully criticized as a 'pan-capitalist theory and practice of explicitly technologized, or "telematic", societies'. He argued that this discourse is primarily 'concerned with legitimating the political and cultural control of individuals, groups, and new social movements' through 'production, promotion, distribution, and consumption of new media technologies' both at the material and ideological levels. The rhetoric of e-governance in the Indian context is also deeply enmeshed in a larger discourse of cyberlibertarian developmentalism that brings together the idea of economic development and neo-liberal discourse of technology arbitrarily to rationalize ICT-based State and civil society interventions for social and economic transformation in the rural setting (Sreekumar, 2003, 2006). Following Armitage (1999) and basing the analysis on the Foucauldian notion of Governmentality, Navarria (2006: 126) argues that 'the concept of e-government does not only signify *efficiency gains* and *economical benefits*—for both the government and its subjects', but, in the long run, 'the overall e-government project, broadly understood as a product of the neo-liberal discourse of technology and the contemporary development of pan-capitalism, could represent a greater and long-lasting threat for citizens' life and freedom'.[3] The mapping of the social and political constraints that marginalized communities and individuals encounter in their interface with e-governance projects, perhaps, has implications for the optimistic political vision of new media technologies as a decolonizing force facilitating development of 'cyborg skills' required for their survival under techno-human conditions theorized in the cyber-feminist approaches to new technologies (Haraway, 1991; Sandoval, 1995).[4] Identifying the structural factors that envelop human–technology interaction in the rural setting in South Asia is thus an inevitable step in understanding social innovations and its impacts either initiated by the State or by civil society or by State–civil society partnerships. This chapter, organized as described next, is an attempt in that direction. The section titled, 'The Advent of E-governance in India', following these introductory remarks, would provide an outline of the recent developments that constitute a default history of e-governance in the country. The section titled 'Gyandoot: Organization and Technology' discusses the technological organization of the Gyandoot project and the section on 'Gyandoot Kiosks: The Social Dynamics' takes up the question of the social dynamics of the Gyandoot cyber kiosks, which provides access to the Intranet. The section titled

'E-governance and the Network Society: Deciphering the Narratives of Success' would take a closer look at the meanings of the seemingly hyped anecdotes and narratives of success along with the exposition of a realistic tale that explains the interplay of technology and power enmeshed in the local dynamics of absorption and appropriation of Gyandoot's Intranet technology. The section titled 'QUANGOs, Civil Society and Private Sector' would outline the contours of the emerging model of e-governance. The final section will offer some concluding observations.

The Advent of E-governance in India

One of the few early responses to the possibilities of using ICTs in development administration in the developing world has been from the Indian State. It may be noted that the Indian State began to design and execute rural development programmes with a relatively visible ICT content in the 1970s, while international attention on the potential of harnessing ICTs for developmental activities is a relatively late phenomenon. The early experiments are, however, examples of attempts to use ICTs for improvising development planning, a key area of State action in the pre-liberalization era. The Dharampur Sub-District Infrastructure Planning for Development 1977 (in Bhatnagar (1990)) is one such early example of attempts to use computer applications for cost optimization and decision-making. The Karwar Rural Development Information System (1984) was yet another initiative formulated with a focus on reducing delay and curbing corruption through a monitoring programme based on computer applications (Kaul, M. et al., quoted in Bhatnagar [1990: 7]).

'Electronic governance' became a key word in the Indian State's efforts to provide a facelift to its administrative machinery in the late 1990s, now geared for 'good governance' as part of the structural adjustment strategies dictated by the World Bank and other international agencies.[5] As noted by a scholar,

> In the second phase, the implementation of the national IT Task Force and State Government IT policies symbolized a paradigm shift in e-governance policies towards using IT for a wider range of sectoral applications reaching out to a large number of people in rural as well as urban areas. Moreover, there has been a movement towards a greater input of NGOs and private sector organizations in providing services to the public. These projects

> have been influenced by the increasing focus of international agencies such as DFID, G-8, the UNDP and the World Bank under the banner of 'E-Governance for Development'. (Madon, 2002: 2)

The Ministry of Information Technology was constituted and the Central Government initiated some tentative projects aimed at testing the potential of e-governance. Besides the Central Government, many state governments also responded, seeking the possibilities of improving administrative functions by introducing e-governance at different levels of the bureaucracy. Specialized agencies have also come up within the government to initiate innovative experiments. In 1999, a national conference was organized in Bangalore attended by 75 senior bureaucrats including IT secretaries of 32 states and union territories in India.[6] The conference affirmed a resolve to create 'one-stop, non-stop, efficient, effective, responsive, transparent citizen governance through the use of information technology' (Katakam, 1999: 78). The conference came up with a declaration on the intent and content of e-governance programme in India, although it was criticized as bereft of specificity and the plans lacking time-frames (ibid.). The declaration emphasized the need for shifting the focus of governance from government centric to citizen centric in the wake of the opportunities opened by information technology for large-scale delivery of quality services. The use of IT was expected to facilitate efficient delivery of government services to citizens and business, 'to anyone, anytime, anywhere through a variety of channels at a reasonable cost' (ibid.: 79). The declaration recognized the need for re-engineering the process of government to achieve synergy with technology. A critical factor identified was the centrality of upgrading skills of the existing workforce, while also underscoring the necessity of industry–community–State partnerships in e-delivery of services. A sound communication infrastructure for ubiquitous access, a conscious effort to harmonize IT with regional requirements and innovative use of IT to prevent possible social exclusion were considered essential components of the future e-governance strategy. The role of the Central Government in supporting capacity-building efforts of state governments was also underlined.

Many projects were announced at the conference although many of them were eventually shelved or totally abandoned due to various reasons.[7] However, since 1999, the proliferation of e-governance projects in India has been phenomenal. Surging numbers of what is identified as e-governance projects indicate that most of the state governments and union territories in India claim to have accepted the need for undertaking e-governance initiatives.

Although the initial efforts on the part of the Indian State to link ICTs with development projects have been limited to localized projects and district-level planning, we can also observe that it soon advanced to centralized projects for connectivity and regulation. There are clear signs of a realization that the ICTs while providing a potential for reassuring the State's commitments to the developmental agenda, opens a gateway for strengthening the arms of governance. Nevertheless, most e-governance initiatives have focussed on development of infrastructure, capacity building and policy changes as well as participation of the private sector while showing a growing reluctance towards integrating Civil Society into its fold (Sreekumar, 2002a, 2002b).

This is true of relatively successful initiatives such as Gyandoot. The Gyandoot Project has emerged as a benchmark for innovation in e-governance and e-commerce according to commentators (Sood, 2001). The local bodies in collaboration with government officials have started ICT kiosks operated by unemployed youth who were selected and trained by the Gyandoot Samiti to run these kiosks. It aimed to cater to the everyday needs of a wide section of rural consumers. The project, which sets its objective as social engineering and development through ICTs, has marked a paradigm shift in the way government functionaries relate themselves to the needs of the poor. This major ICT project was first launched in a region that is largely tribal and impoverished. Nevertheless, it remains an administered programme with little relationship with the civil society. No major social organization has been made a partner in its implementation.[8]

Gyandoot: Organization and Technology

Gyandoot, an Intranet-based Government-to-Citizen (G2C) service delivery portal in Dhar district of Madhya Pradesh was commissioned in January 2000 with an objective of creating cost-effective, replicable, economically sustainable and financially viable model for e-governance. Rajesh Rajora, one of the main architects of the project and District Collector of Dhar when the project was implemented, has claimed that it was envisaged to enhance participation by citizens/government in community affairs through creative uses of ICT and also ensure equal access to emerging technologies for the oppressed and exploited segments of the society (Rajora, 2002: 66–67). Gyandoot, managed by a society called 'Gyandoot Samiti', is registered under Madhya Pradesh Societies Registration Act with the District Collector as President. The CEO of District Panchayat functions as Secretary and various departmental heads are members of the Samiti. Gyandoot operates with a team consisting of a Project Manager, an Assistant Project Officer, a Technical Head and a few computer operators. The promised services offered by Gyandoot encompass a wide range of government departments. These services can be accessed from any Gyandoot Information kiosk called *soochanalaya*, on payment of a nominal fee. There are about 40 kiosks located in different parts of the district run by local managers called *soochaks*.

These *soochanalaya*s would be equipped with PC, Printer and UPS, with facilities to provide e-governance services, commercial Internet and voice connectivity in rural areas. A host of government information and application forms are loaded on the net, and villagers can access these to submit applications to the government departments on the net. Inter-village communication is possible and relevant software has been developed and ported. The *soochak* is in a sense a rural entrepreneur and earns a living primarily from the income that the kiosk can generate. They are trained to run the kiosk by the Gyandoot Samiti. The Dhar Internet project is one of the largest rural Internet/ Intranet projects in India. It was chosen for the Stockholm Challenge Award 2000.[9]

Initially, the kiosks were provided Intranet communication using the telephone lines provided by the government-owned Bharat Sanchar Nigam Limited (BSNL). At that point, communication on telephone lines being poor and unreliable and the dial-up connection charges relatively expensive, the project faced serious limitations of coverage. Later, corDECT WLL (Wireless-in-Local-Loop) developed by IIT, Madras and Midas Communications which provides a wireless access solution for expanding telecommunication networks integrating of both voice and Internet services was adopted with reasonable success. n-Logue Communications also offered a business model to enable rural connectivity using corDECT Wireless terminal, telephone instrument, 100 MHz Pentium PC (with colour monitor, local language word processor, browsing and e-mail software) with a 16-hour power back-up for telephone and 4-hour back-up for PC.[10] Backbone Internet connectivity is supplied by Satyam Infoway. In the first phase, villages within 25 km around Dhar town, the district headquarters, are covered. An Access Centre is set up in the District Panchayat Computer Room in the Civil Station office premises.

The kiosks are offering a wide set of facilities and services such as gathering and disseminating agricultural prices, online registration of applications, online public grievance redressal, rural e-mail, village auction sites, online matrimonial site, information regarding governmental programmes, career counselling for students and facility to ask questions to experts, technical advice channel between experts and villagers, online application formats, village newspaper, and so on. It can be seen that if functioning properly, the centres can be of very high utility to the villagers.[11]

Nevertheless, setting up a kiosk in a village does not necessarily mean that these services are provided or that they are availed by people. More importantly, the question of who avails them and who are excluded either intentionally or voluntarily remains a significant poser in the social context within which the project is implemented. In order to understand this, we may need to go deeper into the actual practices of these kiosks in the rural setting.

Gyandoot Kiosks: The Social Dynamics

In this sub-section, an attempt would be made to identify some of the key features of the *soochanalaya*s and take a closer look at their everyday practices. The analysis is based on the information collected from the five Gyandoot centres that were visited during fieldwork as well as our visit to the Gyandoot Samiti headquarters housed in the Dhar Civil Station, Dhar. The clumping of e-governance initiatives as inherently good from a macro perspective is a major assumption in the literature on e-governance. The basic problem with this rationale is not only the hype or even the 'sweeping grandiloquence' of its rhetoric, as some writers would call it (Woolgar, 2002: 5). But more importantly, our attention to macro-level impacts often ignores what really happens on the ground. These technologies are actually used and experienced in everyday practice quite differently from the way their potential uses and benefits are configured for public consumption by their progenitors as well as commentators.

The locale-specific dynamics of the operation of Gyandoot kiosks provide a key to the understanding of e-governance initiatives with a social content and significantly based on a G2C interface. In this section, we would attempt to discuss four major issues that provide a reasonable assessment of the actual and potential abilities of the kiosks and the model of ICT diffusion it stands for. The issues we map out here are: (*a*) entrepreneurship and employment; (*b*) local infrastructure; (*c*) user perceptions and (*d*) sustainability. We would also look into the major tensions and contradictions that envelop the project from a micro-level perspective.

Entrepreneurship and Employment

The entrepreneurship model of launching *soochanalaya*s is at the centre of the strategy of Gyandoot Samiti, the QUANGO (Quasi Autonomous Non-governmental Organization) sponsored by the District administration of Dhar.[12] The *soochak*s who run the kiosks are chosen from the educated youth of the village who are willing to explore the potential of self-employment by launching a *soochanalaya* in their own village. The minimum qualification requirement of a *soochak*, as suggested by Gyandoot Samiti, is education upto the 12th standard. They should also be capable of self-financing the basic infrastructure for running the kiosk either through a Bank loan or using their own resources.

The user fees from the G2C services offered by Gyandoot and additional services using the same infrastructure that the *soochak* can provide are to form the basis of the earning portfolios of the entrepreneurs. The actual earnings however depend on a variety of factors. This includes the relative prosperity of the village which determines the demand for both the Gyandoot and non-Gyandoot services that the kiosk can offer, the demand for computer training from students and the number of such students willing to undertake the training for the payment of a monthly fee.

The employment potential of the kiosks is in fact very limited. The income from the kiosk cannot in most cases provide for the subsistence wages of more than one manager. The services either belonging to the G2C categories or Government to Business (G2B) categories cannot be identified as providing employment opportunities in the villages. The *soochak*s supplement the services from Gyandoot Samiti with other activities, an effort also encouraged by the Samiti. Activities such as data entry operations, small DTP jobs, Screen Printing, photocopying, computer training, horoscopes and matchmaking, and so on are some of the most prominent supplementary activities that the *soochaks* undertake using the facilities available in the kiosk.[13]

One interesting aspect that deserves mention is the social meaning of being a *soochak*. The advantages of a being a *soochak* are obviously much more than being a small entrepreneur in a village. This stems from the nature of the services offered by the kiosks, which requires the *soochaks* to be constantly in touch with the government officials. Many *soochaks* understand that this has an empowering element ingrained in it. As one *soochak* would put it:

> I am happy that I can mediate between people and government officers. The close contact that I can enjoy with senior officers is really very important. Similarly, fellow villagers also understand the fact that I have contacts with officers. This is definitely some kind of recognition.[14]

However, the complex layers of social power that characterize the village society limit the empowerment of the *soochaks*. Political parties and caste elites have a major role to play in the everyday life of the communities. The caste and social status of the *soochaks* become enormously important in defining their domain of influence. The untold story of the three Gyandoot centres of Badnaver is an illustration of this important aspect of village life in rural India.[15]

A majority of the *soochaks* belong to the educated unemployment category of the village population. Although they have been given reasonably advanced technical training in running the kiosks, they have never been exposed to any useful training on entrepreneurship or management of small enterprises. It is surprising that this aspect has been largely ignored when the model was conceived and eventually implemented. It is certainly a fact that some of the *soochaks* do show innate talents as successful entrepreneurs by trying to supplement the earnings from Gyandoot services with numerous other small-scale activities that earn them a moderate additional income, although the market for such services in the vicinity of the centre determines the potential of such efforts. But in many cases, a shrinking stream of earnings from the Gyandoot services remains the major component of the gross earnings of the *soochaks*.

Moreover, the possibility for increasing the number of kiosks and thereby enhancing the chances of educated youth becoming small entrepreneurs in places where Gyandoot kiosks do not exist is limited by both social as well as technical and infrastructure-related problems. Apart from the uncertainties of adequate demand for these services, unstable supply of electricity in rural India poses a major challenge for the operation of the kiosks. The *soochaks* invariably reported that the availability of electricity to the kiosks averages only at 4–5 hours in a day, which strictly curtails the running time. As one *soochak* puts it:

> How can I run the Kiosk with power supply limited to only 4 to 5 hours in a day? Even this is not steady. I don't think the problem will be solved soon. We have to live with this. Working with low voltage is also a serious problem. The villagers understand these problems but when I can't deliver, the credibility is affected.[16]

Tapping non-conventional sources of energy appears to be a remote alternative to the power supply problem encountered by the entrepreneurs. There has been an attempt to deploy solar panels harnessing solar energy for running the kiosks. It has been set up on an experimental basis in one of the three Gyandoot kiosks in Badnaver. The *soochak* of the kiosk was not able to comment on its utility, as it was yet to become operational.[17]

Local Infrastructure

We do not find a huge disparity in terms of kiosk-wise infrastructure facilities. Most kiosks have a minimum set of facilities that help them to connect with the Intranet services that Gyandoot offers and also supplementary equipments such as photocopier printer and UPS. While some kiosks have procured scanners with own resources, it is unaffordable for many smaller operators unless they take loans from the bank. This is not considered as a viable option by entrepreneurs in the rural kiosks who are doubtful whether the demand for the scanner services would be sufficient to cover the repayment costs. Solar panel is set up in one of the kiosks in Badnaver as we noted in the previous section after paying a safety deposit of Rs 9,000. It is surprising that the pool of equipments in some of the kiosks does not include a printer as is seen in the case of the Amjera Kiosk. Of the five centres we visited, only one (Nagda) did not have a phone connection and Internet facility. Nonetheless, the Intranet of Gyandoot is operational at the Nagda kiosk through WLL.

User Perceptions

Each *soochanalaya* presumably caters to 20–30 villages once it is set up, although in practice the visits from people beyond a 5 km radius to the kiosk is rather rare, according to some *soochak*s and local informants. These villages vary drastically in terms of population ranging from tiny hamlets with less than 500 inhabitants to large villages with nearly 8,000 inhabitants. The actual coverage of each centre is very limited. The experience of many kiosk managers is that the services of the kiosk are mainly availed by a small group of people residing in the villages surrounding the kiosk.[18]

Not surprisingly, people belonging to different social groups have different perceptions about Gyandoot although the villagers as a rule try to see positive elements of the project as outweighing the failures. However, many of the regular customers are dissatisfied with the range of services offered by Gyandoot and believe that it can do much better. Many of them would attempt to articulate in clear terms how it could be improved.

The common complaints about the operation of kiosks can be classified into five groups: (*a*) connectivity-related problems; (*b*) failure to adhere to the stipulated time-frames for redress of grievances as well as delivery of routine services; (*c*) huge disparity between the range of promised and actual services available with the kiosks; (*d*) failures in delivering G2B services and (*e*) attitudinal factors as well as poor performance of the *soochak*s. The connectivity problem, as we pointed out in the previous section, has been crucial in creating an environment of apathy among the villagers about the project.

Although many centres have been connected with optical fibre cables in order to overcome this problem, the instability in connectivity remains a major cause of user dissatisfaction with the kiosks. It is also widely felt that the departments responsible for delivering the services do not adhere to time-frames set by Gyandoot Samiti although the officials at the Gyandoot headquarters inform that these timelines have been fixed after discussions with the concerned departments.[19]

The media reports on the delays in delivering services have also pointed to the lethargy and unresponsiveness of government employees as a major reason for the decline in the credibility of the Gyandoot project.[22] While some farmers complain that timely updating of market rates are exceptions rather than the rule, officials concerned point out the slow process of computerization of government departments as a possible reason for the delays.[23] Several case studies have pointed to the complaints from users that the kiosks remain closed during office hours. *Soochak*s invariably tried to deny this and said that such occasions are rare and probably during times of prolonged failures in power supply.[24]

TABLE 6.1 **Infrastructure, Ownership, Earnings and Expenses: A Comparative Picture of Selected Gyandoot Centres**

Centre/Facilities	*Tirla*	*Amjera*	*Nagda*	*Badnaver 1*	*Badnaver 2*
Number of PCs	2	1	2	2	1
PC purchase	Own resources	Own resources (Loan from Bank of India)	Own resources + Panchayat	Own resources (PMRY)	Own resources
Phone	Yes	Yes	WLL	Yes	Yes
Internet	Yes	Yes	No	Yes	Yes
Printer	Yes	No	Yes	Yes	Yes
Ownership of the centre	Community	Self	Panchayat	Self	Self
Photocopier	1 (Through PMRY)	No	No	No	No
Education of *Soochak* UPS	BA (Political Sciences) and DTP	12th	12th, DCOA, DTP	12th, DTP, MS Office and AutoCAD	BE (Electronics)
Solar Panel	No	No	No	Yes	No
UPS	Yes	Yes	Yes	Yes (Nos. 2)	Yes
Scanner	No	No	No	Yes	
Generator				Yes (Hire)	Yes
Average number of visitors/day	10–15	5–10	5–10	2–5	2–5
Computer training for students	Yes	No	No	No	Yes
Gross earnings (Rupees)	6,500–7,000[1]	2,800–3,200	3,000–3,500	2,500–3,000[2]	8,000–10,000[3]
Earnings from Gyandoot services (Rupees)	1,200–1,500	1,500–2,000	1,500–2,000	500–700	1,500–2,000

Centre-wise Expenses (Rupees)					
Rent	Nil	Nil	Nil	Nil	Nil
Phone	300	300	Nil	300	400
Electricity	300	400	300	300	500
Loan repayment	2,500	1,400	Nil	1,550	Nil
Stationery and other expenses	200	200	300	300	400
Total	3,300	2,300	600	2,750	1,300
Net earnings (Rupees, App.)	3,200–3,700	500–900	2,400–2,900	≥300	6,700–8,700

Source: Based on fieldwork.

Notes: 1. The *soochak* reports that he could earn Rs 3,000 or more every month from photocopying. He explains that since it is the block headquarters, there is a high demand for copying.[20]

2. His income from screen-printing an activity that predates his Gyandoot franchise, is not included. He maintained that he would not probably renew his Gyandoot franchise.

3. The high income is from the education programmes conducted in the centre. The *soochak* in this centre is an engineering graduate and his brother is a franchisee of AISECE with 25–30 students enrolled.[21]

An exploratory study conducted by Center for Electronic Governance (CEG), Indian Institute of Management, Ahmedabad, India (IIMA) came to the conclusion that centres closer to the district headquarters attract more visitors, but the visitors are generally the elite of middle level farmers:

> Generally, awareness of Gyandoot exists among the literate and middle-income group families. The poor laborer or landless farmer is not aware or even interested, as he sees no value addition, in it, for him. All the 16 daily wage laborers interviewed felt this way about G2C services. (CEG-IIMA, 2002: 10)

The report further notes that:

> There seems to be varied understanding, among the people, of what services are available through Gyandoot. At the remote *soochanalayas* (and away from the central hub of Dhar), confusion is more widespread and awareness levels are very low or nil. Even where a board enumerating the services exists, people are unsure of the nature of services and how to avail of them from the *soochanalaya* (ibid.).

It is also widely felt that it is the big farms and merchants who mostly use the services. Poorer farmers and landless agricultural labourers fail to see any benefit for them through the kiosks. Children from richer families are able to avail the facilities in the kiosks at a subsidized rate compared to market rates for similar services. The new technology has not faced antipathy from the rural elites. Instead, they have welcomed it and in a certain sense exercise control over its diffusion.

Sustainability

The question of sustainability has to be addressed on two counts: at the macro as well as micro levels. The macro question is related to the issue of sustaining the Gyandoot Samiti as a QUANGO with the revenue from the kiosks. The entrepreneurs have to pay a lump sum of Rs 5,000 every year as licence fee to the Gyandoot Samiti to renew their franchise. However, this is a paltry amount, given the fact that there are only 30–35 kiosks operational at a time and that the proliferation of new kiosks in Dhar appears to have halted. The District Council, an elected local body at the district level, which funded the initial investments for launching Gyandoot Samiti, would have to continue channelling resources for the everyday operation of the Gyandoot infrastructure.[25]

The sustainability of the *soochanalaya*s depends on the net earnings. We can see from Table 6.1 that of the five cases we have examined, three break even, whereas the other two seem to be earning much less than a respectable monthly income for an initial investment of about Rs 75,000.[26] The Gyandoot Samiti had a modest expectation of a net income of Rs 36,000 per annum for each kiosk manager.[27] It can be seen from Table 6.1 that three out of the five centres do not fulfil this expectation. In the case of the Nagda centre, the kiosk is owned by the panchayat and it pays the electricity bill.

In the case of those kiosks which appear to be financially viable, the major source of income is from non-Gyandoot services. Many of the *soochak*s believe that the income from Gyandoot services is shrinking and in order to survive, they need to expand their operations. The prospects of introducing additional services depend on the local demand for such services. In Dhar District, which has 60 per cent of its population below poverty line, the chances of an immediate surge in demand for information-related services to the scale of bridging the income gap for the kiosks seems to be doubtful.

One of the most important questions regarding the operational dynamics of the kiosks is about the linkages it has with the local economy. In many cases, the Panchayat is financing the infrastructure of the kiosks and this involves channelling local resources towards the establishment costs on a monthly basis. Since the resources of the Panchayat is drawn from the tax pool as well as the fund allotted to them by the state government, the opportunity costs of financing the everyday working of the kiosks without palpable and effective benefits for the community could be very high.

E-governance and the Network Society: Deciphering the Narratives of Success

The most interesting aspect of the ICT-based development initiatives is the accent on stories from the field that finds their way into the media and public at large to highlight the achievements of the projects in a way that differs drastically from the narratives of successes of other developmental programmes which depend primarily on statistical data (whether engineered or real) on their performance and achievements. These stories about the multitude of ways in which absorption and use of ICT by the rural people are facilitated by the initiatives are instructive in constructing the idea of a rural network society as it is emerging in developing regions. But in this section, based on stories told and untold, we argue that the contested nature of the emerging network society and its limitations can be fully understood only when these success stories are deconstructed in their own locale contexts and further, by telling the suppressed stories of conflicts and tension that mar the implementation and evolution of these initiatives. This, in effect, would also open a new pathway to understand the social and political structure of the nascent rural network society. In this section, we make a provisional attempt to demystify the discourse on individual achievements that seek to glorify the relevance and role of the initiative. Further, we also examine the entrenched social factors that over-determine the delivery mechanisms institutionalized by Gyandoot. The stories that are already showcased and a story we gathered during our own fieldwork narrated by a local political activist of Shiv Sena (the Hindu fundamentalist outfit) are examples of the way in which technology–society interaction could be interpreted in a deterministic paradigm while the social shaping aspect is relegated to the background or completely ignored. The case summaries reported in Table 6.2 invariably point to the emergence of a rural network society. Nevertheless, the problematic nature of the network society is also brought to light with the aid of hindsight and locale-specific information on social factors that should underlie any interpretation of these narratives generally considered as examples of the project's success and good performance. The anecdotes analyzed in Table 6.2 have some common threads. They invariably point to layers of hyped narratives that envelope the descriptions of the achievements of the project. As the first five cases in Table 6.2 indicate, the strategy in discourses on projects benefits have been to identify isolated cases of successful use of the project instead of using standard indices of success and achievements in terms of the overall changes it had brought to area covered under the project. But a close look at the anecdotes would reveal that these narratives fail to account for any meaningful qualitative change in the life of the people concerned. They attempt a superficial glorification of the project's benefits in particular and of ICTs in general. These hyped narratives fail to impress when they are juxtaposed against the hidden social and economic backdrop of rural realities in India. In Table 6.2, we have made an attempt to show that most of the inferences drawn from the anecdotes with regard to the benefits of the projects are either misplaced or over-hyped.

In case 1, for example, an e-mail is presented as a superior medium of communication than direct or voice contact. It glorifies the fact that a bureaucrat responded to an e-mail complaint at the expense of

TABLE 6.2 Narratives in Contexts: Understanding Tales from Gyandoot Villages

The Story	*Source*	*Thrust*	*Remarks/Questions*
1. An e-mail complaint for Rs 10 brought drinking water to a tribal hamlet of 39 households. The villagers' previous complaint to local authorities had not yielded results for six months. The complaint filed through the kiosk brought a hand pump mechanic to the hamlet within two days, and he repaired the hand pump within 3 hours.	Bhatnagar and Vyas (2001); Jaffri et al. (2002)	Efficiency and reliability	Is technology capable of changing the mindsets of government employees? Why did the mechanic fail to respond when the complaint was registered through conventional channels? Is there a production of a fear or reverence for technology in the use of ICTs as a medium of communication that prompts people to act differently?
2. Farmers in Bagadi village were quoted a rate of Rs 300 per quintal from local traders for their potato crop. The kiosk was used to get the prevailing market rate in a town 100 miles away, which paid Rs 100 more. Consequently, their potato produce was sold in the distant town. The prices paid to farmers have increased approximately 3–5 per cent, keeping about Rs 200 million from the pockets of middlemen and traders*.	Ibid.	E-business, Liberation from middlemen	Why do middlemen get an upper hand in their dealings with farmers? In most cases, it is not because of the lack of awareness of the farmer regarding the market price. On the other hand, there exist deep inter-linkages between the credit and product market in rural India that forces the farmer to sell his produce to the middlemen and traders from whom they have taken credit for cultivation. Moreover, the perishable nature of some of the agricultural products and lack of warehousing facilities compel framers to sell at a price lower than the market rates. Only the big farmers actually benefit from the information in this particular case.
3. 'I asked for the price of apples at the Dhar wholesale market. The co-ordinator pressed some buttons, and there it was on the screen. I cannot read, but he told me it was 50 rupees cheaper per crate than the rate in the village market. Next morning, I travelled to Dhar to buy fruits'.	Chatterjee (2000)	Consumer freedom	It would be a surprise if a locale retailer or consumer does not know that the wholesale prices are lower in urban areas than in rural areas where incremental transportation costs might lead to a certain percentage of mark-up. If everyone travels to Dhar from this village to buy apples, or any other commodity they require since the price would be invariably lower there, the retail business in the village would come to a standstill!

(*Table 6.2 continued*)

(*Table 6.2 continued*)

The Story	*Source*	*Thrust*	*Remarks/Questions*
4. Kalsingh, a milk farmer, wanted to sell his cow. He registered with the auction facility of Gyandoot (which enables trading of commodities like milch animals, cultivable land, tractors, agricultural tools etc.). He received four trade enquiries and finally sold his cow to the highest bidder for Rs 3,000.	Bhatnagar and Vyas (2001); Jaffri et al. (2002)	Rural e-business	Farmers will not hesitate to use Intranet services of Gyandoot for selling agricultural implements and livestock if the coverage and usage of kiosks for this purpose is high. But the limitations far more outweigh the potential in this respect.
5. 256 milch animals vaccinated in one day: Upon receiving an e-mail from a kiosk that an epidemic had broken out amongst the milch cattle of the village Kot Bhidota, a veterinary rescue team was dispatched the same day. The disease *haemorrhage septicemia* was detected; the team promptly started curative treatment and vaccinated the rest of the animals against the disease. They also conducted a search in the neighbouring villages for signs of the disease and carried out preventive vaccinations. No deaths were reported.	Bhatnagar and Vyas (2001)	Efficiency and reliability	The superiority of e-mailing technology in speeding up the government machinery is not clear in this example also. The alternative of dialling the office would also have worked even better.
6. Closing illicit liquor shop: In Nagda, some miscreants ran an illicit liquor shop neatr the Panchayat office. The devotees of a temple and children attending the Girls School in its vicinity faced bullying by visitors to the liquor shop. Since everyone feared a nexus between the police and the mafia running the shop, no one dared to complain. An activist of the local unit of Shiv Sena, a Hindu fundamentalist outfit used the e-mail facility of the Gyandoot centre to send an anonymous petition to the District Collector. Three days later, the shop was raided and the perpetrators arrested.	Interview with Mohan Jat, Nagda on 14 August 2002	Anonymity, efficiency	Anonymity was important in reporting the matter and a self-styled local custodian of value (Shiv Sena is an organization that protests celebration of Valentines day and similar icons of 'western culture' in India) could make use of the facility and avoid direct confrontation with the mafia. Shiv Sena, incidentally, had 186 volunteers in the village and they mainly work to stop selling of cows and oxen for slaughtering. Mohan Jat is the manager of a Goshaala (place to keep cows rescued from being slaughtered). The person who sent the anonymous mail keeps the ***soochak*** in good humour because if he leaks his identity to the mafia, his life will be in danger.

Source: Various sources including fieldwork.

Note: * There are different versions of this story, see for example http://www.sustainableicts.org/Gyandoot%20F.pdf

other channels of communication which are in fact more effective, such as explaining over a phone which is even cheaper than sending the e-mail spending Rs 10. The anecdote, the way it is narrated, amounts to admitting that technology involved in communication is more important than the content of the communication for eliciting responses. In the second anecdote, the whole set of issues relating to the economic realities and market inter-linkages that characterize rural transactions are squarely ignored in order to glorify isolated incidents of a profitable transaction facilitated by the technology without explaining if the same 'knowledge' could have come only through the Internet. In the case of anecdotes 3 and 4 also, we could see callous assumptions about the Internet having emerged as the only medium of information for facilitating rural transactions. The narratives lack credibility if the points that they try to drive home are placed in the contexts of the socio-economic realities of the rural areas in India. The case of the e-mail-led vaccination of the milch animals that an anecdote describes in Bhatnagar and Vyas (2001) is also not dissimilar. We have ourselves asked people for such stories during fieldwork and rarely did people have any important incidents relating to their use of the Internet to narrate. One of such cases of using the Intranet for a social purpose is given as case 6 in the table. Here, the anonymity of the sender played a crucial role. The cases point to the social embedment of technology, which is neglected while glorifying them as 'success' stories. The value of information that the new technology could generate depends heavily on contexts and it does not flow from any inherent virtues of the technology itself. A detailed case-study that helps examine these issues is provided in the following sub-section highlighting the entrenched social factors that determine trajectories of technological innovations in the Indian rural setting, challenging the futuristic e-topia of the narratives of success.

A Tale of Three Kiosks

A visitor to Badnaver will be surprised to see three Gyandoot kiosks functioning just within a 3 km radius of the block headquarters if s/he knows that the stated position of Gyandoot Samiti is to encourage not more than one kiosk in one locality, since the number of villages not covered by Gyandoot is very high in the district.[28] Nevertheless, it could be too early to conclude that Gyandoot's popularity and utilization could be high enough in Badnaver to warrant the operation of three kiosks.

The first kiosk is operating within the premises of the Block office. It was not open when our visit was made. A villager volunteered to find the *soochak*, but after nearly one hour he returned to announce that the *soochak* was unfortunately not traceable. Some villagers complained that it is not unusual that the kiosk remained closed during office hours. When asked if they had any business with the kiosks, they said they had none. They have just come to see the staff of the Block Development Office. Randomly quizzing a few villagers around, it was found that none of them had used the *soochanalaya* for any of their requirements. An employee of one of the government offices housed in that campus tipped that the other two *soochanalaya*s are quite nearby. We could perhaps walk down.

He was right and it was only 5 minutes' walk from the first *soochanalaya* to the next (Badnaver 1 [B1] in Table 6.1). The *soochak* of B1, which started at a relatively recent period, sounded extremely resentful about Gyandoot Samiti. He said that the income from Gyandoot services was practically inconsequential and that he was running the kiosk on loss.

> I cannot even recover my licence fee of Rs 5,000 that I have already deposited with the Gyandoot Samiti. I survive because I earn moderately from this computer center which I started before Gyandoot franchise was taken and from a parallel screen printing service.[29]

One of the major complaints he had was regarding the unstable connectivity of the Gyandoot Intranet that he experiences. However, his problem was different from the usual troubled connectivity that *soochak*s in other kiosks were also pointing out. He said that he fails to connect when once the results of the 8th, 10th, and 12th standard public examination results and mark lists are made available through the Gyandoot Intranet for Rs 10 per copy. This is regarded as the most lucrative of the services demanded by the public. When the Board publishes the results, Gyandoot would immediately access it and send it to kiosks. Students visit the centre with their respective roll numbers and get the result.

> I don't even have a problem during load shedding. I hire a generator. But I can't connect when the results are published. I complained to the Collector. He said 'I will take care'. But nothing happened. I still cannot access the results. On the other hand, another *soochanalaya* here has no problem with the connectivity. I have a problem and the one housed in the Panchayat office campus also has problems. I suspect corruption. The Project Manager is a very honest man. But I can't say the same about operators at the Gyandoot office.

He said that people in fact demand no other Gyandoot service. 'The Mandi is in Badnaver itself. So why should they visit the kiosk for price list? All offices are housed here. What service can we offer?'[30] This was getting nowhere. So I asked him a pertinent question: When there were two *soochanalayas* in Badnaver, why did he apply for a third one?

> The *Blockwala* (the one in the Block) never opens. The other one is little interior (sic–just two furlongs from his kiosk). When I first approached the Collector he refused. He wanted me open the kiosk in another village where no *soochanalaya* has been opened. But I was not interested. I wanted to take the franchise since I had already had the computer center here. In another village I would have had to rent in a new room. But when the CEO of the District Council visited the block I met him and later he recommended that as an youngster I deserve encouragement and the kiosk was allotted to me.[31]

We could see the familiar stories of corruption, favouritism and arbitrariness in decision-making emerging as the rural network society was taking shape. Is the technology that promised transparency and responsiveness itself getting entangled in the labyrinths of corruption and nepotism? The narratives and counter-narratives pointed to the social tensions created by the high-profile technology-driven project in rural Dhar. It was with these allegations of manipulation at the back of my mind that I visited the third *soochanalaya* in Badnaver (Badnaver 2) in Table 6.1 [B2].

B2 was run by an upper caste, engineering graduate from a relatively wealthy family. Before taking the Gyandoot franchise, the family had already launched an Internet centre under the Bharat Sanchar Nigam Limited (BSNL—the government owned telecommunication company) scheme of 1 Internet dabha (shop) per Block along with an e-education programme under the AISEC licence. The BSNL scheme envisages giving 25 per cent commission to the licensee with a free telephone by which only the Internet could be accessed. Further, he had also acquired contracts for computerizing land records in a few village Patwaris, the lowest level offices of the Revenue Administration. His response was cautious when asked about his motivation to take the Gyandoot franchise when there was another *soochanalaya* operating within walking distance from his centre:

> Gyandoot Soochanalaya was opened in 2000 in the block office premises. But the *soochak* was not a trained computer operator. He got the franchise due to political connections. I met the collector and the CEO of the

> District council when they visited the Block during the inauguration of the *soochanalaya*. I personally felicitated the Collector, presented him a memento and invited him for dinner. During dinner, he asked me to oversee the working of the kiosks since the operator was not properly trained. Removing him was not possible due to his political connection. Further, it could have led to some bad publicity. We were advised to send in a proposal to the Janpath Panchayat requesting them to remove him. But I thought it was unfair. I said: 'I get my Dhal Roti (Food). Why should I prevent him from getting his'. The Collector was impressed by my answer. So he himself took the initiative to give me Gyandoot franchise. Thus this is the first private Gyandoot center in Dhar.

He was soon nominated to the Gyandoot Samiti as one of the two representatives of the *soochak*s in the QUANGO. He said that computers are being installed everywhere—in schools, offices and the hospital in Badnaver and he has played a major role in the installation of the machines in all these places. I asked him if he experienced any problems in connectivity. I told him that in B1, the *soochak* had complained about connectivity problems when public examination results are published. His response was quite characteristic:

> I have no connectivity problem. Others have problems because they don't know how to operate. They are not skilled enough. They have no proper training. When private people are given Gyandoot franchise it should be ensured that they have adequate experience in handling the machine. [B1] was started after I took Gyandoot franchise. It came up because the CEO of the Janpath Panchayat who was the tenant of the soochak recommended. In fact I was consulted before it was given. I tried to stop it. But the project manager told me that whoever gives Rs 5000 can start a Gyandoot kiosks and anybody who is efficient can make money for it.

The narratives differ. But the story is a familiar one of bureaucratic muddle, political interference and mutual distrust. The local elite has been effectively controlling the new technology facilitated by the interplay of cross-linked forces such as political power, influence in bureaucratic circles and caste structure. The narratives of poor farmers and agricultural labourers benefiting from the technology are often exaggerated from the contexts in which such exceptions are made possible.

QUANGOs, Civil Society and Private Sector

This leads us to the larger question of Civil Society participation in e-governance projects initiated by the state governments. As indicated in the beginning, civil society organizations appear to be kept out of the institutional structures of e-governance. Even in cases where such participation is sought, the linkages are either weak or notional. While the State-led projects in e-governance show a relative reluctance to work with Civil Society Organizations (CSOs), it has attempted to follow the QUANGO Model wherever relevant and possible. QUANGOs, like Gyandoot Samiti, form the backbone of the projects launched by several state governments. QUANGOs are defined as organizations which essentially undertake the responsibility of implementing State-sponsored programmes or public policies funded by the State but operating at arms length of the executive without an immediate hierarchical relation with it (Van Thiel, 2001: 5). The formation of QUANGOs is part of the general strategy adopted by the States, informed by the logic of Civil Society mediation. As pointed out in the Pliatzky Report (quoted in Flinders, 1999: 29), QUANGOs are created since,

> The work is more effectively carried out by a single-purpose organization rather than by a government department with a wide range of functions; in order to involve people from outside of government in the direction of the organization; in order to place the performance of a function outside the party political arena.

Nevertheless, QUANGOs are not real substitutes for CSOs and most often degenerate into behemoth bureaucratic entities. Further, this leads to the incorporation of individuals and organizations, not accountable to any constituency even notionally, into the governing structures of developmental and e-governance programmes. Figure 6.1 tries to provide a conceptual model of the emerging e-governance programmes being carried out in India. This is a complex institutional model where sustainability issues are embedded in the policy prerogatives that mould the State's interest in the programme.[32]

FIGURE 6.1 **Emerging Model of E-governance Projects**

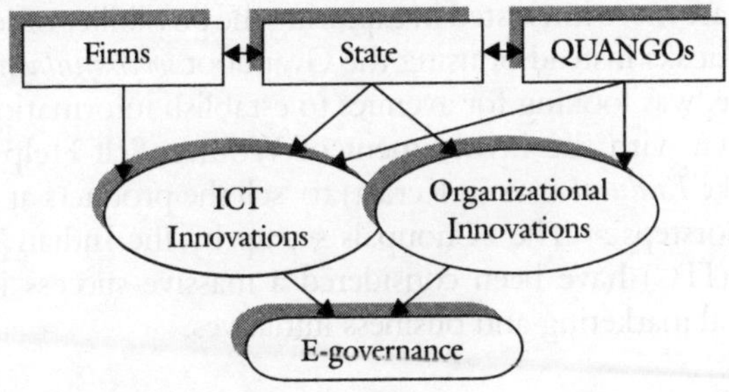

While the civil society involvement in e-governance projects is either notional or limited, private sector participation is seen as essential to the successful implementation as well as sustainability of the project. Our visit to Gyandoot kiosks also led to a chance meeting with a representative of Hindustan Lever Limited (HLL), the Indian subsidiary of Unilever, who was interested in learning more about Gyandoot kiosks.[33] A presentation by Project Manager, Naveen Prakash was aimed at exploring possibilities of future partnership with HLL. The District Collector was hoping that the private sector involvement would help overcome the financial crunch and make the project viable in the long run.[34] Gyandoot's offers to HLL can be listed as follows:[35]

1. HLL can test the success of its new products in rural areas through Gyandoot.
2. Gyandoot kiosks can be used for collecting feedback from HLL customers.
3. HLL can profile rural consumers using information available at the kiosks. Moreover, the *soochaks* could be used as surveyors for HLL for a fixed remuneration.
4. HLL could also use the Gyandoot portal for marketing and advertizing its products.
5. *Soochaks* could display HLL posters, banners and so on in the *soochanalayas*.
6. Since many of the *soochanalayas* are located at bus stands, haat bazaars, Block headquarters and so on, HLL can use the kiosks for selling its products. The brand identity of Gyandoot as a provider of quality services would be useful for HLL in selling its products. The kiosks managers need to be paid the normal commission HLL pays to its retailers.

Although private sector participation in paid technical services is visible in the case of most projects, strategic partnerships as envisaged by Gyandoot are not forthcoming due to a variety of reasons. Private companies are more interested in exploring the possibility of setting up their own kiosks instead of using the Gyandoot *soochanalayas*. HLL, for example, was looking for avenues to establish information kiosks of their own with the involvement of Women Self-Help Groups (SHGs—like *Kudumbasree* in Kerala) to 'sell the products at the consumer's doorsteps'.[36] The eChoupals set up by the Indian Tobacco Company (ITC) have been considered a massive success in kiosk-centred rural marketing and business initiatives.

Conclusions and Discussion

The central themes addressed in this chapter relate to the critique of the notion of e-governance as an essentially administrative innovation facilitated by ICTs and recognition of e-governance as a social process which involves not only attitudinal change and transformation of traditional forms of governmentality, but also as a contested arena of social forces shaping the trajectory of the evolution of this technocratic innovation. E-governance delivered simply as an improvement in the pragmatics of governance exemplified in the efforts to make service delivery quicker or more accessible, would probably end up in reproducing technological practices, which hinges for its existence on the crucial technology component rather than its social dimensions and would help the consolidation of and centralization of power in the hands of those who directly or by proxy own, control or manipulate the technology. We find that the relationship between technology and governance is an area that needs closer scrutiny than is often made possible by perspectives of governmentality and developmentalism. One of the central issues that emerge in the context of exploring the inter-relationship between technology and governance is the questioning of the assumption of the neutrality of technological processes enmeshed in the notion of e-governance. Technical processes defining the contours of e-governance are embedded in the structures of power which reinforce the power relations that e-governance, according to the developmental perspective, is expected to eliminate. This is particularly obvious when we closely scrutinize the consequences of the introduction of ICT-based administrative projects in villages where the local elite play key roles in its implementation and admittedly benefit from this public good. We have seen that the projects are often implemented with the active support and participation of the village elites and their collaboration is a major aspect of the survival of the project.

We have seen that automation on the one hand and projects with a social content on the other form the key strategy of defining the pace of development of e-governance as identified by the State. The Indian state had shown an early interest in automation during the 1970s and the 1980s and took a leading role in the implementation of innovative e-governance projects in the late 1990s. Our focus in this chapter is to understand the projects in the social contexts in which they are introduced, taking exception to an implied essentialism of both these perspectives. The computerization of government departments and the launching of projects with a social content emerged as a near-universal pattern for e-governance processes in India at the state level as well as at the national level. Even states with relatively poor performance in terms of social and economic indicators surged ahead with innovative projects in e-governance. Nevertheless, many of these projects are floundering and have been unable to break the initial inertia. Even when the programmes are able to make modest successes in terms of expansion and sustainability, the conflicts generated in the domain of technology–society interaction in these projects are enormous and deserve closer attention than developmental perspective can offer.

One of the crucial aspects of networked governance is its potential for creating a network society in rural areas. Gyandoot's Intranet is an example of an emerging rural network society with its scattered nodes and decentralized delivery system. However, the sociological aspects such as power relations and technological aspects such as connectivity are important in defining its contours of effectiveness and success. Rural network society developed as an offshoot of networked governance could at best be considered as akin to a techno-social network with a potential to increase both citizen-to-government and person-to-person communication in a specific geographical unit. However, if we consider the contemporary history of projects such as Gyandoot as evolving techno-social networks, their potential for reproducing traditional lines of social inequalities and reinforcing rural power hierarchies rather than eliminating them cannot be overlooked. The idea that ICT is inherently a liberating technology, and hence e-governance, is a new way of transcending inept and inefficient bureaucratic systems, which empowers 'end users', appears to be completely inaccurate in the rural societal setting. Moreover, despite the claims of active networking of people in rural Dhar made on behalf of Gyandoot, its ability to connect to multiple social and economic domains is extremely limited and ostensibly mediated by the social power equations that envelop its institutional setting.

Acknowledgements

Thanks are due to Erik Baark, Greg Felker and Agnus Ku for their comments and suggestions on earlier drafts of this paper. I may also thank Govindan Parayil for useful discussions and Milagros Rivera Sanchez for detailed comments. I have also benefited from critical comments by Ashwini Saith, Steve Fuller, Antony D'Costa, Tojo Thachenkery, Alakh N Sharma, Subhash Bhatnagar Narelle George, and Yanji Bian at seminars and workshops held in Bangalore, Lisbon, Tunis, Singapore, Seoul and Hong Kong, where earlier versions or certain sections of this paper have been presented. The usual disclaimers apply.

Notes

1. For a discussion of the term e-topia, see Mitchell (2000).
2. For a discussion, see Feenberg (1999).
3. See Baddeley (1997) for a discussion of Foucault's concept of Governmentality in the context of ICTs.
4. A host of commentators have provided uncritical evaluation of the ICT-based rural social innovations as providing liberating opportunities that members of the marginalized communities and groups can use successfully for 'empowering' themselves. See, for example, Bhatnagar and Sechware (2000); Hafkin and Taggart (2001); Rajora (2002); Hafkin (2002); Gajjala (2002); Sharma (2003); Pringle and Subramanian (2004); Peizer (2005); Ng and Mitter (2005).
5. For a critique of the technocratic dimensions of the good governance programmes of the Indian State, see Joseph, S. (2001).
6. Also see www.ciol.com/content/services/egov. The conference was one of the highlights of the Bangalore IT.com 1999 event.
7. One such example is the project announced by IT secretary, Kerala for computerizing and networking relevant applications for 1,214 local bodies which claimed development of software for the payment of welfare pensions and a database which can be used for procuring building licences, licence fees and taxes. The secretary was confident that the system would help people living in rural areas to 'find out what quotas they are entitled to, what schemes are applicable to them, or in the case of land records, where they are located and whom to contact' (Katakam, 1999: 79). Five years after the announcement, the project has not yet taken shape!
8. The Warna wired project in Maharashtra is perhaps a significant counter example. It covers 70 contiguous villages, providing information and knowledge to increase efficiency and productivity of the Sugar Cooperatives. It aims to support the Cooperatives with agricultural and medical information besides setting up a constellation of facilitating centres for continued enskilling and education. The national informatics centre (NIC) of Government of India and state's education department collaborated with the Cooperatives to launch a strong network of fibre-optic cables, V-SATs, PCs and other ICT equipments.
9. Gyandoot was a joint winner of Stockholm Challenge Award in the category 'Public Services & Democracy' along with the Australian initiative 'Technology for Social Justice'.
10. Discussion with Sanjay Inamdar, Regional Manager–Sales (Western Region) of n-Logue Communications Pvt. Ltd, Chennai on 13 August 2004 at the headquarters of the Gyandoot Samiti. He considers the collaboration as being mutually beneficial. See also Chapter 5 for a detailed discussion of the corDECT WLL technology.
11. For discussion of the uses of the services often offered by these types of village kiosks see Saith (2002). The author attempts an outsider evaluation of the utility of the services to farmers in the case of MSSRF based on the list of services they have given in their website. He wonders how some of these could be relevant and resulting in any material gains for the poor people.
12. For a discussion of the popularity of QUANGOs among bureaucrats, see Flinders (1999). In India, it is preferred because of a variety of reasons. First, it gives the façade of an NGO for an otherwise bureaucratic set-up run by civil servants deputed by government. Second, it provides certain leeway with budgeting since

their accounting falls outside the annual governmental budgets. Third, it provides a suitable channel to receive foreign funding which can be routed only through NGOs. See the section on 'QUANGOs, Civil Society and Private Sector' for further observations on the theme.

13. In many Kiosks, additional software for preparing astral birth charts and horoscopes based on the Indian system of Astrology is available. The standard rate per copy of the birth chart is Rs 50. Akhil Jain, *soochak* in the Nagda centre, reported that he has purchased the software programme called 'Kundli 2000' from Indore and rates the demand for this service which is essential for match-making as relatively high. Interview with Akhil Jain was conducted on 14 August 2002.
14. Interview with Akhil Jain on 14 August 2002.
15. We would be discussing this case subsequently in this section.
16. Interview with Devendra Chauhan, *soochak* at the Amjera Kiosk, on 12 August 2004.
17. Interview with Manish Sharma, *soochak* at the Badnaver (1) kiosk.
18. Authors often equate the total population in the villages with people who avail the services of the kiosks. For example, many case studies on Gyandoot labour the point that Gyandoot kiosks in a locality may be serving thousands of people in the nearby villages. The statement from an earlier case-study on Gyandoot is typical in this regard: '20 kiosks (*soochanalayas*) were initially set up in various rural centres, with each kiosk typically serving a population of 20,000–30,000 villagers' (Sanjay and Gupta, undated). A similar uncritical observation can be seen in Prahalad and Hammond (c.2002: 11): 'Aggregating demand makes the system highly cost-effective: each kiosk serves 25–30 surrounding villages, while the network as a whole covers over 600 villages and more than half a million people'.
19. Interview with Naveen Prakash (Project Manager) on 12 August 2004.
20. The CEO of the Tribal Block Development Office, S C Sharma verified the statement by the *soochak* and added that the Block takes services form the Kiosk and pays for it. Interview on 12 August 2002.
21. One of the students at the kiosk however said he does not know about AISECE and that they award the degree. Interview with Rithu Raj Sing on 14 August 2002.
22. *Hindustan Times*, for example, reported that: 'the entire plan can get stuck if the downslide in its back-end operation continues. Largely attributed to the lethargic and ignorant government employees, the project has been witnessing snags of late. 'Against the promised reply within a week of a complaint being lodged through Gyandoot, delay has become the order of the day', says an official of the District Rural Development Authority (DRDA), the government agency handling the nitty-gritty of its operations (retrieved from http://www1.worldbank.org/publicsector/egov/gyandootarticle.pdf). Another report puts the scenario as follows: 'The operator at the information centre in Nalcha Block in Dhar has no clients. He has no electricity for hours and his information kiosks are deserted… It would appear as though Gyandoot has not been able to provide all the 44 services it was set up to deliver'. NDTV 27 January 2003. [Archived at http://www.apnic.net/mailing-lists/s-asia-it/archive/2003/02/msg00013.html].
23. Media has been critical of Gyandoot on this front as well: 'The mandi-rates, supposed to be updated daily, are not changed for two–three days on end. As the rates change many times in a day, the facility is more or less useless. Moreover, a farmer has a cheaper option of a telephone call at 80 paisa (Re 0.80) to Dhar mandi to

ask for rates instead of paying Rs 5 user charge at the cyber centre. Land records of only three out of seven tehsils in the district are available on the network. There has also been duplication of work. The Land Records Bureau was doing the job for the last many years while it was undertaken simultaneously by the district administration, only to abandon it later' (ibid.).

24. During our visits, one of the three centres in Badnaver was closed during office hours. Incidentally, this one was housed in Block headquarters itself!
25. The initial investment for the government was Rs 2,500,000, which was collected from the Local Bodies (Interview with Sanjay Dubey, District Collector, Dhar District, on 12 August 2002). The District Council established the Server at a cost of Rs 5 lakhs, spending Rs 1.5 lakhs on the machine and another Rs 1.5 lakhs for software. The District Council also financed the training of the *soochaks* who were selected to run the kiosks. Moreover, the subsidy of Rs 16,000 for each entrepreneur is also borne by the District Council.
26. The *soochaks* have to pay Rs 60,000 for procuring the machine and setting up the centre in addition to a licence fee of Rs 5,000 payable to the Gyandoot Samiti. It is likely that they incur an amount of Rs 8,000–10,000 as other expenses including purchase of stationery to set the kiosk operational (interview with Naveen Prakash (Project Manager) and Deepak Sharma (*soochak*, Tirla Centre) on 12 August 2002).
27. Interview with Naveen Prakash on 12 August 2002.
28. Interview with Naveen Prakash (Project Manager) on 12 August 2002.
29. Interview with Manish Sharma on 14 August 2002.
30. Ibid.
31. Ibid.
32. This model originally identified in Sreekumar (2002a), (2002b) based on the Indian experience has been found to be relevant in the contexts of understanding e-governance initiatives elsewhere (Sokolova, 2006).
33. The potential of using the Gyandoot kiosks for market expansion has been explored by corporate organizations like, HLL, Tata Trust, Mahindra Tractors, Jain Irrigation and S. Kumars.
34. Interview with Sanjay Dubey (District Collector Dhar) on 12 August 2003.
35. Based on Naveen Prakash's presentation 'Gyandoot and HLL: Looking for Strategic Partnership' at the Gyandoot headquarters on 12 August 2003.
36. Discussion with Sarath Dhall (Marketing Manager, Rural–New Ventures, Hindustan Lever, Mumbai) on 12 August 2002.

References

Armitage, J. 1999. 'Resisting the Neoliberal Discourse of Technology: The Politics of Cyberculture in the Age of the Virtual Class, CTheory.net'. Article a068, archived at www.ctheory.net/articles.aspx?id=111.

Baddeley, S. 1997. 'Governmentality', in B. Loader (ed.), *The Governance of Cyber Space: Politics, Technology and Global Restructuring* (pp. 64–96). New York and London: Routledge.

Bhatnagar, S. 1990. 'Computers in Developing Countries', in S. Bhatnagar and N. Bjorn Anderson (eds), *Information Technology in Developing Countries*. Amsterdam: North-Holland.

Bhatnagar, S. and N. Bjorn Anderson. (eds). 1990. *Information Technology in Developing Countries*. Amsterdam: North-Holland.

Bhatnagar, S. and N. Vyas. 2001. 'Gyandoot: Community-Owned Rural Internet Kiosks'. Retrievable at http://www1.worldbank.org/publicsector/egov/gyandootcs.htm.

Bhatnagar, S. and R. Sechware. (eds). 2000. *Information and Communication Technology in Development: Cases from India*. New Delhi, Thousand Oakes and London: Sage Publications.

CEG-IIMA. (Center for Electronic Governance, Indian Institute of Management, Ahmedabad). 2002. *An Evaluation of Gyandoot*. Washington: World Bank.

Chatterjee, P. 2000. 'Internet Takes Roots'. MSNBC, February 27 (quoted in the Project description of Gyandoot for the Stockholm Challenge Award, archived at: http://www.stockholmchallenge.se/projects.asp?ProjectId=585).

Feenberg, A. 1999. *Questioning Technology*. London and New York: Routledge.

Flinders, M.V. 1999. 'Quangos: Why Do Governments Love Them?', in Flinders, M.V and M.J. Smith (eds), *Quangos, Accountability and Reform* (pp. 26–39). Great Britton: Macmillan Press.

Gajjala, R. 2002. 'Cyber-Feminist Technological Practices: Exploring Possibilities for a Woman-Centered Design of Technological Environments'. Background paper prepared for the 'INSTRAW Virtual Seminar Series on Gender and ICTs. Retrieved from www.uninstrw.org.

Ghatak, M. 2002. 'Use of Information Technology in Agriculture'. PGPABM-2002 MANAGE. Hyderabad, archived at www.indiainfoline.com/bisc/itin.pdf.

Hafkin, N. and N. Taggart. 2001. 'Gender, Information Technology, and Developing Countries: An Analytic Study.' USAID. Retrieved from www.usaid.gov/wid/pubs/hafnoph.pdf.

Hafkin, N. 2002. 'Are ICTs Gender Neutral? A Gender Analysis of Six Case Studies of Multi Donor ICT Projects', background paper prepared for the 'INSTRAW Virtual Seminar Series on Gender and ICTs. Retrieved from www.uninstrw.org.

Haraway, D.J. 1991. *Simians, Cyborgs and Women: The Reinvention of Nature*. New York: Routledge.

Jafri, A., A. Dongre, V.N. Tripathi, A. Aggrawal and S. Shrivastava. 2002. 'Information Communication Technologies and Governance: The Gyandoot Experiment in Dhar District of Madhya Pradesh, India', Working Paper 160. London: Overseas Development Institute.

Joseph, S. 2001. 'Democratic Good Governance: New Agenda for Change', *Economic and Political Weekly*, March 24:1011–14.

Katakam, A. 1999. 'Towards E-Governance', *Frontline*, 16(25): 78–80.

Kaul, M. [et al.]. 1989. 'New Information Technology Applications for Local Development in Asian and Pacific Countries', *Information Technology for Development*, 4(1).

Madon, S. 2002. 'The Development Impact of E-Governance', paper presented at the International Seminar on ICTs and Indian Development, Bangalore, 9 November 2002, jointly organized by Institute of Social Studies, The Hague and Institute for Human Development, New Delhi.

Mitchell, W.J. 2000. *E-topia: Urban Life, Jim—But Not As We Know It*. Cambridge, Massachusetts: MIT Press.

Navarria, G. 2006. 'The Three Faces of Government in the Age of the Internet and the Future of Activism Within a Condition of Shared Weakness', *EastBound*, 1(1): 124–52.

Ng, C. and S. Mitter (eds). 2005. *Gender and the Digital Economy: Perspectives from the Developing World*. New Delhi: Sage Publications.

Piezer, J. 2005. *The Dynamics of Technology for Social Change*. Lincoln, NE: iUniverse.

Prahalad, C.K. and A. Hammond, 2002. 'What Works: Serving the World's Poor, Profitably'. World Resource Institute: Digital Dividend. Archived at http://www.digitaldividend.org/pdf/serving_profitably.pdf.

Pringle, I. and S. Subramanian (ed.). 2004. *Projects and Experiences in ICT Innovation for Poverty Reduction*. New Delhi: UNESCO.

Rajora, R. 2002. *Bridging the Digital Divide: Gyandoot Model for Community Networks*. New Delhi: Tata McGraw-Hill Publishing Company.

Saith, A. 2002. 'ICTs and Poverty Alleviation: Hope or Hype?', paper presented at the 'International Seminar on ICTs and Indian Development', Bangalore, 9 November 2002, jointly organized by Institute of Social Studies, The Hague and Institute for Human Development, New Delhi.

Sandoval, C. 1995. 'New Sciences: Cyborg Feminism and the Methodology of the Oppressed', in Chris Hables Gray (ed.), *The Cyborg Handbook* (pp. 407–22). New York: Routledge.

Sanjay, A.K. and V. Gupta. Undated. 'Gyandoot: Trying to Improve Government Services for Rural Citizens in India', eGovernment for Development eTransparency Case Study No. 11. Archived at http://www.egov4dev.org/gyandoot.htm.

Sharma, U. 2003. *Women Empowerment through Information Technologies*. New Delhi: Authors Press.

Sokolova, M. 2006. 'Improving Quality of eGov Strategies in Belarus, Ukraine and Lithuania through Citizens'. Participation in 'eGov Planning Open Society Institute', Center for Policy Studies, Central European University. Archived at http://pdc.ceu.hu/archive/00002840.

Sood, A.D. 2001. *A Social Investor's Guide to ICTs for Development*. Bangalore: Centre for Knowledge societies and Charities Aid Foundation India.

Sreekumar, T.T. 2002a. 'Civil Society and ICT based Models of Rural Change: History, Rhetoric and Practice', paper presented at the International Seminar on ICTs and Indian Development, Bangalore, 9 November 2002, jointly organized by Institute of Social Studies, The Hague and Institute for Human Development, New Delhi.

———. 2002b. 'Civil Society and The State-led Initiatives in ICTs: The Case of Kerala', *Information Technology in Developing Countries,* 13(2)December. Archived at http://www.iimahd.ernet.in/egov/ifip/dec2002/Dec2002.pdf.

Sreekumar, T.T. 2003. 'De-hyping ICTs: ICT Innovations by Civil Society Organizations in Rural India', *i4d-Information for Development*, 1(1): 22–27.

Sreekumar, T.T. 2006. 'ICTs for Rural Poor: Civil Society and Cyber-libertarian Developmentalism in India', in G. Parayil (ed.), *Political Economy and Information Capitalism in India: Digital Divide, Development Divide and Equity* (pp. 61–87). UK: Palgrave Macmillan.

Van Thiel, S. 2001. *Quangos: Trends, Causes and Consequences*. Aldershot: Ashgate.

Woolgar, S. (ed.). 2002. *Virtual Society? Technology, Cyberbole, Reality*. Oxford: Oxford University Press.

7

Crossing a Chasm: Technologies, Institutions and Policies for Developing a Regional IT Industry[1]

Nirvikar Singh

Introduction

What does it take to build a successful local IT industry? This chapter examines the joint role of technology trends and infrastructure, along with the policy environment and institutions, in developing a robust IT industry. We use the experience of the Indian state of Punjab as a case study, since it is a relatively advanced state that has so far failed to develop a robust IT industry despite substantial efforts over the last few years. We also evaluate the connection of the IT industry to the broader structure of the economy and society. For example, it can be argued that Punjab's strength in agriculture and some manufacturing industries must be built on, for a local IT industry to succeed. We suggest that, in fact, information and communication technologies (ICTs) are complementary to existing economic activity in Punjab, and can be viewed in the context of an overall economic strategy of diversifying and increasing the efficiency of the existing economic base. We discuss policies that are emerging, and prospects for success in the light of current technological and economic trends. We also place the Punjab experience in the overall Indian context of building ICT capabilities.

Punjab is a particularly interesting case study, because it is a high-income state with a very good infrastructure as measured by broad-based indices (Table 7.1), yet it has been relatively unsuccessful after India's economic liberalization, including in its efforts to develop an IT industry within the state. It is best known as the centre and vanguard of India's green revolution, the use of high-yielding varieties of food grains, along with complementary modern inputs (tractors, fertilizers, irrigation, and so on) to substantially increase crop yields. In recent years, however, this growth engine has lost steam and Punjab has also been beset by environmental concerns associated with its current pattern of cropping and resource use.

One alternative that has presented itself as a growth area is the IT industry, which has been an engine of growth for several of the states of southern India and captured the imagination of policy makers, academics and entrepreneurs. The Punjab government, following this trend, has made substantial efforts to develop an IT industry in the state. Despite these efforts, the state's success in IT has been very limited. Given Punjab's apparently favourable pre-conditions, this relative failure is paradoxical at first sight. We resolve this paradox by arguing that Punjab's policies have not gone far enough in the right direction. We can use the familiar metaphor of needing to cross a chasm in one big leap, not a series of small hops. We therefore discuss what would constitute 'far enough' in terms of institutions and policies to cross the chasm.

Our answer to the paradox also focuses attention on the relationship of IT to the rest of the economy, and its role in broader economic development. Punjab has a range of manufacturing firms in sectors such as textiles, apparel and transportation equipment. What is the actual and potential role of IT in helping these firms to compete more effectively in a liberalized economy? Similarly, what kinds of roles can IT play in improving agricultural efficiency, through better production decisions or improved market access? Even more broadly, how can IT impact rural development, through improvements in access to local government, delivery of services such as education and health, and general attitudes to social change? Are these benefits significant, and how can they be realized in a sustainable manner?

The structure of the chapter is as follows. In the section on 'The IT Industry in Punjab and Government Policies', we briefly review the

TABLE 7.1 **Relative Infrastructure Development Indices, 14 Major States**

	1980–81	*1991–92*	*1996–97*
Bihar	83.5	81.7	77.8
Rajasthan	74.4	82.6	83.9
Uttar Pradesh	97.7	102.3	103.8
Orissa	81.5	95.0	98.9
Madhya Pradesh	62.1	71.5	74.1
Andhra Pradesh	98.1	96.8	93.1
Tamil Nadu	158.6	145.9	138.9
Kerala	158.1	158.0	155.4
Karnataka	94.8	96.5	94.3
West Bengal	110.6	92.1	90.8
Gujarat	123.0	122.9	121.8
Haryana	145.0	143.0	137.2
Maharashtra	120.1	109.6	111.3
Punjab	207.3	193.4	185.6
All India	100	100	100

Source: CMIE index, taken from Ahluwalia (2002).

state of Punjab's core IT industry and some of the recent policy efforts and contrast its experience with that of more successful states. In the sections on 'Telecommunications Infrastructure' and 'Human Capital', we look at two key inputs for the IT sector, telecommunications infrastructure and human capital and discuss trends as well as policy efforts along these fronts. In the section on 'Forward Linkages and Development Impacts', we examine the possible and actual linkages from the IT industry to sectors that can use IT to improve efficiency or competitiveness and discuss potential and actual development impacts in industry, agriculture and services. In the section on 'Governance and Policy Environment', we look in particular at the role of IT in governance in the state, as well as discussing the current policy environment and new policy proposals. The section 'Conclusions' concludes with an assessment of Punjab's experience, its prospects and the lessons for India's overall prospects in building ICT capabilities.

The IT Industry in Punjab and Government Policies

The success of India's software exports, especially from places such as Bangalore, led other states to try to jumpstart local IT industry clusters. The main focus was on software development, but there were also niches in hardware design. However, despite substantial policy efforts in the late 1990s, Punjab's IT industry remains small. This result is somewhat paradoxical, for several reasons. Chandigarh (the Union Territory that serves as Punjab's capital) and Mohali appeared to be extremely attractive sites for an IT industry. They are close to the nation's capital, the area is pleasant to live in, land is available, and the people of the region have a reputation for being energetic and enterprising. The Punjab government allocated land, a software technology park (STP) was set up, and firms started to move in, from abroad and from other parts of India: major examples included Quark, Infosys and Tata Interactive. Other firms looked seriously at the region as a place for new investment.

However, in 2002, the total size of the IT industry in the region was still tiny, and much of the activity came from struggling small firms. Some firms had even left, and the main new development project for the STP had not got beyond laying of the foundation stone. What went wrong? Interviews with policy makers and people in the industry suggest that, aside from the problems created by the global slowdown, as well as Chandigarh's lack of an international airport and a few other amenities, the policy environment was just not right. As we discuss later in the chapter, Punjab was able to overcome some initial constraints in terms of local human capital and telecommunications infrastructure. However, other, less direct or tangible components of the policy did not fall into place.

The issues were not ones of lack of direct financial incentives, which were indeed offered for greenfield investments in the state. Given the parlous state of Punjab's government finances, such financial incentives can be too costly. They may also be counterbalanced by higher costs down the road. For example, the huge subsidies provided to agricultural users of electric power have been one factor in high industrial power costs (less the case Punjab), as well as poor quality power supplies. The latter, in particular, was mentioned by firms as a serious constraint on operations. To some extent, the power problem is severe all across India, but Punjab has been distinguished by the degree of deterioration in its power situation over the last decade. A survey of 10 states (Goswami et al., 2002) certainly found a correlation between the power situation and the perceived investment climate in those states. The electric power example illustrates a more general point that the IT sector cannot be looked at in isolation of the rest of the economy. We will provide other illustrations later in the chapter.

Even given the constraints on electric power supplies, the government might have compensated by doing more to create a hassle-free environment for business, both in starting up and in continuing operations. Field interviews yielded stories of government-as-usual, including delays, corruption and policy inconsistencies. When there are other location options, such an environment can be fatal in getting the momentum going. In particular, the National Capital Region, comprising Delhi, Gurgaon (Haryana) and NOIDA (Uttar Pradesh), has presented itself as a formidable rival for being an IT destination in North India. It also appears that the Union Territory of Chandigarh did not initially approach the issue of attracting IT investment with much alacrity. As a result of these factors, not a single firm made a big enough investment in the region to get the ball rolling, and there was never a critical mass in Punjab's IT industry. Employment directly associated with IT remains tiny. We also found that there was limited interest in IT at the northern region headquarters of the Confederation of Indian Industry (CII) in Chandigarh.

Having noted the negatives, we also note that several things seem to have changed since eight years ago. First, there are some firms that are based in Mohali and doing reasonably well; so there is a base that did not exist earlier. Second, the telecommunications infrastructure for the STP has improved considerably, so that bandwidth is much less of a problem than it was a few years ago. Third, as we discuss in greater detail in the section on 'Human Capital', the local availability of technically trained manpower has improved, though there are some remaining gaps. Fourth, higher costs in and around Delhi, plus some recent efforts on the part of the Chandigarh administration, may improve Punjab and Mohali's relative attractiveness as an IT destination. Finally, government policy appears to have progressed, at least in the conceptual stage. Whereas earlier efforts were somewhat *ad hoc*, there is now an effort to devise a comprehensive plan for the IT industry. As we discuss in the section on 'Governance and Policy Environment', if new policies make it to the stage of effective implementation, Punjab may be able to do better in the future in developing a local software industry.

Telecommunications Infrastructure

As discussed in the introduction, Punjab is a relatively advanced state, with one of the highest per capita incomes and also the highest infrastructure index among the states. Nevertheless, in practice, we found that the state of the telecommunications infrastructure was inadequate, especially for Internet use in the rural areas.

Initially, in the 1990s, there were problems with the bandwidth required for software exports. Policy initiatives at the state and national levels have gone a long way towards reducing the constraints of bandwidth through new investment in infrastructure, though firms suggested to us in interviews that there were still occasional problems with availability and that bandwidth costs are significantly higher than in some other parts of India. From the state's leading ISP, we also learned about overcharges for leasing lines and other operational obstacles that were consistent with descriptions of the state's policy environment in other areas such as electric power provision and land use regulation.

The problems of rural Internet access were of the kinds that do not show up in official statistics. An immediate problem (based on observations in Bhatinda district) faced by semi-commercial efforts to provide rural Internet access and services was the poor quality of telephone connectivity. While phone lines might be available or installed specifically for an Internet kiosk, the quality of service turned out to be so poor in most cases that potential Internet-based services were undermined. While IT-based services that do not rely on the Internet or telecommunications are valuable, their financial sustainability is much harder to achieve without good telecommunication links.

Many of the problems with telecommunications infrastructure have to do with central government regulatory policies and so a state government has limited room for manoeuvre. Nevertheless, awareness of the constraints is important in order for the state government to be able to lobby the centre to relax unnecessary constraints or to formulate its own policies that work around the constraints.

In particular, India's telecom infrastructure needs must be understood not just as providing high-quality, high-speed international data links for its premier software firms and independently providing basic local voice calling capabilities for poor villagers. While clearly what is provided is closely linked to the ability to pay, the value of being part of a well-functioning network must not be underestimated, even for the bottom rung. For example, the average annual revenue per line of village public telephones (VPTs) is estimated at US$ 16 without long distance, and US$ 760 with that capability.[2] The essence of a network is connectivity and the domestic network and international gateways, fixed and wireless service, and voice and data must all be built out in a coordinated manner to maximize the value of the network to its users. The government's role should be to ensure this coordination, without stifling competition and innovation. We next discuss innovations that potentially provide cost-effective voice and data telecom access to rural and semi-rural populations. To the extent that such access can stimulate demand for IT products and services geared to the domestic market, there is a positive link between widespread telecom access and the domestic IT industry.[3]

Some of the key work on innovation for Indian telecoms has been done by teams led by Ashok Jhunjhunwala of IIT, Chennai.[4] He realistically frames innovation needs in the context of economics. Affordability is critical to making widespread provision of telecom services economically viable. Jhunjhunwala gives the example of cable services in India, which are priced at US$ 2 to US$ 4 per month, and have 35–40 million subscribers. At this kind of price point, however, a telecom operator in India cannot recover set-up costs for access, which are about US$ 800 using conventional technologies.

The economics of widely available telecom services in India is therefore very different from the US, where revenue per connection will be several hundred dollars per year: innovation in the US focuses on increasing revenue through upgrading services, rather than reducing the cost of providing access. The goal of innovations by Jhunjhunwala's team, therefore, has been to bring the cost of access down below US$ 300 per line, and as close to US$ 200 as possible. The latter figure would make access affordable to 50 per cent of Indian households at current income levels. On the other hand, without such innovations, targets of increasing India's teledensity four-fold (from 4 to 15 per hundred), or Internet access 10-fold are empty rhetoric.

With the cost of fibre-based backbones falling rapidly, it is the access component of the network that accounts for as much as two-thirds of the per-line cost. The IIT Chennai group and spin-offs started by alumni have developed several key innovations that can dramatically bring down the cost of access. These innovations include developments in hardware as well as software and address issues of network management and deployment as well as pure access issues.[5] They help bring both affordable voice and Internet access to the rural areas of India.

The benefits of this suite of technologies are not restricted to rural users, but also extend to middle class and working class urban users. Current access costs using these new technologies are estimated at US$ 400 per line, but are likely to fall with further innovation. Pilot projects in rural and urban areas appear to have been very successful, and adoption is finally gaining some traction, despite bureaucratic and policy hurdles.

An alternative technology, which bypasses the local network entirely and is particularly suited for rural Intranets connecting nearby villages is that of high-powered Wi-Fi networks that travel long distances.[6] These versions of wireless networks using the Wi-Fi, or 802.11b, standard create a wireless zone of up to 12 miles long, far beyond the usual 300-foot-radius range that Wi-Fi typically achieves. The new products can achieve long distances by boosting the power inside access points—the radios that create the network, and by adding additional antennas to the access points so that signals can be beamed directly to a user's location, rather than creating a cloud of access. While the technology is in its infancy, it is already being implemented in the US and the main issue for use in India will again be cost and regulatory hurdles.

The bottom line is that bringing down the cost of access through innovation targeted at the domestic market is a critical component of any dramatic increase in telecoms connectivity in India. Economically combining Internet and voice access also has the benefit of increasing the value of connecting to the network. The benefits accrue not just to the poor, but also to the tens of millions of lower middle class households who are currently outside the affordability radius.

Human Capital

Advanced technical and managerial education is an essential input into the development of an IT industry. From the perspective of an individual state, it might be argued that local educational facilities are not important, since labour, especially skilled labour, is mobile within the country. However, the evidence suggests that there are mobility barriers, in the form of language, culture and substantial costs of moving. Even in the US, which is culturally and linguistically much more homogeneous than India, proximity to educational facilities (universities in particular) has been an important factor in developing the IT industry in various local regions—Silicon Valley being the most prominent of them, of course.

The case of Punjab is interesting since, while it has been a relatively prosperous state for some time, the extent of technical education it provided tended to lag behind its ostensible level of development, and behind several of the southern states in India.[7] To some extent, this reflected educational policy decisions taken at the centre, but also substantial periods of political turmoil, as well as a political focus on agriculture as the dominant sector of the state's economy.

In the late 1990s, some catching-up began to take place, as part of a nationwide effort by the central and state governments to ride the wave of the IT boom. Punjab has significantly increased the number of engineering colleges, as well as non-degree granting technical institutes. A significant step was the establishment by the state government of the Punjab Technical University (PTU) in 1996–97. The PTU acts as an independent provider of technical education, as well as an umbrella for technical education in pre-existing universities and colleges in the state. Its decentralized model, including the use of small entrepreneurs at the rural or small town level, is well-suited to making a basic level of IT education available to a broad cross-section of the population. In particular, the costs of shifting to a city for education can often be a substantial barrier to the less-well-off rural youth. Education closer to home is also particularly attractive for girls, where social norms might otherwise not permit them to go to a city for further education.

Thus, the past five years have seen a significant broadening of Punjab's local supply of technical education geared towards IT and related knowledge industries. What is probably missing from Punjab is a top-notch institute at the level of the Indian Institutes of Technology (IITs). While the IITs are elitist in some respects and a significant proportion of their graduates contribute to the brain-drain, they also can act as a source of high-level technical skills that are important for innovation. The IIT Chennai group, discussed in the previous section, is an example of the possible positive role that more advanced institutes can serve. IITs are also important in enabling faculty of regional engineering colleges to upgrade their skills by acquiring doctorates.

A similar quality gap lies in the field of management education, where Punjab, despite its long tradition of entrepreneurship, has no world-class management institute. Rankings differ, but the only management institute in the region that appears regularly in such lists is the University Business School, Chandigarh, which had ranks of 56, 18 and 28 in three different lists.[8] Of course, Chandigarh, while the shared capital of the state, is not in Punjab. In some respects, this gap may be more serious than in the case of engineering education, where the best regional engineering colleges are close in terms of quality.

Despite the increase in technical training within Punjab, it seems that the vast majority of engineering and other technical graduates go beyond the state's boundaries for jobs. To some extent, this is to be expected, since places such as Bangalore are well established and offer the most attractive opportunities. In general, however, the problem is that the number of IT jobs in Punjab is tiny, as we have noted earlier. Nevertheless, a supply of high-quality local graduates can be an important factor in the future, since firms seeking to invest in the state will be less concerned about having to attract workers from elsewhere in the country.

Forward Linkages and Development Impacts

In addition to requiring educated manpower as an input in production, the IT sector itself can play an enabling role in education delivery: this is an example of what is termed a 'forward linkage'.[9] Low-cost digital processing, storage, transmission and printing together lower the costs of developing and delivering educational materials. There is no reduction in the role of human capital, except to the extent that interactive materials can substitute partially for human teachers. On the contrary, there is a need for precisely the kind of human capital that can be produced by technical institutes, since the preparation of IT-based educational materials now requires IT skills as well as knowledge in the discipline in which the educational content lies. In the case of IT-education, of course, the two overlap.

There is evidence that broad-based delivery of non-traditional education can be enhanced by the use of IT. For example, fieldwork in Bhatinda in studying TARAhaat suggests several general points. First, in all attempts to introduce IT to rural India in a manner that promotes development, financial sustainability is a key issue. The TARAhaat franchisee model offers important promise in this regard with respect to incentives and scalability, though there have been difficulties in implementation. Second, the experiment validates the idea that IT costs have come down sufficiently to make rural IT services financially viable. This is an example of the force of technological change, since a basic desktop system now costs less than US$ 700.

Third, there is the issue of complementarities, both technological and pecuniary. To elucidate this point, note that one major roadblock for TARAhaat has been the poor quality of existing telecoms infrastructure. This has severely limited the scope of services that its franchisees could offer. On the other hand, the provision of complementary inputs such as financing and physical infrastructure, through subsidized loans from nationalized banks and the use of local government buildings, have been important in reducing start-up as well as operating costs. The most important complementarity emerged when the PTU quickly piggybacked on TARAhaat's efforts, enhancing the franchisees' initial financial viability through its own offerings of college-level IT education.[10]

Fourth, the scope of potential educational services is quite broad. While ventures such as this cannot be financially self-sustaining in providing basic education or adult literacy classes, they fill several gaps in the educational offerings available in rural and small town India. For example, TARAhaat offers classes in Tally accounting software and some local small business owners have taken these. Going beyond IT education, it is also developing classes in rural retailing and in personality development. Again, these are not 'basic needs', but they raise the level of human capital outside the big cities and without the costs of travel to those cities.

The key inputs provided by TARAhaat are educational content and organizational capital. Our observations suggest that both these are scarce in rural areas and any efforts to create and leverage such resources are likely to have a positive impact.

The use of IT in agriculture, industry and services creates forward linkages from IT to those sectors. To the extent that IT can have significant effects on the efficiency of operations in other sectors, there are complementarities between the IT sector and the rest of the economy. Examples of areas where increased efficiency may be possible include accounting, procurement, inventory management and production operations. These are all examples of forward linkages, since IT adoption has positive impacts on operations. This is, of course, the standard argument in developed countries such as the US for the virtues of the 'new economy' based on IT. In a developing country such as India, the use of IT is still scattered, but falling costs and homegrown solutions may provide some promise. We would argue that, ultimately, the linkages between IT and the rest of the economy are critical for the success of a regional IT industry.

Agriculture

The forward linkages from IT to agriculture are varied in nature. In the case of farming itself, good and cheap telecommunications access can enable farmers to access market information. Interestingly, in the case of Punjab, the role of the Internet appears to be currently limited by the existing institutional structure and physical infrastructure. In other parts of India, where infrastructure is less well developed, information about market prices and conditions allows individual farmers to make more beneficial decisions on where to sell their crops. In Punjab, the good network of roads and market towns makes the benefits of the Internet more marginal. Better-off farmers, for example, use cell phones to gather market information as needed. In other cases, the existence of dominant intermediaries, who, in addition to providing credit, have pre-purchased the farmer's crop, make the Internet redundant.

One aspect of agriculture where there may be potential for use of IT is in providing information to farmers. The Punjab Agricultural University, Ludhiana, is well known for its research and extension services. These include training programmes, farmers' fairs and telephone helplines. However, the use of IT does not appear to go beyond the posting on the Web of some basic information about scheduled events and locations and contact information of district offices. All this information is in English. E-mail contact possibilities are restricted to a single, general address. Hence, the potential of the Internet and Web as a medium for farmers to pull in useful information about inputs and techniques has not been tested.

To summarize the discussion so far, the benefits that could flow from linkages from IT to agriculture include improvements in the working of input and output markets and improvements in the quality of decision-making in productive activities. Interestingly, in Punjab, immediate and obvious benefits from improved market information via the Internet did not appear to be available. In general, this could be understandable as a consequence of the relatively good infrastructure of markets and roads (and to some extent, voice telecommunications) in Punjab, in contrast to more backward areas such as Dhar (Madhya Pradesh) and Bundelkhand (Uttar Pradesh), where Internet access seemed to make a dramatic improvement relative to the status quo.

However, further analysis also suggests that existing market institutions may have a lock-in effect, preventing the benefits of improved information from being realized. Fieldwork in Bhatinda district suggested the importance of 'commission agents', intermediaries who provided credit as well as pre-purchasing farmers' crops. In such circumstances, information of spot market opportunities that the Internet could provide is made irrelevant.

On the input market side, our interviews with government officials suggested that input traders constitute a strong political lobby. Hence, direct purchases of inputs from manufacturers, which might be aided by the Internet, in a straightforward example of disintermediation through improved information, were precluded by lobbying-influenced policies.

In terms of using the Internet to reach new markets, especially abroad, we did not find any strong evidence for this taking place, though it was being discussed in policy-making circles. To some extent, the problem was again lock-in to an existing institutional structure, which determines with some rigidity the cropping patterns, market channels and prices paid. It is also true that the complementary physical infrastructure is available for the existing institutions, but not for newer efforts that might take more of IT.

The lack of any noticeable efforts or impacts with respect to IT and Punjab agriculture is, to some extent, the consequence of needing to change a set of complementary institutions and policies in tandem. However, large players, such as the state government or corporations with agricultural interests (for example, agro-processing), can make a difference through coordinated implementation, as illustrated by the ITC effort in several southern states. Their annual report (ITC, 2002) states:

> Project 'e-Choupal'...links the Indian farmer with domestic and international markets... It already reaches out to more than half a million farmers to provide web-enabled real-time information on the weather, best farming practices and commodity prices. Through virtual clustering, these 'e-Choupals' are conferring the power of scale on even the smallest of individual farmers. This...e-infrastructure will dramatically enhance efficiency in the purchase and sale of agri-inputs and farm produce, with direct benefits to the farmer. 770 'e-Choupals' are already operational, covering 4,500 villages across four states in India.

Of course, the Punjab case requires some major changes in the thrust of future agricultural development in the state for the benefits of IT use to be realized, and this presents a larger challenge than the ITC example.

Industry

In the case of industry, the general picture that emerged from discussions with policy makers, industry group representatives and

IT managers was that the use of IT was not a front-burner issue. The general industrial investment climate in India appears to be poor, with the financial sector carrying a significant amount of non-performing assets and unwilling to engage in much new lending. Equity markets in India remain in their infancy as well. Other hurdles are the difficulty of exit from any industrial activity, over-regulation in general and corruption that accompanies such regulation. It was also suggested to us that the loss of guaranteed markets in the former Soviet Union continued to weigh negatively on Punjab's woollens and light manufacturing industries.

While these problems exist nationwide, they do so to varying extents. For example, Goswami et al. (2002) have examined the quantitative impact of state-level variations in policy on manufacturing productivity. Using a survey of 1,000 manufacturing establishments across 10 Indian states (including the National Capital Territory of Delhi), they find that states that are poor performers and identified by survey respondents as having a 'poor investment climate', have total factor productivity (TFP) that is 26 per cent lower than the high-performing states. About a tenth of this gap is found to be due to a higher regulatory burden (specifically, labour market regulations such as factory inspections) in the worse states. Punjab is in the lower half of the states surveyed, as shown in Figure 7.1.

The situation of Punjab, in the bottom half of the 'league table' of perceptions of investment climate, is most strikingly illustrated by plotting a particular measure of perceptions of investment climate against per capita SDP. This is shown in Figure 7.2, where Punjab is the only clear outlier among the nine states that were surveyed, excluding Delhi. Clearly, Delhi would also be an outlier if it was included and this might suggest that the problem is overstated, since Delhi is a successful IT destination (as is NOIDA in Uttar Pradesh). To some extent, this caveat is warranted, but Delhi is a single high-income urban region, and therefore is distinct from the nine states in the sample.

In any case, there is limited use of IT in Punjab industry. Perhaps, the best example is the case of accounting and inventory software. Two strong domestic competitors exist here. One is Bangalore-based Tally Solutions, a private company that makes Tally accounting software. Fieldwork interviews in Punjab suggested that Tally is a market leader for small business accounting software and that even small

FIGURE 7.1 **Firms' Perceptions of Investment Climate**

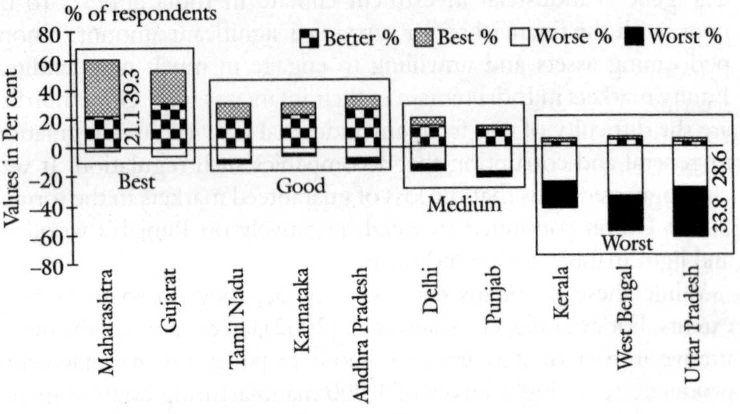

Source: Goswami et al. (2002: 10).

FIGURE 7.2 **Per Capita SDP and Perceptions of Investment Climate**

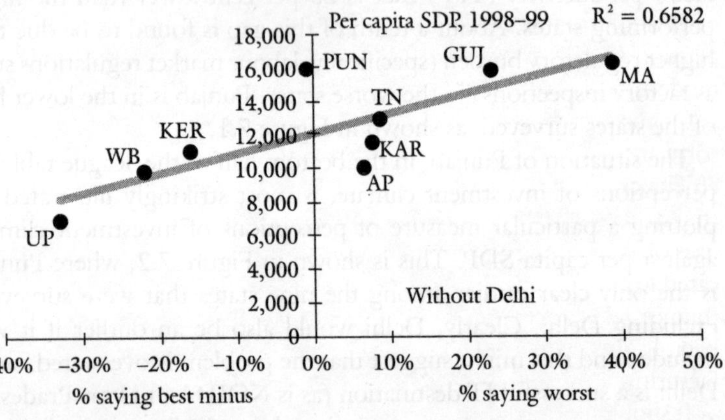

Source: Goswami et al. (2002: 10).

shopkeepers are interested in acquiring it and learning to use it. For larger businesses, Tally has recently introduced a server version, which fits the needs of businesses with multiple locations or offices. The second product is E.X. from Tata Consultancy Services (TCS), which is the oldest of India's IT services companies, currently the largest exporter and a firm with a substantial presence in the domestic business software market. The two companies' products are priced similarly, and in each case one can find all the features of distribution channels, after-sales support, marketing and product versioning that exist in competitive US markets.

Another example of IT use was that of CAD software in manufacturing firms such as Hero and Punjab Tractor. Unlike the case of accounting software, the products here were not domestic, but made by the US and global market leader in design software, Autodesk. The software is distributed through local resellers, but demand is not very high. Companies tend to purchase limited amounts of software, upgrade infrequently and demand little customization or maintenance support. Thus, the market for such software is not very strong as yet. The reseller also identified piracy as somewhat of a problem. It may be noted that the small local market also means that the manufacturer appears uninterested in a pricing strategy that would build volume in small country markets, thus perpetuating limited adoption.

In the case of design and other higher-end software, it appears that the structure of Punjab industry may also act as a barrier to IT adoption. Smaller firms are less likely to be able to afford the fixed costs of IT adoption and this problem is compounded by the lack of competition (partly due to over-regulation), which reduces incentives to modernize production.

On the positive side, as we have noted, accounting software and CAD software were two examples of use of software to enhance productive efficiency, and these are natural ones in the context of Punjab's industrial structure, with many smaller manufacturing firms. More complex uses of software, including networked applications, or more sophisticated logistics or production management, seem to be far away. The thinness of IT use suggests that its development impacts in practice have been limited. The situation reflects the weakness of Punjab's manufacturing industry, as discussed earlier.

It should be emphasized, however, that even at India's level of development, there is tremendous scope for the use of IT. For example, Miller (2001), who surveys the potential for B2B e-commerce in India, gives the example of Reliance Industries, which, though still quite diversified, is now heavily into production and distribution of chemicals. In this area, of the company's 20,000-plus customers in India, about 3,000 are major buyers, accounting for over three-quarters of total sales. These major customers are electronically linked to a Reliance-controlled Internet-based market exchange. Using leased lines, customers can process orders, and Reliance can communicate dispatching details, better manage inventory, carry out invoicing and provide customer support. Using this system, Reliance has reduced receivables from 310 days to 90 days. General cost improvements have come from an overall tightening and acceleration of processing within the company and between the firm and its customers. The speed of order delivery has greatly improved and inventories reduced. A shift from leased-lines to the Internet will provide further cost savings.

Reliance is one of the largest firms in India, as well as being one of the most dynamic. The absence of firms like this from Punjab may be partly responsible for the lack of any discernible impact of IT on the functioning of firms in Punjab. Specifically, both the size distribution of Punjab firms (skewed towards smaller firms) and their focus in more traditional industries may reduce the adoption of IT in Punjab industry. One emerging exception may be in pharmaceuticals, where a more knowledge-based industry cluster might be developed. Otherwise, in general, the sophistication and scope of Punjab's industry does not seem to provide an attractive place for IT use to make an impact, despite Punjab's relatively high per capita income levels and good traditional infrastructure.

Services

One area where higher-end software may be more likely to be adopted is in the financial sector, where the inherent possibilities for digitizing financial information are stronger. This takes us over to the services sector, however, in exploring forward linkages. Large financial institutions, such as ICICI, State Bank of India and Bank of Punjab appear to be adopting IT for internal business processes as well as customer interfaces, but this is a nationwide phenomenon and Punjab is neither a leader nor a laggard in this respect.

The financial sector in India in general is not in good shape, with the banking sector in particular being overstaffed and subject to political compulsions in making loans. New private banks are somewhat free of these problems, but their role in the sector is still limited and is overshadowed by the continued dominance of public sector banks and other lending institutions. Government-controlled lending institutions, because of their own financial problems, are unlikely to be leaders in IT adoption, though some progress is being made.

Interestingly, in the case of finance, there is also an important opportunity at the other end of the spectrum. The use of IT in rural banking and micro-finance can impact a much broader cross-section of the population. The evidence of pilot schemes such as the SKS InfoTech Smart Card project is encouraging. Hand-held computers and smart cards can substantially reduce the costs of making loans, as well as monitoring them. Reducing these transactions costs may turn out to be critical for the scalability and sustainability of micro-finance schemes. On the other hand, as in the case of market information for farmers, the relative benefits for Punjab may be lower, since it has entrenched credit institutions that serve the majority of farmers, and the scope for micro-finance may be more limited.

The best hope for Punjab in terms of the use of IT in the services sector is the opportunity that is currently exciting many Indian firms. IT-enabled services (ITES) have shown the strongest growth among IT-related sectors in the last two years. They include a variety of types of service, ranging from customer call centres, to accounting services and other business process outsourcing (BPO), to GIS and engineering services. Thus, the required degree of technical sophistication of the workforce and the level of use of IT can vary widely. In fact, the three categories mentioned make up most of India's ITES exports, with the first two showing high growth and representing over 60 per cent of the total of Rs 71 billion. The list of ITES segments constructed by NASSCOM displays ample scope for specialization within the category (Table 7.2), and it is clear that individual firms are already trying to capture niches. For example, Daksh has a 90 per cent concentration on customer service, whereas HCL Frontline (a division of HCL) is 100 per cent focussed on technical support (*Dataquest*, 2002: p. 134).

In the case of Punjab, the draft government IT policy makes ITES a priority and lists the strengths of the state in this respect. These are said

TABLE 7.2 **ITES Types**

- Customer Interaction Services
- BPO/Management; Back Office Operations
- Insurance Claims Processing
- Medical Transcription
- Legal Databases
- Digital Content
- Online Education
- Data Digitization/GIS
- Payroll/HR Services
- Website Services

Source: http://www.nasscom.org/it_industry/spectrum.asp

to include a large and technically competent English-speaking workforce and high quality infrastructure. We have already discussed some of these issues earlier. We will also return to a discussion of the policy environment in the section on 'Governance and Policy Environment'. However, here we note that a concerted effort will be required if Punjab is not to miss the boat again. For example, the data on the location of ITES companies from NASSCOM shows that Punjab (and Chandigarh) is nowhere in the current ITES picture (Table 7.3). Furthermore, the region did not even make the survey of promising ITES cities, in which the cities were assessed on factors such as manpower availability, real estate, telecom infrastructure, policy initiatives, power infrastructure, city perception and entrepreneurial history. Thus, Punjab still faces an uphill battle in establishing a niche in this sector.

TABLE 7.3 **ITES Penetration and Ranking of Cities**

City/Region	*Number of companies*	*Survey rank*
National Capital Region	53	8
Mumbai	45	7
Bangalore	35	6
Chennai	35	3
Kolkata	29	4
Hyderabad	24	1
Kochi	10	2
Ahmedabad	9	5
Pune	6	9
Others	32	
Total	278	

Source: NASSCOM (2002).

Even if Punjab is successful in establishing a niche in ITES, the overall employment impacts may be relatively small, in the order of tens of thousands additional people employed (the official target is 50,000 jobs over the next two years). Nevertheless, given the severity of youth employment problems in the state, even marginal impacts will be useful.

Governance and Policy Environment

Our discussion so far suggests that the quality of governance in Punjab has been one factor in its failure to take off as a destination for IT investment. Governance is well-established as an area where IT can lead to substantial improvements in functioning. There are two broad uses of IT for improved government functioning. First, back-office procedures can be made more efficient, so that internal record-keeping, flows of information and tracking of decisions and performance can be improved. Second, when some basic information is stored in digital form, it provides the opportunity for easier access to that information by citizens. The simplest examples are e-mailing requests or complaints, checking regulations on a web page, or printing out forms from the web so that a trip to pick up the forms from a physical office can be avoided. More complicated possibilities are checking actual records, such as land ownership or transactions. Still more complicated are cases where information is submitted electronically by the citizen, for government action or response.

As in the broader case of using the Internet for communications and transactions, sustainability of e-governance initiatives is a significant issue. Since governments at all levels are financially strapped, the initial investments and ongoing expenditures for IT-based service delivery may act as a barrier to adoption as well as to long-run sustainability. However, a franchise model can be successful here. Low-cost rural Internet kiosks, a tiered franchising model and a suite of basic government access services for which users are willing to pay are key components of what Drishtee, an outgrowth of the Gyandoot project in Madhya Pradesh, is implementing in several parts of India.[11] Co-operation of local governments and subsidized financing have been important elements for Drishtee, as in the case of TARAhaat, with the former being obviously critical in the case of Drishtee. It is important to note that once Internet access is available, its benefits are not restricted to e-governance. Individuals can obtain market information, training, job information, advice on farming techniques, and so on, as discussed earlier in this section.

The Punjab government's implementation of e-governance at the rural level appears to have been very slow, both for reasons of funding and prioritization. For example, TARAhaat explicitly promised e-governance services when it began operations in Bhatinda district, but has since abandoned this direction, partly because it perceived a lack of anything to offer. That, in turn, was apparently the result of a lack of progress on the front of computerizing land records. On the other hand, the state government's own sponsored e-governance initiative in Sangrur district appears to have had some success, based on our conversations with state officials in charge of IT policy. Another pilot project that has received attention is at Fatehgarh Sahib, with a Web-based Citizen–IT Interface for or services offered by district administration (for example, see http://informatics. nic.in/archive/inf2002apr/e_governance.htm).

IT policy makers in Punjab have developed an elaborate analysis and plan for e-governance and in some ways have been conceptual leaders among the states, but implementation, as noted, has lagged. To some extent, the elaborateness of the plan, including a statewide identity scheme to manage digital access and exchange of information between citizens and government, has also contributed to delay by substantially raising the set-up costs.[12]

Areas where some implementation has occurred are in tax collection and information availability. According to Nirmaljeet Singh Kalsi, who has directed much of the effort in this direction: 'The results have been very encouraging so far with the improvement in revenue collection by the excise and taxation department. Also, the punjabsewa.gov.in portal offers almost 300 services to citizens (see note 11). Our own examination of this 'citizen services portal' suggested limited functionality and available information, although the framework and interface are both attractive and comprehensive. One obvious positive feature of the portal is that its offerings include complete options in the Punjabi language: this contrasted, for example, with the lack of such options on the PAU web-based information for farmers (the section on 'Forward Linkages and Development Impacts').

Turning to the broader policy environment, we have noted that Punjab has been a successful agricultural state, and evolved an infrastructure and a set of institutions that provided excellent support for its agricultural economy. At the same time, Punjab's development was very much in line with India's economic policies that emphasized heavy government involvement in all aspects of the economy. In some respects, Punjab's success with agriculture and its relative success with government intervention in the state economy have together made it harder to develop the potential of IT in the state. In particular, policy-making remains geared towards traditional crops and the interest groups that benefit from them, including farmers, distributors, input suppliers and other middlemen.

In the late 1990s, as we have discussed in the section on 'The IT Industry in Punjab and Government Policies', there were policy efforts towards developing an IT industry in Punjab. From conversations with people in government and in industry, it appears that implementation lagged substantially behind intentions. Much of the implementation activity within government appears to have been entrepreneurial efforts on the part of a few individuals, with the bulk of the government machinery unchanged in its orientation and approach, and therefore failing to provide an environment for business that was attractive enough to bring in firms with several locational alternatives to choose from within India. The closeness of the Delhi–Gurgaon–NOIDA IT cluster, with better international access and other infrastructure, has been a major factor in this respect. While a few firms did enter Chandigarh and Mohali, major impediments in the allocation of land and the provision of other infrastructure deterred any substantial build-out of a local IT industry.

In 2002, the policy environment appears to have improved in several respects. The government has accumulated substantial experience (its own and that of other states) to help it understand what initiatives are crucial to success. The current policy documents—including the policy on IT and ITES, as well as on industrial policy—are more comprehensive and detailed than earlier ones with respect to the different dimensions of government policy. In particular, there is more of an emphasis on creating a hassle-free business environment, without the roadblocks to start-up (for example, clearances for getting started) and continuing operations (for example, various kinds of health, safety and environmental inspections) that have continued to exist. There is also considerable attention being paid to issues of complementary infrastructure, such as housing, social amenities and transportation.

On the input side, technological and institutional changes have worked together to relax the severe bandwidth constraints that earlier hampered growth of the IT sector, though there are still issues of relative cost and reliability. Finally, there is a larger local pool of human capital available than was the case a few years ago, though again quality is an issue and people with the right technical and organizational skills are still going to be scarce. In order to attract or retain such people, the social infrastructure is crucial.[13]

Of course, it remains true that implementation will be the test. Two constraining factors have not changed since the 1990s. The first is what we alluded to earlier in this section, namely, the strength of the societal groups that have a stake in the traditional agricultural economy of Punjab. The second is the continued poor condition of the state government's finances. These two factors are closely connected, of course, through the provision of lavish subsidies for agricultural inputs such as water and electric power.

The state government will have to make several changes in its policy approach in order to overcome its resource constraints. The first is to reallocate its resources towards developing infrastructure for a 'new economy' in Punjab. This new economy is not necessarily restricted to software for export, ITES, or even IT alone. It includes IT products and services that can be geared to the needs of the regional economy more broadly. The second is to rely more on the private sector in developing infrastructure. The third is to avoid narrowly targeted financial incentives in an attempt to build up a local IT and ITES sector. Such incentives are typically a strain on the budget, without necessarily overcoming key roadblocks to industry growth.

Conclusion

We offer three conclusions in this chapter. First, we offer some understanding of the paradox of Punjab's lagging in the development of IT capabilities, despite its being one of the richest states in India in terms of per capita income levels. We have argued that the policy environment was not sufficiently supportive of what was, in fact, a substantially new form of economic activity in the state. In turn, this policy environment has been a function of the state economy's dominance by particular interest groups tied to traditional agriculture.

Second, we have suggested that some policy shifts are both feasible and can support the regional development of IT and ITES. In particular, policies towards industry that are broadly more favourable and reduce the costs of starting up and doing business are likely to be more beneficial than narrowly targeted financial incentives. Furthermore, while the direct employment benefits, even with ITES added to IT proper, are likely to be small, the spillover effects can be substantial. We were struck by how little impact the use of IT was having on industry in Punjab, though it is ostensibly an 'advanced' state. To some extent, this reflects policies that do not allow successful small firms to expand easily—this is something that has been a problem throughout Indian industry, we believe, since it is the result of policies determined at the national level.

Third, at the micro level, we have been struck by how easily small collaborative entrepreneurial efforts in IT can have an impact on local development. This was most clearly illustrated by fieldwork in Bhatinda district. While TARAhaat's efforts in Bhatinda district have been fraught with several mis-steps and implementation problems, they have validated a model of local entrepreneurship that uses IT as an input for a variety of educational purposes. The impacts included new employment created, new services generated, lower cost and broader education delivery, and most importantly, the economic testing of the use of IT in rural areas. Thus, despite the failure of local e-governance to take off (as has been true throughout the state), and mistakes by TARAhaat in strategy and execution, the Bhatinda example is indicative of what is possible in terms of developmental impacts.

If anything, the Punjab experience illustrates the importance of policy, but in a non-obvious way. A few years ago, one might have expected that Punjab would easily establish itself as a destination for IT-related investment. It appeared to have very favourable pre-conditions, in terms of high incomes, good infrastructure, and an enterprising population. Instead, other states and locations have been the ones to thrive. In a sense, they have leapfrogged Punjab in developing ICT industries.[14]

This is not surprising if one believes that local human capital (with technical and organizational skills) by itself is necessary and sufficient to explain the development of ICT industry locations in India, since the southern states had an advantage in this respect. However, it is clear that provision of complementary physical inputs was also critical. Most of all, generally favourable government policies for doing business in a particular state or location have been the crucial ingredient for success. The survey we referenced in this chapter was not restricted to IT, but the states that are seen as attractive places to invest in general are also mostly states that have succeeded on the IT front. While specific targeting of policy to remove critical infrastructure bottlenecks is certainly necessary, our analysis suggests that a general policy shift is also needed. This is where Punjab took only a short hop, rather than a leap that would be necessary to cross the chasm to a new economic future. This is what we would suggest is our main conclusion from examining the case of Punjab and comparing it with other Indian states.

Notes

1. This chapter is a revised version of a paper prepared for the International Seminar on ICTs and Indian Development, to be held in Bangalore on 9–11 December 2002. It draws partly on research funded by the Rajiv Gandhi Institute for Contemporary Studies and the Academic Senate of the University of California, Santa Cruz. I am indebted to Dipankar Bhanot, A.S. Chatha, Bibek Debroy, Nirmaljeet Singh Kalsi,

Pritam Singh Kohli, P.D. Kaushik, Gurnihal Singh Pirzada, and numerous individuals in IT firms and industry organizations for help at various stages of this research. The views expressed here, as well as any errors or shortcomings, are solely those of the author.

2. See Telecom Regulatory Authority of India (2000). Yale Braunstein has pointed out to me that this wide spread partly reflects an inefficient tariff structure, with domestic long distance and international calling being priced significantly above cost.
3. An example from field research in Bhatinda district of Punjab in December 2001 can illustrate: a farmer told us that he had taken computer lessons, bought a home computer and signed an Internet service contract so that he could exchange e-mail with his brother in Toronto, Canada. All three IT-related products and services depended on basic telecom availability. See also Prahalad and Hart (2002).
4. See Jhunjhunwala (2000), as well as numerous presentations available on his website, at www.tenet.res.in/ashok.html. The overall group is called the Telecommunications and Computer Network (TeNeT).
5. The innovations are described in more detail in Jhunjhunwala (2000). They include some wireless components, combined access to voice and Internet connections, and low-cost access devices.
6. See the news article, 'Wi-Fi stretches its boundaries', By Ben Charny, Staff Writer, CNET News.com, 27 September 2002.
7. This was partly a North–South imbalance. For example, the four southern states of Andhra Pradesh, Kerala, Karnataka and Tamil Nadu, together with the Union Territory of Pondicherry, accounted in 1997 for about half of engineering college seats, though they had only about a quarter of the population of India. See Palit (1998).
8. The Punjab School of Management Studies, Punjabi University, Patiala, appeared at rank 49 in one list, at www.rbeindia.com/topbschools.htm. Other lists were available at www.geocities.com/shanmugavelbabu/50bschool.htm, and www.namasthenri.com/Education/b_schools.htm.
9. In the case of technical education for employment in the IT industry there is a 'backward linkage' from IT to its inputs. A similar backward linkage exists in the case of telecommunications. These linkages are among the most obvious and national and state government policies all over India have been geared toward relaxing the constraints of telecoms infrastructure and IT skills.
10. Subsequently, TARAhaat has chosen to forbid its franchisees from offering PTU courses, in an attempt to solidify its brand and quality control. Whether this is the optimal strategy requires a separate analysis.
11. Further details of Drishtee's efforts are in Kaushik and Singh (2002). Drishtee has a pilot effort in Jalandhar, but we did not get much sense of it going forward with any speed. In Jaipur, Rajasthan, it is moving rapidly by piggybacking on the firm Aksh's expansion of a fibre-optic cable network.
12. The 'citizens' database scheme' is described in detail by Nirmaljeet Kalsi as follows: 'Almost all transactions with the government require a citizen to prove his identity, the ownership of property as well as his professional credentials. The documents that verify these are typically pre-defined transactions with agencies like the state electricity board, municipal corporation and transport department. If the process of acquiring these documents is automated, the delays in getting such work done will be cut down by 80 per cent. Key services include dissemination of information,

billing, handling complaints and grievances, licences and approvals, certificates, financial transactions or procurements. Therefore the citizens' database, property database and business database along with the departmental databases like Punjab State Electricity Board, old age pensions, food and civil supplies or transport would cover almost all services being provided by the government to the citizens and businesses. The creation of these three authenticated databases will be at the core of our e-governance strategy. While Punjab already has the business database in the form of master dealer files (MDF) with the department of excise and taxation, the citizens' database would take some time. We first need to complete the pilot project at Fatehgarh Sahib, the pro forma for which has already been finalized. Funds for the door-to-door survey have also been allocated to the deputy commissioner (DC).' (From a 24 June 2002 interview with Dataquest, available at www.dqindia.com/content/special/102062403.asp.)

13. While these factors are hard to quantify, they came up again and again, in different forms, depending on the profiles of the individuals raising the issues. Among the aspects brought up were clubs for IT professionals with families, entertainment for young single adults and branded restaurant chains—all these were found to be deficient in Chandigarh. Interestingly, the new IT and ITES policy of the government recognizes such factors explicitly.
14. On the efforts of the government of Andhra Pradesh, see Eischen (2000), and for Tamil Nadu, see Bajpai and Radjou (1999) and Bajpai and Dokeniya (1999). On Karnataka, there are numerous references: see, in particular, the works cited in Saxenian (2001).

References

Ahluwalia, M.S. 2002. 'State Level Performance under Economic Reforms in India', in A. Krueger (ed.), *Economic Policy Reforms and the Indian Economy*. Chicago: University of Chicago Press.

Bajpai, N. and A. Dokeniya. 1999. 'Information Technology-Led Growth Policies: A Case Study of Tamil Nadu', Development Discussion Paper No. 729, October, Harvard Institute for International Development.

Bajpai, N. and N. Radjou. 1999. 'Raising the Global Competitiveness of Tamil Nadu's Information Technology Industry', Development Discussion Paper No. 728, October, Harvard Institute for International Development.

Dataquest. 2002. 'SW INDUSTRY: Working around the Slowdown, 14 February'. Available at <http://dqweek.ciol.com/content/search/showarticle.asp?artid=21244>.

Eischen, K. 2000. *National Legacies, Software Technology Clusters and Institutional Innovation: The Dichotomy of Regional Development in Andhra Pradesh, India*. Santa Cruz, USA: University of California, Department of Sociology.

Goswami, O., A.K. Arun, S. Gantakolla, V. More, A. Mookherjee, D. Dollar, T. Mengistae, M. Hallward-Driemier and G. Iarossi. 2002. *Competitiveness of Indian Manufacturing: Results from a Firm-Level Survey*. Confederation of Indian Industry and The World Bank, January. Available from www.ciionline.org.

ITC. 2002. *Annual Report*.

Jhunjhunwala, A. 2000. 'Unleashing Telecom and Internet in India', paper presented at the Conference on Telecommunications Reform in India, Asia/Pacific Research Center, Stanford University, November 9 and 10.

Kaushik, P.D. and N. Singh. 2002. 'Information Technology and Broad-Based Development: Preliminary Lessons from North India', UC Santa Cruz Working Paper.

Miller, R.R. 2001. 'Leapfrogging? India's Information Technology Industry and the Internet', IFC Discussion Paper No. 42, May. Washington, DC: The World Bank.

NASSCOM (National Association of Software and Service Companies). 2002. 'FY02: Software and Service Industry Performance, 18 July'. Available from www.nasscom.org/download/FYo2-results.pdf.

Palit, S.K. 1998. 'The Development of Engineering and Technical Education in India', *Global Journal of Engineering Education*, 2(3): 317–26.

Prahalad, C.K. and S.L. Hart. 2002. 'The Fortune at the Bottom of the Pyramid'. Available from http://www.strategy-business.com/media/pdf/02106.pdf.

Saxenian, AnnaLee. 2001. 'Bangalore: The Silicon Valley of Asia?', Working Paper No. 91, Center for Research on Economic Development and Policy Reform, Stanford University, February.

Telecom Regulatory Authority of India. 2000. *Consultation Paper on Issues Relating to Universal Service Obligations, 3 July*, New Delhi: TRAI.

8

The Electronic Grassrooting of Democracy: A Case for Community Radio[1]

Vinod Pavarala

> As for the radio's object, I don't think it can consist simply in prettifying public life.... The radio would be the finest possible communication apparatus in public life, a vast network of pipes. Whatever the radio sets out to do it must strive to combat that lack of consequences which makes such asses of almost all our public institutions....
>
> —Bertolt Brecht, 1932

> When radio fosters the participation of citizens and defends their interests; when it reflects the tastes of the majority...; when it truly informs; when it helps resolve the thousand and one problems of daily life; when all ideas are debated in its programs and all opinions are respected; when cultural diversity is stimulated over commercial homogeneity; when women are main players in communication and not simply a pretty voice...; when no type of dictatorship is tolerated...; when everyone's words fly without discrimination or censorship, that is *community radio*.
>
> —José Ignacio López, 1997

The Brechtian mandate to use radio for social change seems never more pertinent than the present times when media globalization has led to the predominance of commodity logic in the functioning of media institutions and to a stultifying monotony in its programme content. As firms grow larger, through mergers and acquisitions, they become reluctant to take risks on innovative or progressive information, or cultural forms that could potentially challenge the status quo or strive towards even limited structural change. It is then left to civil society organizations (CSOs) to struggle for access to information and communication technologies (ICTs) and mobilize them for autonomous self-expression as well as for development at the level of local community. This chapter seeks to provide an overview of community radio in India and elsewhere, against the background of recent insights on participatory development, the inequities of the ICT revolution, the right to communicate and the democratization of airwaves.

Participatory Development

Participatory Development Communication envisages, among other things, democratized and decentralized information and communication structures as key agents of empowerment to those who have been traditionally marginalized (Nair and White, 1993; Servaes et al., 1996; White et al., 1994). The mid-1970s saw disenchantment with the postulates underlying the dominant paradigm of development, as they did not correspond to the social realities in Third World countries (Escobar, 1995). The dominant paradigm promoted a top-down approach to planning and growth which treated people as objects. Information was expected to be an agent of diffusion of new ideas (Rogers, 1962) and help in transforming traditional individuals into modern citizen-subjects. Media were used as persuasion agents for 'domesticating' people through information, therapy and manipulation (Melkote and Steeves, 2001).

The Third World response to the dominant models of development came not so much as a well-defined approach. It was more a critique by scholars in the 1980s, who disapproved universal application of development models. These alternative approaches were stimulated by Freire's (1970) influential work on the pedagogy of the oppressed through conscientization and Schumacher's advocacy of appropriate technologies (Schumacher, 1973). Issues articulated through the rise of new social movements emphasize self-determination and democratization and generative power, rather than distributive

(Kothari, 1984). The primary issue in participation is the facilitation of dialogue among members of a community so that they become the ultimate arbiters of the development that is suitable for them (White, 1999). A democratic, independent communication infrastructure at the community level can play a constructive role in enabling marginalized people to manage their own development and to acquire a sense of control over its course through self-management and self-representation.

ICT Inequities

The post-industrial society, heralded by scholars like Daniel Bell (1974), was supposed to put an end to centralization and standardization and bring about decentralized and diversified production processes and democratic decision-making. However, the advent of IT, as Hamelink (1986) pointed out, has belied these expectations. Control over technologies became more centralized and the power structure between the information-haves and the information-have-nots has been reinforced. In the face of the increased volume of information on specialized subjects, where more and more is known about less and less, where access and skills required to exploit complex technologies are unevenly distributed in the society, information will become a source of power only under specific conditions of equity (Lyon, 1988; Wresch, 1996). As long as ICTs cater only to the interests of the most powerful sectors of society, that is, those who control and manage information, the information revolution remains only a myth (Traber, 1986).

This early critique of the role of IT in perpetuating information inequalities takes on a new dimension with the emergence, over the last decade, of what Manuel Castells (1996) calls 'the network society'. Painting a broad historical canvas, Castells suggests that while networking as a form of social organization has existed previously, the IT paradigm provides 'the material basis for its pervasive expansion throughout the entire social structure' (Castells, 1996: 469). In a stratified social structure, the spread of networks cannot but have differential effects on different segments of society. However, Castells is optimistic about the potential for reconstructing democracy through mobilization of information and communication technologies (ICTs). These technologies, he thinks, could possibly lead to greater decentralization and citizen participation and enhance horizontal communication among individuals and communities. The 'electronic grassrooting' of democracy could possibly help reconstruct a new kind of civil society (Castells, 1997: 352).

In the context of development discourse in non-Western countries, States and civil society organizations (CSOs) now routinely speak of information technologies enabling improved access to governmental and quasi-governmental resources and services, new credit and financial services available through palmtops and information kiosks, new opportunities to design, manufacture and market their products through the Internet and Intranet systems, more and better education through computers or about computers, and superior medical advice, diagnosis or knowledge in their own locality (Sood, 2002). The applications of ICTs in development have broadly been categorized as: those that provide decision support to public administrators for improving planning and monitoring of developmental programmes; those that improve services to citizens and bring in transparency; and those that empower citizens through access to information and knowledge (Bhatnagar and Schware, 2000). Focussing on the third role of ICTs within the perspective of alternative approaches to development, this chapter seeks to explore the relationship between ICTs and democratization of the process of development, with specific reference to community radio.

Media Globalization and the Public Sphere

The concerns about the inequities of the so-called information society have further intensified by what recent scholarship has come to call 'globalization'. The term has been used to include an array of processes such as: time–space compression (enabled by expanded transportation and communication infrastructure); technological, commercial and cultural homogenization; and a general 'intensification of the consciousness' of the world as a whole (Robertson, 1992; Pieterse, 1994; Appadurai, 1996). Wilson and Dissanayake (1996) underline the dialectics of globalization by directing our attention to a new terrain of cultural production and national representation that is simultaneously becoming more globalized (around the dynamics of capital) and more localized (into contestatory enclaves of difference, coalition and resistance).

The worldwide trend towards economic liberalization since the 1980s and the ongoing process of convergence in ICTs have hastened globalization of media and other channels of communication. Deregulation of government controls and concentration of ownership of ICTs in a handful of conglomerates based in the US, Europe and Japan have led to an oligopolistic control of these gigantic corporations over the terms of public debate and discourse (Pavarala and Kumar, 2002). This has raised questions about the free flow of information to and from citizens, which is an essential prerequisite if the realm of civil society must include 'organized and substantial capacity for people to enter into public discourse about the nature and course of their lives together' (Calhoun, 1994). As globalization of ICTs diminishes freedom of information, erodes the diversity and quality of information in the public sphere, civil society becomes increasingly ineffective in the cultural and information marketplace.

Popular resistance to globalization and the widely shared anxiety that ICTs would ultimately cater only to State and transnational commercial and political interests are clearly discernible in the countervailing public demand for greater decentralization of political power and distribution of the benefits and control of media and communication. So the question is: How can ICTs be used to stimulate a kind of publicness, which is neither part of the State nor wholly dependent on the processes of the market? Oskar Negt and Alexander Kluge (1972) raised the possibilities for an alternative, proletarian, autonomous 'counter' public sphere that would enable dissonant experience and knowledge of the working classes to be freely articulated, exchanged, debated and developed. Gradually, this new alternative public sphere could be expanded to a point at which it might replace the processes and structures of Habermasian bourgeois public sphere (Habermas, 1962).

Nancy Fraser (1992) critiques the silences in Habermas's public sphere and argues that it claimed to be open to all, but could not promote participatory parity as bracketing of social inequalities in discursive arenas is not possible in stratified societies. She claims that subordinate social groups have found it advantageous to constitute alternative publics. She proposes the forging of what she calls subaltern counterpublics, spheres parallel to that of the dominant social category where 'members of subordinate social groups invent and circulate counter-discourses to formulate oppositional interpretations of their identities, interests and needs' (Fraser, 1992: 123).

The Right to Communicate

It is against this background that ICTs may be seen as providing to the marginalized an arena that could be used for democratic deliberations and negotiations that would, in turn, help forge a more responsible and responsive alternative public sphere. Organizations experimenting with participatory communication for social change across the world, citizen groups, community organizations and media activists have been pushing for an institutional space that can be effectively used for revitalizing the civil society (Dagron, 2001). A necessary condition for the creation of this alternative institutional environment is progressive State policy that facilitates the development of new centres of symbolic power, which lie outside the spheres of influence of the State and the market, respectively (Pavarala and Kumar, 2001a).

A number of CSOs in India and elsewhere in the South Asian region are articulating the need for popular and community-based ICTs—small local newspapers, community radio, alternative video, and so on—that seek to subvert both State and market monopolies over people's minds. Community radio stations in Nepal and Sri Lanka and the campaign for an independent community broadcasting sector in India and Bangladesh (Noronha, 2003), and participatory video projects in Andhra Pradesh and Gujarat in India are significant efforts to hand over control of media technologies to the poor and non-literate (Satheesh, 1999/2000; Stuart and Bery, 1996). More recently, the introduction of computers and the Internet in a major way into the communication strategies of CSOs has widened the range of available technologies for empowerment at the grassroots level. In this sector, especially in India, one can witness a range of innovative experiments in the use of IT for development, initiated and supported by the State, the market, as well as the non-governmental organizations (Bhatnagar and Schware, 2000; Sood, 2002).

Throughout the world, the right to communicate and to have affordable access to the means of communication is increasingly being acknowledged as a fundamental human right. The citizens' communication rights[2] proclaim that people and communities must be the principal beneficiaries of the developments in ICTs. While stating that the right to communicate is a universal human right, the 7th World Congress of the World Association of Community Radio Broadcasters (AMARC), held in Milan, Italy, in 1998 emphasized, among other things, the following:

- The market economy is not the only model for shaping the communications infrastructure. People must be seen as producers and contributors of information and not be defined solely as consumers.
- Community media can play an important role in strengthening cultural rights, and in particular, the rights of linguistic and cultural minorities and indigenous peoples by providing access to the means of communication.
- Access to the means of communication must be supported by education and training to assist a critical understanding of the media and to enable people to develop their media and communication skills.
- The democratic participation of women in communications media should be guaranteed at all levels (AMARC, 1999).

The *Toronto Platform* put together at the International Symposium on Women and the Media: Access to Expression and Decision-Making, held in Toronto, Canada in 1995 underlined the need to:

- Recognize the rights of all women to have access to expression and participation in the media, in particular those from discriminated groups such as other-abled, indigenous, women of colour and women of diverse sexual orientation.
- Introduce, support and extend community media as a way of increasing women's participation and contribution to the media and local economic development, especially in areas of high illiteracy rates.
- Develop educational and training methodologies to enable women's organizations and community groups to effectively communicate their own messages and concerns and access to existing media.

Radio for Social Change

Radio has historically been the medium of choice for those committed to democratization of media and participatory modes of communication in development. Experiences from Latin America, dating as far back as the mid-1940s, have demonstrated the potential of community radio for social change. A recent Rockefeller Foundation report asserts that community radio is 'one of the best ways to reach excluded or marginalized communities in targeted, useful ways', and in giving them a 'voice' that matters most in development communication (Dagron, 2001).

The reach and access of radio to deprived, rural areas makes it an appropriate and powerful tool for development and education in developing countries. The ownership of radio sets among the rural poor is relatively high as compared to other forms of media. Recent studies have shown that in far-flung and isolated areas, where illiteracy renders inert the print media and where television has made little inroads, there is an embedded culture of radio listening and radio becomes a lifeline to the outside world (Pavarala, 2003). Moreover, the nature of the medium is such that it can take advantage of the auditory senses of humans to dramatize messages. It is a flexible medium that allows for a range of content formats, ranging from song and drama to news and talk to documentary. The costs of setting up a radio station as well as the costs of production are relatively low. Low-power FM frequencies provide the technological options for geographically-specific programming that can potentially include locally relevant information and promote local talent.

Many of the liberal democracies in the world have enacted laws that promote independent and pluralist media, India being the conspicuous exception (Pavarala and Kumar, 2001b). In Australia, community broadcasting has been allowed for nearly two decades on the condition that stations licenced under that category must be: (*a*) non-profit enterprises that reflect community purposes and interests and (*b*) operated with the participation of community members. The objectives of the Australian policy include: promoting the diversity of broadcasting services, developing and reflecting Australian identity, character and cultural diversity, and providing programming material that is locally significant. Similarly, the US has a long history of community radio thriving alongside commercial and institutional broadcasting sectors. The Federal Communications Commission gave a further boost to the community radio movement in the country by announcing, in recent years, a new low-power FM radio licensing policy. This has enabled even smaller players to enter the fray.

The policy objective for establishing community radio stations in Canada is the provision of public access to the airwaves with the aim of offering a unique type of grassroots-level local programming not generally available on the radio dial. Organized as a three-tier broadcasting system (public, private and community), Canada also provides for a simplified process for licensing low-power 'developmental' community stations that could serve as a training ground for the future establishment of higher-power community stations. The new Irish broadcasting policy formulated in the mid-1990s recognizes the 'surge in community and voluntary activity over the past 20 years', especially in areas such as education and health, and identifies the third sector of broadcasting clearly with community and voluntary organizations. The Independent Radio and Television Commission of Ireland has also started a Community Radio Support scheme to provide financial support for licenced community radio stations. The post-apartheid broadcasting policy in South Africa spearheaded by the Independent Broadcasting Authority articulates a three-tier system based on universal access, cultural diversity, democratization of the airwaves and nation building. Being hailed as one of the most comprehensive and progressive communication policies in recent times, the South African policy seeks to address historical inequalities in: access for the disadvantaged, resource allocation (frequencies, human resources and finance), and language, cultural and educational programming.

In the UK, the government published a communications White Paper in December 2000 and is currently assessing the viability of a not-for profit, 'third tier' of access radio sector (Davies and Tacchi, 2001). Even more pertinent for the Indian context are the experiences of the less-developed countries that are allowing community radio stations to function as autonomous media spaces for people.

Radio Sagarmatha is Nepal's first independent community broadcasting station run by a body of environmental journalists, Nepal Forum of Environmental Journalists (NEFEJ), and represents South Asia's first major effort at independent community radio broadcasting (Noronha, 2003). Today, Nepal has about five community radio stations, some in remote rural areas, and each established and operated on a different model of ownership, ranging from NGOs to village government and co-operative societies. In 1998, UNESCO set up a pilot project in Sri Lanka to assess the prospects of converging community radio and information from the Internet to determine its possible impact on development efforts in rural communities. The Kothmale Community Radio, one among the community radio stations under the Mahaweli Community Radio project, was established in May 1999, and has become a global yardstick for participatory media projects. In the Philippines, the Tambuli Community Radio Project set up a management and training team that co-operates with communities to organize independent community radio stations in less-developed rural areas (UNESCO website). Bangladesh too is expected to licence non-profit radio for community groups soon.

Policy Ferment in India

Satellite transmission and the Internet have no doubt broken the State monopoly in broadcasting in India too, but only in the direction of making it highly commercialized and entertainment-oriented. The Government announced, in mid-November 1999, the auction for licences to set up 140 FM (frequency modulation) radio stations in 40 cities.[3] However, a group of CSOs, communication scholars and media activists, which met at Hyderabad for a UNESCO-sponsored workshop in July 2000, urged that

> [the government should] take the current government policy of freeing broadcasting from State monopoly to its logical conclusion by expanding the available media space and permitting communities and organizations representing them to run their own radio stations. (The Pastapur Initiative on Community Radio Broadcasting, 2000)

The spirit of the 1995 judgement by the Supreme Court was to spell the end of government monopoly in broadcasting and to underline the rights of listeners and viewers rather than those of the broadcasters.[4] Implicit in the Court's ruling is the need to provide a more democratic broadcasting space for plurality of views, opinions and ideas (Price and Verhulst, 1998). The judgement was clearly not in favour of total deregulation of airwaves for use by the private business firms. It pointed out that the Indian Constitution forbids any private individual or institution or the government from claiming exclusive rights over the airwaves. Broadcasting media as a whole should promote freedom of expression and speech and, therefore, should be able to enjoy freedom from government monopoly and control. The judgement further orders the Central Government to 'take immediate steps to establish an autonomous public authority to control and regulate the use of the airwaves'.

An articulation of several of these vital concerns is also discernible in the reports of a number of committees periodically appointed by the Government of India to evolve a comprehensive media policy for the country (Kumar, 2003a). The demands from CSOs for an appropriate public policy framework led to a revival of some of the critical questions concerning broadcasting in India: What should be the structure of regulatory system that responds suitably to the new media organizations and technologies? What has to be the fate of national (read political) objectives? How to support non-profit access to broadcasting and public communication?

The working paper on national media policy, formulated by a parliamentary subcommittee in March 1996, states that one of the key objectives of the policy should be to project the developmental needs and the social, cultural and economic aspirations of the people, particularly in the rural areas.[5] It also underscores the need for non-commercial broadcasting stations to be run by universities, educational institutions, panchayats, local bodies and state governments. The draft Broadcasting Bill 1997 is seen as a confluence of such major concerns, but remains a major piece of unimplemented good intentions till today. The report of the Prasar Bharati Review Committee submitted in May 2000 reinforced the faith in broadcasting by NGOs and other non-commercial bodies for the benefit of the community.[6] However, none of the recommendations favouring alternative media institutions seem to be materializing in the near future, as the government has now decided to tackle the issue of convergence of telecommunications, IT and electronic media before announcing a comprehensive policy that will cover the entire communications sector.

In the first quarter of 2003, in response to the campaign for community radio in the country, the Ministry of Information and Broadcasting announced a scheme by which 'established' educational institutions, such as universities, IITs and IIMs, could obtain licences to run their own radio stations. Mistakenly labelled 'community radio', the norms laid down for licences include content guidelines which suggest that these campus radio stations air programmes on agriculture, environment, health and other development-related information.[7] Apart from the fact that university campuses are privileged 'communities' with more than adequate access to media resources, it is unrealistic to expect campus radio stations managed by young students to eschew fun and entertainment. There is no apparent fit between the form and the content in this new policy, even as marginalized rural communities continue to wait for real community radio.

Community Radio Initiatives in India

In India, although radio to a large extent still remains centralized, government-controlled and lacking in editorial independence,

All India Radio (AIR) has had some historic experiments in development. It also has a network of local radio stations in the districts for serving the local communities, but the stations have not proven to be locally relevant as they are not community-run (Mukhopadhyay, 2000). In the 1970s, an experimental FM station was set up in Nagercoil district in Tamil Nadu. The project successfully elicited participation of the listeners and could have turned out to be a workable model for development-oriented FM stations. The station functioned with methods that did not conform to the top-down attitude of AIR. K. Anjaneyulu, the initiator of the project, asserted that, 'Local radio should identify itself so completely with the interests of its local population that the heart of the people beats in every pulse of the programme it broadcasts' (Page and Crawley, 2001). This modus operandi could not be replicated in stations administered and staffed by an all-India cadre of broadcasters with only temporary local affiliations.

Several non-profit and development organizations have been campaigning to get permission to set up low-cost local radio broadcasting facilities to support their community development work and to get communities access to means of representation. The Bangalore-based communication campaign group, 'Voices', convened a gathering of radio broadcasters, policy planners, media professionals and not-for-profit associations in September 1996 to study how community radio could be relevant to India and to deliberate on policies appropriate for such an action. A declaration calling for the establishment of community broadcasting was signed. A suggestion to the effect that AIR's local stations should allocate regular airtime for community broadcasting was put forward. Requests were also made for grant of licences to NGOs and other non-profit making organizations for running community radio stations (Voices, 1996).

In 1998, 'Voices' was allotted airtime on the local AIR station in Chitradurga, a rural district in Karnataka. The project proved to be a hands-on learning experience towards the objective of setting up an independently run community radio station. 'Voices' has been working on a project at Kolar, mainly as a centre for training women's credit and savings groups to produce and narrowcast radio programmes. UNESCO made available a portable production and transmission 'briefcase radio station' kit to do experimental broadcasts of programmes produced in Kolar. In September 2001, 'Voices' inaugurated its community radio production centre, 'Namma Dhwani'

(Our Voice) at Budikote in Kolar district. As the matter concerning licences to set up community radio is still under discussion, the programmes are not broadcast but narrowcast, that is, people play back to each other on a tape recorder. According to Ashish Sen, Director, 'Voices', the centre is part of the effort to democratize new information technologies towards community empowerment. The aim is to prepare for the advent of community radio to train the community in scripting, programme-making, recording and other aspects. This station now transmits their radio programmes through the television cable that has networked residents of the village (Sen, 2003).

In Pastapur village of Medak district in Andhra Pradesh, a UNESCO-supported community FM radio centre was set up four years ago, with a recording studio, control room, two transmitters, one transmitting tower and all other related facilities to function as a radio station. This station, managed and run by rural Dalit women, who are members of a local NGO, the Deccan Development Society (DDS), is also awaiting licence to broadcast (Pavarala and Narsamma, 2000). The women record programmes on numerous issues related to women empowerment and local problems, edit them on their editing equipment in the studio and play them back on tape recorders at village *sangams*.[8] More than 300 hours of programming has been done so far and put together in a magazine format that includes talk, interviews, discussions, folk songs and other forms of entertainment available in rural Telangana. P.V. Satheesh, Director, DDS laments, 'The community radio centre is languishing for want of a licence from the Government of India which strangely, on paper, represents the Sovereign People's Republic of India, but has more faith in Rupert Murdoch rather than its own people when it hands over broadcast licences. In this Kafkasque situation, we must continue to operate and create our own small spaces for the real communities of the country' (Satheesh, 2001). One of the few NGOs in the country to apply for an independent licence to broadcast, the DDS had to face rejection from the I&B Ministry.

The Kutch Mahila Vikas Sanghatan, (KMVS) Gujarat, an NGO working with rural women on their concerns in 150 villages of Kutch district, offers a different model for community broadcasting in India. KMVS has been involved in community development work in the Kutch district of Gujarat for over a decade. In December 1999, the Ujjas unit of KMVS started airing a weekly, 30-minute sponsored Kutchi language programme on AIR's Radio Bhuj, for which the team was conferred the prestigious Chameli Devi Award for 2000. The radio programme was entitled *Kunjal Paanje Kutch Ji* (Sarus Crane of Our Kutch). According to Shabnam Virmani, who had directed this radio serial on behalf of Drishti Media Collective, a media NGO based in Ahmedabad, 'the programme raised specific gender-related issues such as the participation of women in political processes, specifically panchayats, girl's right to education, female foeticide, harassment of brides for dowry, unnatural deaths and suicides of women at their in-laws, pressure on women to produce boys, maternal mortality and disregard of mother's health' (The Hoot.org, 2001). The Centre for Alternatives in Education, which is a part of the Indian Institute of Management in Ahmedabad, conducts feedback surveys to ascertain the ground-level impact of the programme.

The second round of the interactive community radio programme by KMVS went on air in March 2001 from the Bhuj radio station, this time specifically aimed at addressing the needs of the earthquake-devastated people of Kutch. The programme titled *Tu Jiriyo Ain* (translated *Tu Zinda Hai* in Hindi) was primarily interview-based and also included songs by local artistes on the earthquake. The programme was meant to facilitate lateral communication between affected communities and vertical communication between the people and structures of power (The Hoot.org, 2001). The organization's current programme titled *Kutch Log Ji Bani* (The Voice of Kutch) is handled wholly by the media unit of KMVS and is aired on AIR Rajkot, which is a medium-wave station.

The community radio reporters of KMVS have become some sort of a model for other community radio efforts in the country. The capacities of these reporters have been strengthened systematically through conscientization, confidence-building and technical-training workshops, as also on the art of seeking information (Kumar, 2003b). With special investigative features and emphasis on local issues such as water, panchayat, literacy, alcoholism, mid-day meal and health-care facilities, the programme has gained immense popularity and has aroused the aspirations of the community.

From August 2001, the National Foundation for India (NFI), in collaboration with a grassroots-level NGO, Alternative for India Development (AID), and their technical partner, Manthan, started airing a community radio programme from Daltongunj, a backward region in the Palamau district of Jharkhand state by using the All India Radio (AIR) slot of the local station on similar terms as the KMVS in Bhuj. The 30-minute programme, *Chala Ho Gaon Mein*, started broadcasting every Sunday on FM band at 7.20 p.m. from the local AIR station in Daltongunj.[9] NFI and its partners conducted participatory workshops for gradual and methodical capacity-building of community volunteers in radio production and in identification of core issues to be addressed by the community radio (NFI, 2001). Made in the local dialect, the half-hour programme concentrates on local issues such as alcoholism, health, literacy and child marriage. The community reporters of AID travel in and around the villages, carrying back reports of local problems and developments, folk songs and folk tales from the region, and record plays on local issues with performers from the villages. The issues discussed in the programme not only strike a chord among the listeners, but also help forge solidarity among those placed in a similar socio-economic situation. The programme motivates people to work for the development of the village and also provides opportunities for 'suppressed artistic talent' in the villages (Pavarala, 2003).

Conclusion

Even as some groups are taking advantage of tiny spaces opened-up within AIR, many other NGOs and media-activist groups are struggling to carve out any autonomous space. In summary, the idea of community radio as it is understood and implemented in India and other parts of the world encapsulates the following:

- a conception of the marginalized as active social agents in communication;
- a conception of communication as a process that helps poor, disadvantaged women reframe histories of oppression and strengthen their resolve to resist further marginalization; and
- a strategy that seeks to anchor rural women's communication experiences and modes of communication in the local, everyday realities of their social environment; and a recognition that grassroots-level participation and horizontal circulation of ideas are necessary pre-conditions to democratization of communication and redistribution of power.

Notes

1. An earlier version of this paper was written as an introduction to the Trainers' Manual on Community Radio brought out by the Manipal Institute of Communication and Ford Foundation India, 2004.
2. Many of the principles of citizens' communication rights have been articulated and developed in the People's Communication Charter, an initiative supported and endorsed by AMARC, Cultural Environment Movement, Third World Network, and many other organizations.
3. By early August 2000, it was announced that some 26 companies received letters of intent, from the Indian government, after bidding to set up FM radio stations in 40 Indian cities. While some private FM stations have since gone on air, the fact that the advertizing pie is not large enough for too many players has deterred many companies from venturing into this sector.
4. Judgement dated 9 February 1995 in Ministry of Information and Broadcasting v. Cricket Association of Bengal states: 'Airwaves constitute public property and must be utilized for advancing public good.'
5. This report formalizes the intentions first set out by the Joshi Committee in 1983 of achieving an 'Indian personality' for Indian broadcasting.
6. Ministry of Information and Broadcasting, Government of India constituted a Committee on 22 November 1999 with N.R. Narayanamurthy as the Chairman to carry out a comprehensive review of Prasar Bharati.
7. See www.mib.nic.in for detailed guidelines and application for licence. At the time of writing this article, only one radio station started by Anna University, Chennai has obtained a licence under this policy and gone on air.
8. See Muppidi (2001) for an account of the DDS *sangams*, that is, women collectives.
9. As a direct consequence of its popularity and listener demand, the programme now airs twice a week.

References

AMARC. 1999. 'Communications Infrastructure for Citizens' and Community Media—An AMARC-Europe Response to the 1999 Communications Review'. Available at www.amarc.org/europe/Publications/Policy/Comm-rev-response.htm.

Appadurai, A. 1996. *Modernity at Large: Cultural Dimensions of Globalization*. Minneapolis, MN: University of Minnesota Press.

Bell, D. 1974. *The Coming of Postindustrial Society: A Venture in Social Forecasting*. Harmondsworth: Penguin.

Bhatnagar, B. and R. Schware. 2000. *Information and Communication Technology in Development*. New Delhi: Sage Publications.

Calhoun, C. 1994. *Social Theory and the Politics of Identity*. London: Basil Blackwell.

Castells, M. 1996. *The Rise of the Network Society (Vol. 1 of The Information Age: Economy, Society and Culture)*. Massachusetts: Blackwell Publishers.

———. 1997. *The Power of Identity (Vol. 2 of The Information Age: Economy, Society and Culture)*. Massachusetts: Blackwell Publishers.

Dagron, A.G. 2001. *Making Waves: Stories of Participatory Communication for Social Change*. New York: The Rockefeller Foundation.

Davies, E.P. and J. Tacchi. 2001. *Community Radio in a Global Context: A Comparative Analysis, Unpublished Report*. London: Community Media Association.

Escobar, A. 1995. *Encountering Development: The Making and Unmaking of The Third World*. New Jersey: Princeton University Press.

Fraser, N. 1992. 'Rethinking the Public Sphere: A Contribution to the Critique of Actually existing Democracy', in Craig Calhoun (ed.), *Habermas and the Public Sphere*. Cambridge: MIT Press.

Freire, P. 1970. *Pedagogy of the Oppressed*. New York: Seabury Press.

Habermas, J. 1962. *The Structural Transformation of the Public Sphere*. Cambridge: MIT Press.

Hamelink, C.J. 1986. 'Is there Life after the Information Revolution', in M. Traber (ed.), *The Myth of Information Revolution*. New Delhi: Sage Publications.

Kothari, R. 1984. 'Communications for Alternative Development: Towards a Paradigm', *Development Dialogue*: 1–2.

Kumar, K. 2003a. 'Mixed Signals: Radio Broadcasting Policy in India', *Economic and Political Weekly*, 38(22) 31 May–6 June 2003: 2,173–182.

———. 2003b. 'Kutch Mahila Vikas Sangathan (KMVS) Community Radio Initiative: A Case Study—Some Preliminary Observations', paper presented at the UNDP Consultation on ICTs and Development, Indian Institute of Management, Bangalore, 11 March.

Lyon, D. 1988. *The Information Society: Issues and Illusions*. Polity Press: Oxford.

Melkote, S.R. and H.L. Steeves. 2001. *Communication for Development in the Third World: Theory and Practice for Empowerment*. New Delhi: Sage Publications.

Mukhopadhyay, B. 2000. 'Why AIR has Completely Ignored Community Radio', in *Humanscape*, June 2000: 24–25.

Muppidi, S.R. 2001. 'Using People's Organizations for Development: Role of Women's Sanghams in Zaheerabad India', in S. R. Melkote and S. Rao (eds), *Critical Issues in Communication*. New Delhi: Sage Publications.

Nair, S.N. and S.A. White. 1993. *Perspectives on Development Communication*. New Delhi: Sage Publications.

National Foundation for India (NFI). 2001. *Process Document for Community Radio Initiative*. New Delhi: NFI.

Negt, O. and A. Kluge. 1972. *Öffentlichkeit Und Erfahrung (Public Realm and Experience)*. Frankfurt: Suhrkamp. A section translated in A. Mattelart and S. Siegelaub (eds). 1983. *Communication and Class Struggle, vol. 2. Liberation/Socialism*. New York: International General.

Noronha, F. 2003. 'Community Radio: Singing New Tunes in South Asia', *Economic and Political Weekly*, 38(22) 31 May-6 June 2003.

Page, D. and W. Crawley. 2001. *Satellites over South Asia*. New Delhi: Sage Publications.

Pavarala, V. 2003. 'Building Solidarities: A Case of Community Radio in Jharkhand', *Economic and Political Weekly*, 38(22) 31 May–6 June 2003: 2,168–172.

Pavarala, V. and 'General' Narsamma. 2000. 'Voices from the Margins: A Case Study of Community Media Initiatives', *Voices for Change: A Journal on Communication for Development*, 4(2) October.

Pavarala, V. and K. Kumar. 2001a. 'Legislating for Community Radio: Policy Framework from Other Countries', *Voices for Change: A Journal on Communication for Development*, 5(1) May: 13–17.

Pavarala, V. and K. Kumar. 2001b. 'Broadcasting in India: New Roles and Regulations', *Vidura*, 38(3) July–September: 18–20: 2,188–197.

———. 2002. 'Civil Society Responses to Media Globalization: A Study of Community Radio Initiatives in India', *Social Action*, 52(1) January–March 2002: 74–88.

Pieterse, J.N. 1994. 'Globalization as Hybridization', *International Sociology*, 9(2): 169–84.

Price, M.E. and S.G. Verhulst. 1998. *Broadcasting Reform in India: Media Law from a Global Perspective*. New Delhi: Oxford University Press.

Robertson, R. 1992. *Globalization: Social Theory and Global Culture*. Newbury Park, CA: Sage Publications.

Rogers, E.M. 1962. *Diffusion of Innovations*. New York: The Free Press.

Satheesh, P.V. 1999/2000. 'An Alternative to Literacy', in *Forests, Trees and People Newsletter No. 40/41*.

———. 2001. 'Inauguration of Community Radio Center'. Posting on cr-india listserv (cr-india@goacom.com), 27 September 2001.

Schumacher, E.F. 1973. *Small is Beautiful*. New York: Harper and Row.

Sen, A. 2003. 'Waiting to be Heard: Bringing Marginalized Voices to the Centre', *Economic and Political Weekly*, 38(22) 31 May 2003–6 June 2003: 2,198–201.

Servaes, J., T.L. Jacobson and S.A. White (eds). 1996. *Participatory Communication for Social Change*. New Delhi: Sage Publications.

Sood, A.D. 2002. *Guide to ICTs for Development*. Bangalore: The Centre for Knowledge Societies.

Stuart, S. and R. Bery. 1996. 'Powerful Grassroots Women Communicators: Participatory Video in Bangladesh', in J. Servaes, T. L. Jacobson and S. A. White (eds), *Participatory Communication for Social Change*. New Delhi: Sage Publications.

Traber, M. (ed.). 1986. *The Myth of Information Revolution*. New Delhi: Sage Publications.

Voices. 1996. *Bangalore Declaration on Radio*. Bangalore.

White, S.A. (ed.). 1999. *The Art of Facilitating Participation*. New Delhi: Sage Publications.

White, S.A., K.S. Nair and J. Ascroft (eds). 1994. *Participatory Communication*. New Delhi: Sage Publications.

Wilson, R. and W. Dissanayake. 1996. *Global/Local Cultural Production and the Transnational Imaginary*. Durham, NC: Duke University Press.

Wresch, W. 1996. *Disconnected: Haves and Have-nots in The Information Age*. NJ: Rutgers University Press.

9

Wired Watchdogs: How ICTs are Aiding Civil Society to Enhance Public Accountability

Gopakumar Krishnan Thampi

Central to this dominant narrative sketching the interactions of the State and civil society is the rapid diffusion of powerful cutting-edge ideas and tools. A good case in point is the integration of information and communication technology (ICT) tools into the traditional structures and processes of civil society action. The proposed paper will explore this theme in the specific context of how civil society organizations (CSOs) are addressing the phenomenon of public accountability. Three case studies will be presented—two from Latin America and one from India—as exemplars to highlight the emergent trends. The paper will also attempt to draw some relevant conclusions and strategic insights for adaptation and replication.

The Backdrop: Evolution and Alienation of the Modern Democratic State

The legitimacy and point of reference for the civil society is the 'civil' part of the term which originates from the Latin word *Civilis* meaning of or relating to the State or its citizenry.[1] Semantics certainly seems to have directed the future trajectory of civil society's conceptual evolution and practical applications. Whether used as a construct to highlight a 'popular' response against the church as it happened in medieval Europe or the contemporary struggles in challenging the monopoly powers of a hegemonic State or the abusive distortions of a rampaging market, civil society has often been positioned as a critical countervailing power.[2]

Given the inextricable links between civil society and citizenship, it is quite logical that we may now turn our attention to the concept of citizenship and its ramifications on the manner in which the concept has influenced and informed the language of praxis in the civil society arena. Historically, citizenship in the contemporary sense—the wide rubric encompassing legal, political, social, cultural and economic rights and duties that regulate the relationship between State and subjects—is the outcome of the rationalizing activity of the rising modern State (McNeill, 1983). Administrative, coercive and extractive procedures were thus homogenized in the name of greater efficiency and State capacity, consequently, was greatly enhanced. The new modes of exercising power required new modes of legitimization. Thus universal consent to be ruled became a crucial factor of politics and popular sovereignty a necessary response to the intensification of State power. However, this transformation brought about radical changes in the attitudes and responses of society, mainly in terms of accepting the rationalization but simultaneously challenging the monopoly power of the State. Rather than play itself out as a static endgame, relations between the State and the civil society has come to represent extremely dynamic and often creative interfaces acted out within the ambit of an interesting canvas highlighted by this bi-polarity of *accepted rationalisation* and *contested hegemony* of an overpowering State.

How did the modern State come to be alienated? The answer lies in the praxis of the State couched under the rubric of public policy. In many sense, the most visible form of existence of the modern State is the domain of public policy. Public policy effectively captures all nuances of democratic governance—ideology, representativeness, legitimacy and accountability. However, in spite of sweeping liberalism and democracy, public policy stands alienated from social aspirations and representativeness. And practice of democratic tenets has unfortunately failed to address this gap effectively. The concept of 'democracy' though unchallenged in principle is increasingly being seen as flawed in practice (Sartori, 1991). Especially so, when one looks at worrying issues like rampant corruption, abuse of power and dwindling voter participation.

The evolutionary history of the modern State entered an interesting epoch following the tumultous events of the early 1990s. An increasingly resurgent civil society across the world started pitching in a startling resistance to the hitherto unchallenged power of a bellicose State. Far from being comfortably playing the role of passive recipients of public policy actions, civil society movements, organized and reactive, began to actively contest and challenge the State in various spheres, mostly within the democratic framework. Contemporary narratives on governance provide an excellent visual of this engaging dialectics between a powerful State and an equally potent civil society.

The Public Accountability Conundrum

> In framing a government which is to be administered by men over men, the great difficulty lies in this: you must first enable the government to control the governed; and in the next place oblige it to control itself.
>
> —Thomas Jefferson (Rossiter, 1961)

Even after two centuries, the prophetic tone of this statement rings loud. The dominant narrative in current debates on governance hinges around the theme of power and the need to control it. Public policy design and governance have come to represent insensitivity, narrow vision, opaqueness and non-responsiveness. The State in its obligation to impose and validate its own rationality in the exercise of power tends towards one-sidedness, absence of feedback and a dominant bureaucracy, which co-opts the political system. Though the State has made many impressive strides along the development trajectory in many developing countries, lack of public accountability has severely undermined the fruitful realization of development goals. Though, contemporary reforms seek to bring in wider access and greater transparency, governments will become more accountable and perform better only when citizens play a watchdog role and challenge abuses.

It is at this juncture that contemporary civil society is playing a decisive role by actively focussing on the theme of public accountability by illustrating a vivid spectrum of advocacy strategies ranging from contestation to partnerships.

Civil society is increasingly being conceptualized as a 'space' between the State (government) and the Market (business) to challenge, contest, confront and collaborate for improving the quality of governance and safeguarding individual rights and ensuring social justice. And advocacy is a key strategy to use that space effectively. Most actors in the civil society arena have sometime or other, used advocacy as a means to make their interventions more effective and focussed. The emerging challenge in the context of the new economy is to tap evolving technological frontiers to enhance the potency of existing advocacy strategies and more important, to explore new and innovative modes of advocacy. Already, the responses to this challenge in the form of increased use of IT to enhance social capabilities can be discerned.

ICTs and Civil Society—Can Activism Blend with Technology?

It is now a fairly accepted fact that breakthroughs and advancements in communication have been largely instrumental in the spread of democracy worldwide. The ongoing ICT revolution has raised the expectations on the transformative potential of communication technologies to extremely high (and as some would argue, worrying) levels. Consider this statement from a recent work extolling the virtues of ICTs in transcending the seemingly insurmountable development challenges: 'Earlier, the question was *what technology could do* for better information organization and communication. With Internet-based ICTs, we speak of *what there is left to be done*' (Bedi et al., 2001).

However, in the New Economy, the powerful have shaped and dictated the discourse on transparency, accountability and openness, leaving little space for civil society to define these concepts, to protect their stakes as citizens. Nevertheless, in the past decade, we have seen inspiring global campaigns, aided by IT and the net, in civil society efforts to develop a critique of, and challenge hegemonic anti-people and anti-environment aspects of the new economy. More important, as a recent briefing note highlights:

> The vision of ICTs as a support to local empowerment is continuing however, in part through the variety of community-based ICT initiatives such as local public access, telecentres, community networking, and community informatics programmes and community-based and often self-organized E-health, E-culture, and E-participation initiatives. These activities, often combined with the use of Open Source Software, place the control over ICT deployment and implementation into local hands

and provide tools locally for adapting the productive, commercial and service delivery capabilities of ICTs to local needs and local resources. (Gurstein, 2003)

Unfortunately, the contours of the sweeping changes ushered in by the ongoing information revolution have, however, largely bypassed the non-profit sector or the civil society. The failure of civil society organi isations (CSOs) to participate in the ongoing technology-led information revolution will enfeeble them in their socio-political struggles and result in their inevitable marginalization in the long run.[3]

However, increasing evidences from across the globe point to the enabling potency of ICT-led strategies to give a new thrust to civil society actions, especially in the realm of public accountability. The rest of this chapter will sketch three emergent examples—two, from the universe of Transparency International, the global coalition against corruption and one enabling Indian initiative from Public Affairs Centre, Bangalore to contextualize the narratives and facilitate learning and adaptation.

ICT-led Interventions as an Aid for Monitoring Public Accountability: Three Case Studies[4]

The following illustrative examples show how civil society groups are successfully adopting cutting-edge and traditional ICT techniques to empower their endeavours in fighting corruption.

Anti-corruption Radio Spots, Brazil

As the medium for public service messages par excellence, and the only such medium with a constant audience throughout the day, radio provides an excellent vehicle for short, repetitive messages aimed at promoting awareness about the problems of corruption and to empower citizens to make a difference in the fight against it.

This initiative was pioneered by Transparência Brasil (TBrasil), the Brazillian Chapter of Transparency International (TI). TBrasil approached a São Paulo-based radio station (owned by the O Estado de S. Paulo media group, Eldorado plays exclusively in the city of São Paulo and primarily to an upper-middle-class audience) with the suggestion that it adopt as a permanent fixture the regular airing of short anti-corruption spots aimed at awareness-raising and behavioural change.

The initiative, called 'Say no to corruption', began broadcasting in August 2001. Eldorado airs the radio spots four times per day, two of which are aired during prime-time listening hours (intervals during its noon news programme). Other Brazilian radio stations approached by TBrasil replicated the initiative.

The themes used in the spots are intended to address common questions related to corruption and, whenever possible, current events. All radio spots produced in Portuguese up until the writing of this tool can be read on TBrasil's website (www.transparencia.org.br). The radio spots produced to date have had the following themes:

- Corrupt people gloating over being 'clever'.
- Taking bribes is a crime, so is paying them.
- Feeling guilty about paying bribes.
- Instead of paying a bribe, make a complaint to the person's boss.
- Do not pay bribes to the traffic police.
- Information about money laundering (examining a local case).
- TI's Integrity Award for the Londrina anti-corruption initiative.
- Congressmen who resign to avoid losing their mandates. Do not vote for them anymore.
- Monitoring the procurements of your municipality.
- 'It is not the government's job to see that citizens are corrupted. It is the citizen's job to see that government is not corrupted.' (Robert H. Jackson, American judge in the Nuremberg trials)

Obviously, the tool's effectiveness depends mainly on the number of radio stations that participate in the campaign. With this in mind, the difficulties faced by a country of such enormous territorial size as Brazil are larger than those faced by smaller countries. Hundreds, if not thousands, of participating radio stations would be necessary to reach a significant percentage of the population. In order to get other radio stations to air the spots, TBrasil collected in a database information on over 500 radio stations, and then sent invitations to them via e-mail to join the campaign and radio spots. The first such initiative in 2001 resulted in eight additional radio stations joining the campaign. A second round of invitations in April 2002 resulted in three more radio stations joining the campaign.

To inform an even larger audience about the campaign, TBrasil also asked about 800 people, 200 of whom are journalists who are informed on a weekly basis about TBrasil activities, to tell other people in their communities about the campaign. All spots are reproduced on TBrasil's website. TBrasil also invites people to submit ideas for new spots. Local organizations created to curb corruption in their surroundings are also invited to take part in the initiative. One local organization in the southeastern State of Espírito Santo arranged to air spots with two radio stations in the State's capital, one of which was the State's leading AM/FM station. In this particular case, the local organization (Transparência Capixaba) included a number of additional spots of their own, many of which referred to local cases of corruption. TBrasil reviewed all such 'new' spots and suggested alterations. Both Transparência Capixaba and TBrasil have endorsed these spots.

Feedback is not frequently received and always comes in the form of praise of the initiative. Overall, some spots are praised more than others are. Criticism of spots always addresses the naïveté of expecting people to change their behaviour on the basis of such messages, which is, of course, quite true when looked at exclusively from that point of view, but also misses the point by not considering the awareness-enhancing dimension of the initiative.

Licitenet.com: Making Public Procurement More Transparent in Ecuador

This is an exciting joint initiative between the private sector and civil society to create an online database to inform the public about the process of public procurement in Ecuador, and to harness the power of the Internet to foment greater transparency and competitiveness. Corporación Latinoamericana para el Desarrollo (CLD), Transparency International National Chapter in Ecuador is the civil society 'driver' in this initiative.

For years, business associations and other organizations have expressed concern about the high level of corruption in public procurement in Ecuador. Since 1997, CLD has participated in several initiatives to try to reform the country's legislation but has met with little success. After waiting in vain for Congress to modify the legislation, business associations such as Gente por el Cambio and Ecuador Positivo joined forces with CLD to undertake a new project. The purpose was to create a viable system offering relevant information both to the public and to State contractors, as well as providing a forum for the discussion of public procurement in Ecuador.

One of the most important objectives was to disseminate information to citizens. Anyone can visit the Licitenet.com site and make inquiries about procurement processes that are underway, completed or pending approval. They can find ample information to develop a clear and precise understanding of the procurement process in Ecuador since the system went into operation. Access to this type of information is empowering for civil society. It provides citizens with more knowledge about the processes involved and enables them to demand accountability from the State if anomalies arise. Licitenet.com and CLD are thus seeking to create a forum for discussion and analysis of this issue. First, the project aims to demonstrate that making information available to the public leads to cost savings and transparency for State institutions. It also aims to highlight the different failings that make current systems of procurement bureaucratic and inefficient. Ultimately, it is hoped that this will foster the development of a better public service and prompt the State to acknowledge that legislative reforms are necessary. At the moment, the specific goal is to reduce the discretionary power exercised by officials at different stages of the public procurement process. These include: the submission of specifications or pre-contractual tender documents; the obligatory fee to review the specifications; the requirement of disclosure only for processes that exceed US$ 112,000; as well as the detailed conditions of each purchase, the ambiguity of the deadlines, and so on.

The number of 'hits' on the Licitenet.com site has grown rapidly. The site has received over 13,500 visits since its launch with an average of 15 hits per visit. In total, it has received approximately 202,500 hits. Less than 10 per cent of Ecuador's population is currently online but the government has placed considerable emphasis on increasing connectivity. For this reason we hope that the site will become increasingly popular. Licitenet.com was invited to serve on one of the working groups to develop Ecuador's National Connectivity Agenda. This will oversee progress on connectivity and telecommunications in the country for the next 10 years. The group is considering making Licitenet.com the starting point for automating the entire process of public procurement in Ecuador (that is, the creation of a fully-fledged e-procurement portal).

Agreements with the Public Sector to Extend Information Coverage After signing a contract to receive information from the Attorney General's Office, CLD discovered that the data was available only in print—a format that was nearly impossible for citizens to access easily. Consequently, the Attorney General's Office developed a database specifically to send relevant information to Licitenet.com.

This allowed citizens to view data in a much more efficient manner.

Assistance from Business Associations A number of different business associations worked on the project free of charge, for example:

- The Quito Chamber of Commerce invited Licitenet.com to participate in the 'Ecuador: Business, Opportunities and Franchises' fair at no charge, which helped to promote the initiative.
- The Pichincha Chamber of Construction invited Licitenet.com to participate in its permanent exhibition of different construction services.
- The Guayaquil Chamber of Commerce organized an event to present Licitenet.com to its members at no cost.

In addition, business associations like Gente por el Cambio, Ecuador Positivo and the CLD joined forces with Licitenet.com. The role of these two business associations was very important. Both of them consist of senior-level, young executives at some of the country's most influential enterprises, who are keen to promote social progress and reform in Ecuador.

Media Coverage Media coverage was crucial to the success of the project. Newspapers and magazines have published a series of articles and reviews of the site. Nevertheless, the coverage has fallen short of Licitenet.com's initial hopes because the media has been reluctant to provide it with free publicity, on the grounds that this is a private initiative that seeks to make money.

Citizen Reaction Citizens have been the main users of the site and they tend to view it as an alternative way of accessing public sector information. The questions and comments submitted by citizens who visited the site are very interesting and they change constantly. Frequently, individuals and companies have posed specific questions concerning public procurement, or requested help in understanding how the process works.

Campaigning for Electoral Transparency, Bangalore[5]

The quality of governance can be no better than that of the elected representatives. There is a logical corollary to this in the sense that any improvement in the quality of governance, therefore, has to address the issue of improving the quality of the electoral process. Recognizing this crucial link, Public Affairs Centre (PAC), a not for profit organization in Bangalore has been consistently striving to influence both the supply and demand sides of the electoral process. In an effort to bring in an element of electoral probity and transparency and thereby enable the voters to make an informed choice, PAC in association with other civil society partners have designed and conducted information campaigns on the candidates standing for elections. The other parallel concern addressed by PAC was to engage the disinterested citizenry in the electoral process. Through a series of simultaneous innovative campaigns, PAC attempted to enhance voter participation during elections. An important strategy identified in this context has been the effective scope for blending ICT applications in to the traditional domain of advocacy.

Campaign to Stimulate Bogus Votes Preparations for this campaign began when the Bangalore Mahanagara Palike announced, in August 2001, that the voters lists would be on display from 10th of August to 20th of August 2001 in all the ward offices and the Assistant Revenue Officers offices. The display was for verification in view of the forthcoming elections to Bangalore Mahanagara Palike. As an immediate response to this announcement PAC launched a *Campaign for Verification of Voters List* on the same day to support the initiative. The main focus of the campaign was to inform the public about the verification process and give them more information on how to participate in the effort. In order to achieve this, PAC devised a campaign drawing upon an interesting spectrum of ICT applications—mass e-mails, announcements on Radio City—a popular FM radio channel and cable TV networks.

A major highlight of the campaign was the design of a '*One-day Hotline*' on 18th of August 2001 between 11.00 a.m. and 5.00 p.m. at the PAC office to receive the complaints from citizens facing any difficulty in verifying the list, and to record their complaints. The calls started coming in much before the scheduled time and continued for few weeks. These complaints were forwarded to the concerned officer in the BMP for further actions. This exercise was subsequently scaled-up in eight other cities in Karnataka. After the Summary revisions, PAC decided to contact the people who had called the Bangalore help-line, as a follow-up so that an analysis could be done on the effectiveness of the campaign. About 69 per cent of the people who contacted PAC mentioned that they took further action.

Making an Informed Choice—Voting for Responsive, Capable and Upright Candidates Given the crucial role of elected Councillors in the development of Bangalore and the need to ensure a responsive, efficient and non-corrupt administration, PAC and Swabhimana along with the various Resident Welfare Associations in the city designed a powerful campaign to bring in probity and informed participation to the electoral processes. Candidates contesting in each of these wards were interviewed by the members of the association based on an interview schedule of pertinent questions probing their background—whether they are tax payers, do they reside in the ward from which they are contesting and criminal record; level of awareness of civic issues; commitments and priorities for the ward; past achievements; and willingness to abide by the electoral code of conduct, and so on. The information thus collected was widely disseminated through door-to-door distribution of leaflets and display of posters at strategic locations.

ICE TV network, which reaches to about 30 per cent of population in Bangalore came forward to do a similar exercise in collaboration with PAC and Swabhimana by recording the interviews of candidates and airing the same in about 20 wards covered under its transmission. Though the logistics were set for the task, candidates' response was poor and only about 12 candidates dared to reveal themselves in front of the camera. However, the cable network aired PAC's powerful advocacy video documentary 'Whose Vote is it Anyway?' and the interviews of PAC representatives.

Interestingly, this exercise was carried out by other civil society partners in nine other cities in Karnataka for the 2001 Urban Local Body Elections. The reach and power of cable television was exploited in a powerful manner in many cases: In Mysore, the partner organization—Mysore Grahakara Parishad (MGP) participated in a phone-in programme on SITICABLE (a cable channel which is received in 80 per cent of the homes in Mysore), to discuss on the quality of candidates. In Tumkur, four partner groups carried out their information dissemination largely through the cable network, which covers 70 per cent of the city. They prepared a video filler/clipping to air on the channel and also showed PACs video documentary 'Whose Vote is it Anyway?'. The local TV network blocked all channels on the day of the election and showed messages regarding voter's awareness all day. The groups also carried out the 'Making an Informed Choice' in all the 35 wards in Tumkur—a huge task—and the information collected was disseminated through the cable TV network.

Voter Motivation, Awareness and Communication Programme

- Slides with motivating messages were shown in two major city theatres for a week prior to the election. Similar slides were also aired on cable TV network with the help of ICE TV network that covers 30 per cent of the population in the city.
- Perhaps, the most interesting initiative was that of coxing the public sector undertaking Bangalore Telecom to play recorded messages in several exchanges reminding citizens to cast their votes on the day of the elections. Subscribers of the telephone company had a pleasant surprise on election day when they lifted the telephone receiver to make a call!
- Radio City FM 91 kept citizens informed about the BMP elections and the importance of their participation for a week prior to elections. An interview with the Chairman of PAC, Dr Samuel Paul was also aired to encourage and educate voters about the importance of exercising their franchise. This popular radio channel has supported their campaign from the start.

Conclusion

In the emergent New Economy, the powerful have often shaped and dictated the discourse on transparency, accountability and openness, leaving little space for civil society to define these concepts, to protect their stakes as citizens. Nevertheless, in the past decade, we have seen inspiring global campaigns, aided by ICT, in civil society efforts to develop a critique of, and challenge hegemonic anti-people and anti-environment aspects of the new economy. Unfortunately, the contours of the sweeping changes ushered in by the ongoing ICT information revolution have however, largely bypassed the non-profit sector or the civil society. The failure of CSOs to participate in the ongoing technology-led information revolution will enfeeble them in their socio-political struggles and result in their inevitable marginalization in the long run.

But as the illustrative examples prove, CSOs are fast harnessing the potential of cutting edge ICT tools and techniques and incorporate them to the traditional domains of their functioning.

What then are the key learnings from these examples? Rather than offer 'sure-shot recipies', the following list attempts to summarize some insights and strategies that may prove to be useful pointers for replication and adaptation in the Indian context:

- ICT-led initiatives cannot replace conventional grassroots advocacy. The essence is to blend the potential of technology with the zeal of the advocate.
- The tool has to fit the context. One major fallacy perpetuated by the acolytes of the ICT revolution is the primacy accorded to high-end tools like the Internet. In countries like India, radio and cinema are still powerful in terms of reach and access and these media are largely unexplored in terms of effective use by civil society groups.
- Disaggregating the audience/target group. Technology offers various entry points to launch effective advocacy programmes. As the case of PAC illustrates, the same objective (voter motivation) can be packaged as mass e-mails to educated urban users and also as stills in cinemas and as promos in the cable network to reach effectively into the rural enclaves.
- No need to reinvent the wheel: ICT applications can easily dovetail into existing programmes and initiatives, and thereby calls for minimal demands on the organizations.
- Forming strategic alliances: Since most civil society groups find it difficult to locate competent technical support (primarily due to high costs), it becomes imperative to forge strategic alliances with marketing professionals, software design groups, and so on, as the case study from Ecuador illustrates. The role of the private sector is quite under-utilized in this regard.

Notes

1. Webster's Ninth New Collegiate Dictionary (Springfield, MA: Merriam Webster Inc, 1988).
2. Contemporary interest in civil society could be traced to the early 1990s following the collapse of communism and the subsequent efforts to establish democratic or liberal social and political order. For an interesting early discussion on this theme, see Gellner (1991).
3. This theme was the focal point for an innovative workshop organized by the Public Affairs Centre, Bangalore, IT for Change and Mahiti, Bangalore in January 2001. For more details, visit ITfC's website at http://www.itforchange.net.
4. The first two cases are quoted from The Corruption Fighters' Tool Kit (Berlin: Transparency International, 2002). The Corruption Fighters' Tool Kit, first produced

as an international edition in CD-ROM format in 2001, is a compendium of practical civil society anti-corruption experiences described in concrete terms and accessible language. It presents innovative anti-corruption tools developed and implemented by TI National Chapters (NCs) and other civil society organizations (CSOs) from around the world. The toolkit can be accessed online at http://www.transparency.org/toolkits/index.html

5. Further information on this intervention can be found at www.pacindia.org

References

Bedi, K., P.J. Singh and S. Srivastava. 2001. *Government@net: New Government Opportunities for India*. New Delhi: Sage Publications.

Gellner, E. 1991. 'Civil Society in Historical Context', *International Social Science Journal*, 129 (August): 495–510.

Gurstein, M. 2003. 'Communities: The Hidden Dimension of ICTs', CRIS CAMPAIGN, Issue Paper No. 7.

Mc. Neill, W. H. 1983. *The Pursuit of Power*. Oxford: Blackwell.

Rossiter, C. 1961. *The Federalist Papers*. New York: New American Library.

Sartori, G. 1991. 'Rethinking Democracy: Bad Polity and Bad Politics', *International Social Science Journal*, 129 (August): 437–450.

10

Influence of Software Trade on Development of Web-enabled Governance Initiatives: Comparative Perspectives of Indian and UK Experiences

Amit Mitra
Alwyn Didar Singh

Introduction

Over the last couple of decades, it is clear that gradually India has established itself as a country that is responsible for producing a substantial amount of software that is used throughout the world. More recently, BPO is also becoming more common where large multinational companies in the US, the UK and elsewhere are transferring their call centre management and other basic office operations to a cheaper cost environment like India. Clearly, this has meant that over the years, apart from the role of getting software development assignments, what has gradually emerged is an awareness of the potential for application development in almost all areas of software engineering. Such a process has surely done something in changing mindsets from a tool to an institutional or organizational perspective. Even within the Indian context, additional functionality is no longer the primary issue; use of information and communication technologies (ICTs) today is more about extending support to a wider range of governmental activity. For instance, emphasis has been laid on what was referred to as SMART (acronym for Simple, Moral, Accountable, Responsive, Transparent) governance initiatives, which have long been flagship endeavours of the Indian bureaucracy both within central as well as regional governments. However, despite all these strides in objective terms, not very much has happened in terms of the overall approach to electronic governance. Regional champions driving initiatives continue to be the focal approach to growth in ICT use. Unlike other parts of the world, there is still a higher reliance on the type of services that are provided by government agencies in India.

Governmental activity differs between the developed countries like the UK in comparison to a developing country like India through an important dimension. Whereas, extending provision of facilities to unconnected parts of the geography is important in the context of a developing country like India, the primary role of governance is maintenance of functionality within a developmental context. In recent times however, even governments in the developed world are using ICTs to ensure that among others, difference between the information poor and the information rich is addressed through mediation that is available through resident infrastructure at the local level. On the other hand, emphasis within the developing country context is to firstly develop mechanisms through which delivery of basic services could be delivered as well as to ensure reduction in any disparity between the 'haves' and the 'have-nots'. At this juncture, illustrations to make the difference in requirements between developed and developing economies easy to visualize might be useful.

In the developed world, as a matter of fact, there is hardly any need to interact with government agencies on a regular basis, as many imperative transactions have long been automated through either bank mechanisms or through other regional government capacity. For instance, in the UK, council taxes that residents pay in return for garbage disposal and other facilities provided by the local authority, in the normal course of events are paid off through direct debit arrangements on a fixed-duration cycle. Residents therefore do not need to interact with government agencies except when transactions between the government and their bank payment facilities run into problems.

In recent times, the agenda of local governance in Britain and other parts of the European Union is gradually shifting to using ICTs in targeting social exclusion. Disparity within society in the UK as well as other parts of the EU is changing in such a way that the difference between the rich and the poor is becoming exceedingly large. One of the characteristics of such disparity is the difference between information rich and the information poor within society. Access to Internet provides people with an avenue to not only interact but also to browse relevant information useful for optimally utilizing facilities provided by the local authorities. On the face of the fact that people do not normally interact with local government, it was difficult to envisage a situation where providing services on the Internet was going to be popular. When speaking to officials responsible for e-governance within Derventside, Liverpool, Hampshire and other local authorities, it was suggested that the move of providing facilities was primarily designated to encourage interaction with hard-to-reach groups. At this point, social exclusion in both the developed and the developing worlds tends to relate to the problem of identity and its relationship to marginality. Whereas in the British context, marginality may mean lack of information, the Indian perspective is more connected to the issue of provision and its reach.

Although, identified beacon councils are successful in developing these initiatives, they are yet to translate initiatives to the lowest levels of the needy. As a matter of fact, a new stream of research currently carried out suggests that introduction of electronic governance initiatives are actually distancing groups of people within societies within the UK. While the use of customer relationship management mechanisms by local authorities does have substantial grounds to cover, the endeavour to engage with citizens is very useful. The basis of engagement though is somewhat different in the electronic interface. Whereas face-to-face interaction was the cornerstone of traditional formats of engagement, within the normal course of events distance within communication at the local level is a clear characteristic of Internet-based media.

ICTs and Development

It is axiomatic today to be aware of the wide-ranging benefits and opportunities of ICT usage within enterprises, and indeed for whole economies, in the developing world. Traditional understanding of use of ICTs in governance has advocated the belief that extensive use is likely to lead to both an enabling of citizens to easily transact business with governmental agencies and secondly to change perceptions among people of how government currently operates. In a democracy like India, it is vitally important that the impression of what the government plans to do and how they go about delivering those are brought as close together as possible. From around the early 1990s, the theoretical connection between technology and society has gradually acquired prominence in information systems research (Avgerou, 2002). One could argue, therefore, that there is a correlation between the level of ICT usage and the level of development within government agencies. Figure 10.1 illustrates a close relationship between Internet host density and human development. While such a construct on its own does not prove causality, the connection must nonetheless be significant. The obvious inference is the link between development and penetration of ICT technologies; the other side of the coin could very well be understood to be the linkage between the prosperous nations and their access to new technologies. It would thus lead to the conclusion that as with every other technological advancement, benefits tend to be derived mostly by the resource-rich (Tanburn and Singh, 2001).

The advent of the Internet as an important platform for activities of people across the world has established a new avenue for governments to be able to interact more frequently with citizens. In a world where

FIGURE 10.1 Correlation between Hosts Per Capita and UNDP Human Development Index

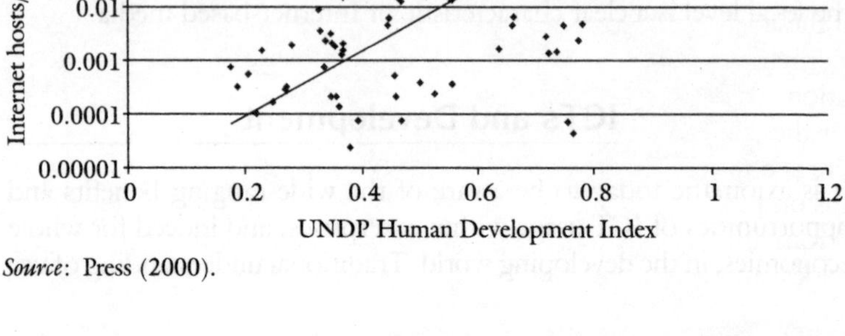

Source: Press (2000).

geographical distances are gradually being broken down through advances in connectivity, it is obvious that people and expectations would gradually become more localized. The ubiquity of the Internet has raised the hope that government initiatives can be made popular and at the same time be communicated to a wider audience. Such interaction may be said to have raised a variety of expectations. While a part of the efforts to use the Internet have been used to publicize information about activities of government, others have been designed for delivery of services. In general, promotional literature on activities of government agencies can be quite easily passed on to the public through electronic interfaces created within the World Wide Web. However, experiences of the commercial world would suggest that promotion on its own does not tend to be effective.

The traditional decision-making model where government decided on behalf of citizens and sought acceptance of such decisions through publicity measures, is gradually becoming obsolete. In democratic structures of governance, it is no longer sufficient to 'decide–announce–defend' policies. Such models of the role of government are being challenged (mostly in developed countries) by the contrary position whereby the citizen is taken to be a partner in the decision-making process. It is not entirely clear how emergent dichotomies as a result of partnerships with citizens may actually be resolved, as well as how an equitable participation of stakeholders may be ensured. In e-governance terms, there is a wide range of issues that are involved. For instance, the management of content, ownership of the site, privacy of information, and extent of risk bearing, are all factors that may be taken care of by different players in a developed economy like the UK. Using an example of spatial data use in the UK local government might help us understand some of the complications about conflict of interests. The Ordnance Survey in the first instance and the local authority, subsequently, are the custodians of all spatial data on resident information in the UK. Dedicated information processing agencies buy these data sets from the local authority, add metadata to it and then sell these value-added data sets to different commercial agencies like supermarkets and insurance brokers. From the perspective of information recipients, the citizens or residents have virtually no control over the commercialization of data that the local authority sells. Yet there is a tacit assumption that the citizen is an equal partner in the decision-making process.

From another perspective, where data is being bought over from the local authority, it is clear that there is no other agency from where such data may be acquired. In which case, devoid of competition, it is unclear as to how quality may be developed through competitive marketing of data sets. So although there is a move towards making the citizen a partner in decision-making as well as local authorities becoming sellers of services rather than providers, it is unlikely that the functionality of the model can be taken to its logical end. In a developing economy like India, these are some lessons that may be useful in creating information-driven public services that may be delivered over the World Wide Web.

Connection between Software Development Success and E-government Initiatives

According to Pratchett (1999), growth and sustenance of local democracy, enabling public policy formulation and facilitating direct service delivery to resident population are three complementary expectations that e-governance could be directed to fulfil. There are subtle variations on the type of support to governance that web-based facilities may be able to offer. Examples within the UK, where local democracy is being primed through the use of electronically mediated services over the last decade or so could be a valid reference point. Increasingly, public participation in decision-making is proving to be an important characteristic of successful governance within the developed world. Local government within the UK functions independently of Central Government. Within such a context, it becomes incumbent upon local government offices to ensure that engagement through providing adequate avenues of debate and information dissemination to local population groups is sufficiently well established. Over the last three decades, development of public opinion within local government in the UK has varied in relation to local government offices. For instance, the study conducted by Mitra (2001) on geographical information system used in British local governments showed that local authorities governed by liberal democrats have tended to support increasing formats of engagement with the local population, whereas, Conservative and Labour party-led authorities have tended to resolve sensitive decisions through traditional ideology or technological means of persuasion.

Although no immediate analogical parallels may be appropriate here between the Indian scenario and that of the UK, as e-governance initiatives have been available within the UK for much longer, yet it is clear that with maturity, similar manifestations of the need to engage with the public might be evident within the Indian context. Like everywhere else, newspapers within India are gradually making available editions on the Internet, providing avenues for feedback through e-mail and even becoming hosts for audio-visual discussions through their websites. Several government policy papers and regulatory authority-proposed initiatives are now being regularly posted on their websites calling for comments. Local-level decision making might develop much more cohesively, as the role of discussion groups becomes more readily established. Within the current scenario, making a petition with the support of a large number of signatures is a kind of traditional mechanism through which public opinion seems to have developed. In recent times, however, the structure of such opinion generation has changed to a more dynamic format. Citizen complaints and procedures have been disseminated and carried forward through the use of discussion lists. In the health sector, the use of second opinions is something that is becoming increasingly common. It is clear within this context that the direction has thus changed from what is made available by the State to what is feasible to be designed by the consumers and their willingness to engage. The latter may be also a characteristic that is common to the nature of local democracies of modern times. For the purposes of the present chapter, development of local democracy may be taken to be a gradual and emergent capability dependent on the extent of Internet usage that the resident population find an immediate need for.

Land parcel data is among a range of archives that have been digitized by different state/provincial[1] governments in different parts of India. Theoretically, such archival material may be put to use in a couple of ways. First, there might be a role whereby operational information like levels of taxes and other levies might be assessed. The second is developmental plan making. When governments develop plans, it is imperative that they have access to a range of land related information that would form the basis of allocation of resources and the projected direction of development. Man–land ratios have occupied an important place in the planning scenario over the last three decades. Given the role of spatial capabilities that information systems can play, it is obvious that service delivery in the Indian context is likely to be significantly related to the use of such archival information.

Over the last decade or so, an overwhelming use of ICTs in government has meant that many primary responsibilities of government are serviced through a substantial dependence on them. However, recent researches by Pratchett (1999), Mitra (2001), among others, show that facilitating service delivery needs are being targeted to a greater degree through dependence on ICTs as compared to other roles. At the same time, traditional understanding of the role of government has usually tended to view impact and dependence on ICTs as a secondary influence. While expectations of the tool, that is, ICTs have gradually risen, the role of institutions has simultaneously altered to that of a less serious mediator. Government, within the modern context, is seen as an enabler rather than a provider of services. The enabling role is also based on a certain amount of selling of services, rather than being generally made available for free.

Methodological Basis for the Present Chapter

For the present chapter, the methodology adopted, besides desk-research, involved questionnaires and semi-structured interviews of software development firms as well as government functionaries responsible for contracting work to the former in India. A questionnaire to determine the perception and experience of software development companies regarding e-governance was circulated to around 50 enterprises mostly in North Indian software hubs at Chandigarh, NOIDA and the capital city of New Delhi in northern India. Both perception and experience were important in the appreciation of e-governance initiatives of the government due to the important role that government tends to play in the minds of the urban population. Unlike the UK where emphasis is primarily maintenance, government in India continues to be involved in a wide range of developmental activity. Such a position makes its capabilities more important in the day-to-day operations of the state.

Responses received from 31 of these enterprises were tabulated for a quantitative and qualitative analysis that was matched with semi-structured, personally administered interviews with over 20 top-level functionaries in the private and public sectors. Information and responses have been utilized in the analysis and recommendations in this chapter. The second author who is based at New Delhi in India primarily carried out the fieldwork for the present chapter. It must be noted here that the views and opinions expressed by the second author in the present chapter are his own. These ought not to be taken to be the official position of the Government of India. It is also probably right to take into account access factors that might have been easily overcome due to the second author's official position as a senior civil servant within the Government of India.

Questionnaires designed for the present chapter were developed on a couple of primary foundations. Extensive involvement of both the authors in public administration involved the logistical elements of governance issues covered in the questionnaires. The first author's doctoral work within British local government enabled understanding of various dimensions used to develop comparative elements that qualified work in India as well as in the UK for the present chapter.

Historically, e-government within either a developed- or a developing-country context has tended to be introduced in three distinct stages. Czarniawska (1997) has argued that there is a gradual emergence of similarity between public and private sector organizations. While government institutions in general have become more 'companized', large public sector autonomous institutions have tended to become more 'politicized'. In other words, the 'companization' of bureaucracies both within Britain and India has contributed to ensuring that outlined processes in Figure 10.2 are valid for both contexts.

FIGURE 10.2 **Traditional Processes of E-governance**

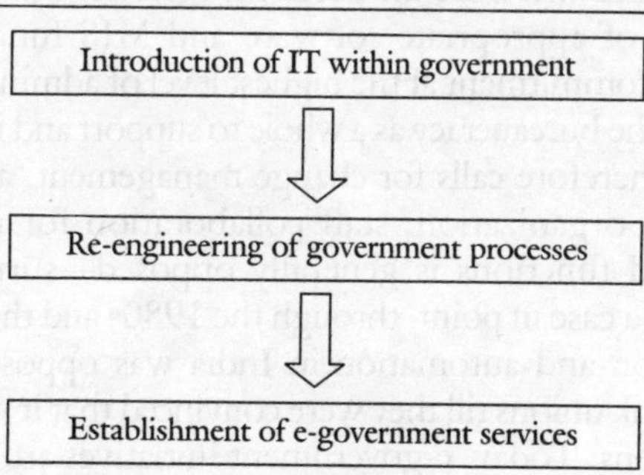

In many ways, the process of introducing computerization in government is similar to doing so in any large private company or organization. Experience in the latter would be useful for the former. However, its results in implementation may not always be the same. Particularly from the point of view of the service providers and therefore as a process of management and implementation, these three steps (as illustrated in Figure 10.3) essentially entail.

FIGURE 10.3 **Implementation Phases of Computerization**

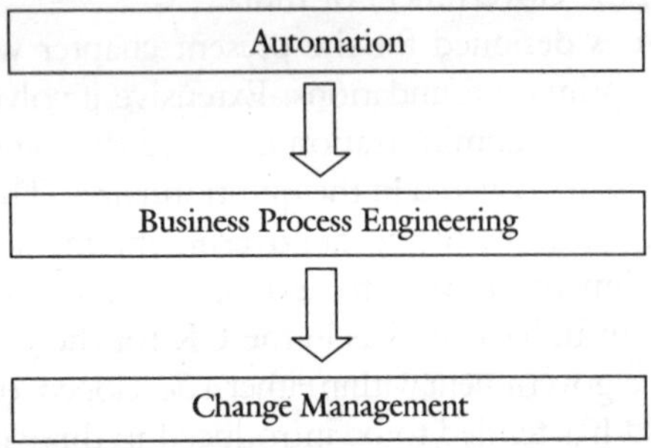

Automation involves the extensive use of ICTs for governance. This besides basic computerization, presupposes the establishment of an 'Intranet' within the government department concerned, if not an online linkage to all or several related departments and organizations for the purpose of automated processing of functions and services. In being able to do this it is important that administrative processes are re-engineered and made compatible with linked and related departments and organizations. Without this e-governance would remain only a partial and therefore unsuccessful exercise. This also involves the designing of appropriate software and MIS functionality and needs not just commitment at the highest level of administration but a willingness of the bureaucracy as a whole to support and implement the system. This therefore calls for change management, as in any large administrative organization, staff collaboration for readapting of their roles and functions is generally opposed—surreptitiously if not openly. As a case in point, through the 1980s and the 1990s, Bank computerization and automation in India was opposed for almost 10 years by bank unions till they were convinced that it would not lead to job reductions. Today, e-government initiatives are facing similar hurdles in many states.

At each of the stages of implementation (as in Table 10.1), the process involves much greater effort and therefore time. This is not just a matter of technology, but really of bringing in a paradigm shift in government departments. Obviously, this is difficult and such experiences have been telling on the reputation of public sector projects for e-governance.

TABLE 10.1 **Statement Outlining the Implementation Process for Computerization between Private and Public Sectors**

Stages of implementation	*Private sector (UK experience)*	*Public sector e-government in India*
Automation	Usually 1st stage of computerization already there	Usually Nil
Business Process Engineering	Fairly advanced business processes	Usually archaic laws and procedures
Change Management	Much easier with top management support	Very difficult

Indian Competitive Advantage

Compared to typical US-loaded costs, India-based software development delivers a huge advantage. This is expected to remain 3:1 for a long time. For web-enabled services like medical transcription and call centres where just English speaking graduates are required (as opposed to engineering graduates for software), the cost differential is nearer 5:1. Besides cost, more and more US and European companies are looking at talent lock-in as a competitive issue, as availability in the long-term basis will be in question. Even in situations like the present US new economy slow-down, outsourcing makes sense in a strategy of cost-cutting to address the increased competitiveness. Some experts have calculated that a typical Western bank, for example, can outsource 17–24 per cent of its cost base, reducing its cost-to-income ratio by 6–9 percentage points, and in many cases doubling its profits.

Listed items in Table 10.2 form a rough guide comparing costs of the five most essential components in the software industry in the US, Europe and India.

Since manpower accounts for some 80 per cent of the production cost in the software sector, overall development of software in India is 40 to 50 per cent cheaper than the US or Europe/Japan. As items (*c*), (*d*) and (*e*) become on par with the US, as they will be in due course (India's telecom costs have declined over the last few years), then Indian salaries could be raised and the competitive advantage still retained.

For IT enabled services (ITES), the competitive advantage is even higher. NASSCOM (2001) in a comparison across the Asia-Pacific region (Table 10.3) of several factors such as workforce, market access, infrastructure and cost, has outlined that, on an average, the best location by far is India.

TABLE 10.2 **Comparison of Essential Cost Components for the Software Industry**

Software development costs	*USA (as base of 100)*	*Europe (and Japan)*	*India*
a) Engineers	100	120 or so	33
b) Land/Rentals	100	100	80–100 (in metros)
c) Cost of connectivity	100	120 or so	200
d) Telephone	100	130 or so	200
e) Allied services (courier, air freight, and so on.)	100	130 or so	150 and slower

Source: Field interviews.

TABLE 10.3 **A Comparison of Competitive Advantage in ITES among Some Countries in Asia-Pacific**

Country	*Workforce*	*Market access*	*Local market*	*Infra-structure*	*Cosmopolitan*	*Cost base*
New Zealand	2	2	–	2	3	2
Malaysia	1	2	–	2	2	2
Japan	1	2	1	3	1	3
Hong Kong	1	2	2	2	2	2
India	3	2	2	2	3	1

Sources: IDC, NASSCOM.
Notes: 1 denotes low; 2 denotes average; 3 denotes high.

From the almost complete concentration on software, Indian exports are evolving into these three distinct growth areas. The present status and projections or targets (of the industry and government) are as follows.

These projections (Table 10.4) are based on estimates as data for domestic e-business and e-governance and cross-border e-commerce is not readily or easily available. The figures therefore are indicative of the expected growth. The importance of these figures and projections is that IT enterprises see and expect large potential from their businesses. They also now see horizontal and varied opportunities as opposed to the specific and export order delivery that they started with. In times of economic slow-down they are thus better prepared to shift business focus and avail of opportunities such as that of e-governance.

TABLE 10.4 **Growth Areas for Indian Software Exports**

	External Sector		*Internal Sector*
Year	*Total software export*	*E-commerce-related software and business*	*E-business (including e-government)*
1999–2000	US$ 5.7 bn	US$ 0.5 bn	Rs 450 cr (US$ 100 m)
2000–01	US$ 9.5 bn	US$ 1.4 bn	Rs 15,000 cr (US$ 330 m)
2004–05	US$ 35 bn	US$ 5 bn	
2007–08	US$ 50 bn software and US$ 20 bn web-enabled services	US$ 10 bn	US$ 50 bn

Source: NASSCOM.

Prevalent Models E-governance within the UK

E-governance is probably one of the very interesting examples of how functionalities found in sites operated and developed by the private sector has gradually gone ahead to be emulated by the public sector. Over the last decade or so, eight different models like virtual storefronts, marketplace aggregator, information broker, online service provider, transaction broker, electronic clearinghouses, reverse auction, digital product delivery and content provider have found recognition as standard platforms on which businesses have been developed. Several of these, as discussed by Laudon and Laudon (2001), may be integrated into usage platforms necessary for the development of government agencies interested in delivering services and information that the public might buy or use. Although the auction model is becoming popular for the private sector, it is still unclear how such a model might be a useful platform for different governmental activity. Various one-stop shop arrangements exist within the UK government's initiatives that might correspond to the marketplace aggregator model.

Several significant e-initiatives like the Royal Mail and the Foreign and Commonwealth Office operate their sites more on the lines of being an information broker rather than providing direct sales outlets. In many cases, standard forms and documents are available in downloadable format to facilitate submission and efficient use of particular governmental facilities. From these examples, it is clear that the information broker and content provider are a couple of models that would best suit the needs that service delivery requires. Moreover, the serious needs of ensuring security on the website *vis-à-vis* the protection of privacy and the prevention of abuse are overwhelming concerns that persuade the use of models. In such a consideration, mere revenue generation unlike the private sector is not the only expectation. Given these different aspects, it is logical to presume that the combination of authentication, authorization, confidentiality and non-repudiation as characteristics do play an important role in the model (Turban et al., 2002) that is going to be best suitable to deliver different advantages for e-governance. A shared activity series where some elements of browsing through availability and then use of some form of identification code would best suit further physical validation in the actual governmental department. Through such an arrangement, some part of the transaction may be carried out using available technology, thus reducing time allocated for the entire validation procedure to be done manually in an office. All these different requirements are likely to demonstrate the appropriateness of a hybrid model incorporating both information broker and the content provider capabilities.

In the context of software outsourcing, suitability of such a hybrid raises the probabilities that there is going to be a close connection between outsourcing and e-governance initiatives. Several firms based in different parts of the country would in all likelihood be involved with the setting up of similar platforms for foreign clients. It is thus logical that similar companies may be able to participate and develop initiatives in partnership with the government.

Towards a Framework

Given the range of initiatives that are currently operational, it is clear that a gradual emergence of a pattern may be discernible. E-governance initiatives have a couple of primary actors. On the one hand, there is the government, which creates the initiative/scheme and, on the other, there is the citizen who is the principal client or user of such an initiative.

With the passage of time, the Indian software scenario clearly demonstrates the continued involvement with software development for government agencies of software houses that are heavily involved in developing outsourced software. Government, in general, is keen to promote its activities and one of the means of ensuring popularity of such measures is by getting companies within the private sector to develop some of the code and processes. Expectations of citizens are mainly embryonic at this stage. However, with the continued growth of capabilities within the private sector and the over-stretching of government offices with increased pressure of numbers, it is clear that experiences of the citizen will change in a way that would add to expectations. In the direct service delivery area of the Indian economy, the cycle of impacts would be like that depicted in the diagram in Figure 10.4. In this context, an extension of the model would suggest that a rise in outsourcing will lead to a rise in development of e-initiatives, although the reverse may not necessarily be true. The advantages of cost and efficiency that business process outsourcing (BPO) offers and the experience that the Indian IT companies are gaining in it would further lead to more aggressive e-initiatives in e-government opportunities, both domestically and abroad.

FIGURE 10.4 **Impact Cycle of E-initiatives**

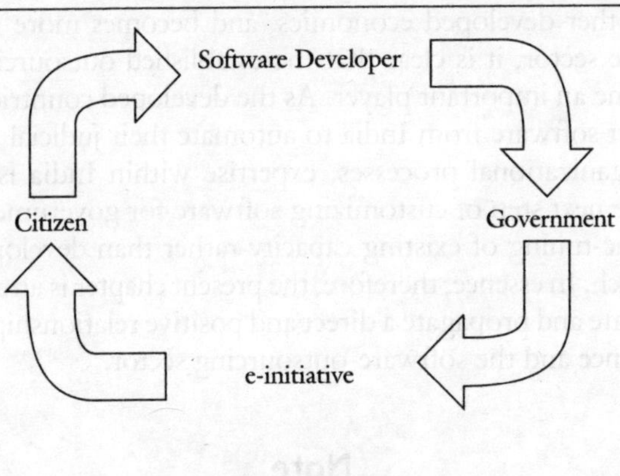

Conclusion

Over the last decade or so, the idea of e-governance has steadily gathered momentum towards obscuring the 'e' part with an 'i' that could interchangeably stand for 'interactive', 'intelligent' or 'informative' governance. In the Indian context, much of the awareness of the potential of such governance has emerged from addressing the initial curiosity that arose from the wondrous developments of 'body shopping' and the lure of 'exponential pay' that the software market offered. However, as in the developed world, it is obvious that outsourcing will gradually emerge to be a more abiding option than employing software professionals in-house. Given such a developing reliance on the software outsourcing market, it is a matter of time before most e-governance initiatives are likely to become reliant on time-tested formats of outsourcing. Like any other newly discovered tool of governance, the establishment of trust in its functional efficacy would need to be steadily built in the minds of its principal clientele, that is, the citizen. Such trust could grow initially out of two different avenues of interaction. The first could address issues that would include improving services to citizens, improving the productivity or efficiency of government agencies and strengthening law enforcement. The second could be more holistic to include improving the quality of life of disadvantaged communities, strengthening good governance by broadening public participation and promoting priority economic sectors.

The benevolent impressions that enabling governance initiatives conjure up could be made more useful through an active assembly of inputs from the flourishing software outsourcing sector within the Indian economy. As the Indian government seeks to emulate the flexibility in other developed economies, and becomes more reliant on the private sector, it is clear that the established outsourcing sector will become an important player. As the developed countries like the US import software from India to automate their judicial and other public organizational processes, expertise within India is likely to grow. The next step of customizing software for government would just be fine-tuning of existing capacity rather than developing code from scratch. In essence, therefore, the present chapter is an attempt to demonstrate and propagate a direct and positive relationship between e-governance and the software-outsourcing sector.

Note

1. A province in the Indian sub-continent is referred to as a 'state'. This term would be used subsequently in this senses.

References

Avgerou, C. 2002. *Information Systems and Global Diversity*. New York: Oxford University Press.

Czarniawska, B. 1997. *Narrating the Organisation: Dramas of Institutional Identity*. Chicago and London: University of Chicago Press.

Fountain, J. 2001. *Building the Virtual State: Information Technology and Institutional Change*. Washington, D.C: Brookings Institution.

Heeks, R.B. (ed.). 2000. *Re-inventing Government in the Information Age: International Practice in IT-enabled Public Sector Reform*. London: Routledge.

Laudon C and J. P. Laudon. 2001. *Management Information Systems: Organization and Technology in the Networked Enterprise*. New Jersey, USA: Prentice Hall.

Mitra, A. 2001. 'An Interpretation of the Organisational Context of Geographic Information System use in British Local Government', unpublished PhD dissertation, UK: University of Birmingham.

Pratchett, L. 1999. 'New Technologies and the Modernisation of Local Government: An Analysis of Biases and Constraints', *Public Administration,* 77(4): 731–50.

Press, L. 2000. 'The State of the Internet: Growth and Gaps', *INET 2000 Proceedings*. Available from www.isoc.org/inet2000/cdproceedings/8e/8e_4.htm.

Tanburn A. and A.D. Singh. 2001. *Innovation and Support in Business Support Services-(FIT)-ICTs and Enterprises in Developing Countries: Hype or Opportunity?* Geneva: International Labour Office.

Turban, E., D. King, J. Lee, M. Warkentin and H.M Chung. 2002. *Electronic Commerce 2002: A Managerial Perspective*. International Edition: Prentice Hall.

11

Evaluating the Developmental Impact of E-governance Initiatives: An Exploratory Framework

Shirin Madon

Introduction

India has made enormous progress since it achieved Independence over 50 years ago. Economic growth since 1980 has been one of the fastest in the world and social indicators have steadily improved. These successes have taken place against a backdrop of India's well-established democratic system, which is the largest in the world and provides the population with a significant degree of political freedom and stability. Despite this, India is faced with a large number of challenges that are typical of developing countries. For the majority of people in India, clean water, sufficient power supply, earning a daily wage as labourers, providing basics for the family and decent health services are priorities (Castro and Chopra, 2000; Jayal and Pai, 2001).

The reform of government administration and the provision of improved services to citizens has long been acknowledged as a major criterion for development and today's drive towards e-governance can be considered a part of this wider developmental goal. The Indian experience in e-government can, broadly speaking, be divided into two main phases. The first from the late 1960s/early 1970s to the late 1990s, and the second from the late 1990s onwards. In the first phase, efforts to develop e-government were concentrated on the use of IT for in-house government applications with a principal focus on Central Government requirements such as defence, research, economic monitoring and planning, and certain data-intensive functions related to elections, conducting of national census and tax administration (GOI, 1985). During this first phase, the introduction of IT in the public sector did not result in automation of many key departmental activities. In the second phase, the implementation of the national IT Task Force and state government IT policies symbolized a paradigm shift in e-governance policies towards using IT for a wider range of sectoral applications reaching out to a large number of people in rural as well as urban areas. Moreover, there has been a movement towards a greater input of non-governmental organizations (NGOs) and private sector organizations in providing services to the public. These projects have been influenced by the increasing focus of international agencies such as DfID, G-8, the UNDP and the World Bank under the banner of 'E-governance for Development'.

The Government of India has set up the Ministry of Information Technology that works with the Department of Electronics to achieve various e-government objectives with a pledge that at least 25 per cent of government services will be delivered electronically by 2005 (GOI, 2000). In India, many different types of e-governance projects are being implemented in parallel as displayed on the website of the World Bank funded e-governance Centre located at the Indian Institute of Management, Ahmedabad in Gujarat state (Centre for Electronic Governance, 2001). Some projects described on the website aim to introduce IT automation in individual government departments. Some aim to improve transparency and accountability within government by introducing electronic file handling and public grievance systems. Others specifically aim to enhance the delivery of government services through IT for a range of high volume routine transactions such as the payment of bills and tax dues to government. E-governance initiatives are also being introduced to help people escape poverty by providing them with vital information on market prices and by helping them to make a living through entrepreneurial activity centred around information and communication technologies (ICTs).

Studies that have described e-governance activity in India have focussed on the extent of wide coverage within many states in India completely networked. For example, Unnithan (2002) explores e-governance trends and drivers in India from an infrastructure and policy viewpoint. However, to date, there has been little serious analysis of these initiatives in terms of their long-term developmental impact, although such a need has been recognized from recent evaluation of projects in which issues of long-term sustainability of the projects have been raised (Harris et al., 2002; Kanungo, 2002).

In this chapter, we argue that a more focussed approach is needed to ensure that citizens derive true benefits from such investments. Many government services need to be re-engineered to benefit from the emerging technology. There is an inherent distrust with citizens on the service delivery mechanisms of government and this image needs to be improved with confidence-building measures (Rama Rao, 2003). A host of factors relating to the ability of individuals and communities within developing countries to avail of e-governance services have hitherto been neglected. These factors depend upon resources, skill-levels, values, beliefs and motivations of a range of actors involved in implementation of the projects, including institutions of local government (such as panchayats), civil society organizations (CSOs) ranging from international NGOs to co-operatives and civic associations, and private corporations.

The next section proposes a framework for the evaluation of e-governance projects in developing countries based on a social constructivist account of how e-governance applications are shaped by local actors over time. Following a brief discussion on our choice of methodology, we describe our three cases of e-governance in India. In the analysis section, we draw on our framework to provide illustrations and raise further questions for research on e-governance for development.

A Framework to Study the Developmental Impact of E-governance Projects

The link between ICT and Development has increasingly been articulated by international agencies in terms of the digital divide raising alarm bells with the concern that developing countries are being deprived of the opportunities for economic growth and social development enjoyed by advanced economies because of scarcity of ICT. As a consequence, many high profile initiatives have been undertaken to create awareness of ICT benefits, raise investment levels and promote ICT policy such as the G-8 DOT Force initiatives and the recent UN World Summit for the Information Society (WSIS). An assumption underlying this discourse is that technology and human development are mutually reinforcing (Avgerou, 2003) and that the market is the mechanism through which ICT is associated with economic development. Another key assumption from modernization literature is that deficiency in knowledge is partly responsible for underdevelopment. This argument derives from modernist development communication literature of the 1960s according to which the communication of scientific and technical knowledge from the developed world to developing countries would result in economic growth. This argument reappears in the conventional development discourse of international agencies, for example, in the World Bank 'Knowledge for Development' report (World Bank, 1999), and more recently in the Bank's Global Development Gateway portal.

On the other hand, a body of literature argues that the 'ICT-for-development' movement is the latest in a series of fads in the history of development thinking since the 1950s (Castells, 1998; Ciborra, 2002; Wade, 2002). Adopting a social deterministic stance, these writers see ICT initiatives as fully inscribed into the political, economic, and social relations of capitalism and argue that they serve to recast development orthodoxy through the lens of modernization theory. Wade, for example, argues that in taking for granted that 'bridging the digital divide' is the central issue of development, literature from the World Bank and other international agencies have neglected details of costs, cost/benefit ratios in development terms and risks (Wade, 2002). Castells (1998) puts forward a troubled vision of the information society in which 'old binaries are reworked' around the ICT revolution such that the poor majority of the developing world becomes irrelevant to the new society. Essentially, these writers tend to recast the arguments of postmodern development theorists who draw on Foucouldian discourse analysis to explore the way in which knowledge, power and action are linked in the work of development agencies and projects around the world (Escobar, 1995 Ferguson, 1990;).

These arguments put forward by development studies scholars focus on providing institutionalist accounts of the constraints that institutions impose on IT innovation projects, but neglect analysis of how and why these constraints change over time. According to recent work by development anthropologists, understanding the impact of development interventions entails a social constructivist account of how they unfold and are shaped by local players over time. This is undertaken by a careful study of documents, manifestos and frameworks which are usually the outcome of complex political and social processes, actual development activities, and the way in which the beliefs, practices and norms of individual and organizational actors at the local level shape the implementation of projects on the ground (Gardner and Lewis, 2002; Grillo and Stirrat, 1997; Schech, 2002). For the study of the developmental impact of ICT projects, a research question of interest would be the extent to which these applications are amenable to further shaping at the implementation stage according to the social context within which they are embedded.

In this chapter we adopt a constructivist perspective to evaluate the development impact of e-governance initiatives. According to this perspective, the interaction between the context within which an e-governance application is implemented and the process of human development that takes place is a central point of study. A process view of development is put forward in Amartya Sen's Capabilities Approach (Sen, 1999, 2000). Sen emphasizes that development should be perceived not only in terms of primary goods and economic resources, but also in terms of the ability or freedom each human being has to lead the life they desire in material, social and spiritual terms. Sen's Capabilities Approach puts forward a relativistic and contextual approach to development according to which individual choices determine the life they desire, but that the freedom to achieve these desires is determined by many different institutions and actors. There are similarities between Sen's capabilities approach and other 'alternative' approaches to viewing development. For example, the 'basic needs approach' has a close proximity to the capabilities approach as it directly focuses on fundamental human needs and at the same time takes into account individual differences among the poor. Similarly, Chamber's notion of powerlessness and isolation is similar to Sen's notion of freedom (Chambers, 1995). Likewise, as de Haan and Maxwell (1998) argue, the social exclusion approach has a close affinity with the capabilities approach to development as both approaches accept that many of the relevant capabilities are relational by definition. For example, being adequately nourished and educated requires participation in effective working market, administrative and social relationships (Sen, 2000). But Sen's approach goes beyond these approaches because the capabilities approach expresses needs in terms of human functionings—that is, the states of being and doing valued by people such as being healthy, obtaining an education, paying government bills without hassle, earning a living, and so on. By expressing needs in this way, Sen places a strong focus on process and human agency. The role of human agency is linked in Sen's analysis to a discussion of the role of institutions as vehicles for capability creation since people's knowledge and beliefs as well as the norms of the organization and society they live in shape their capability set. Although dedicated to an analysis of human development at the local level, Sen's analysis therefore spans various levels of abstraction from individual to community to society to State to international.

In this chapter, we do not attempt to validate or challenge the extent to which Sen's capabilities approach constitutes a comprehensive theory of well-being. Indeed, the discussion around whether Sen's approach is more an economic theory of human capital rather than an all-encompassing theory of human choice and freedom forms part of a lively current debate in development studies (Apthorne, 1997; Gasper, 2001). On the other hand, we do acknowledge that Sen's theory can be further enhanced. For example, Apthorne claims for further inclusion of social anthropological study in human development research in order to provide a 'nearer-the-ground' understanding of social activities and institutions as perceived by local actors themselves. Other writers have argued that Sen's analysis suggests that to become 'developed' is simply a matter of maximizing individual freedoms within an institutional context. They suggest, instead, that empirical work needs to focus on identifying concerted struggles against powers of vested interests at all spatial scales (Chopra and Duraiappah, 2001; Corbridge, 2002).

The impetus to work with elements of Sen's work has been spurred by recent writers who have tried to apply the capabilities approach for the study of ICT access in society raising the important question of whether new options, such as the ability to hold government accountable, the ability to pay bills, the ability to generate income through e-government applications should be added to the capability set of individuals, communities, organizations and States (Garnham, 2000; Mansell, 2001). To date, e-governance evaluation guidelines have tended to focus on the supply-side benefits of ICT infrastructure and on the identification of constraints within the structure of institutions in developing countries. For example, Heeks (2001a,b) bases his analysis of pre-conditions needed at strategic and tactical levels for e-governance initiatives in developing countries on 'networked readiness' criteria. These criteria originally formulated at the Harvard Centre for International Development defines a framework of five inter-related categories with 19 subtopics where a grading of each subtopic into one of four stages provides an indicator of each country's e-readiness in terms of access to ICT infrastructure and in terms of the socio-economic and political institutions which exist to support access and usage of the technology (Kirkman, 2002).

In the context of developing countries, recent IDRC studies have indicated a distinction between two classes of evaluation criteria. The first is input measures which are not immediately associated with benefits to the user community, but which are regarded as essential pre-conditions for those benefits to emerge such as the provision of ICT infrastructure and the role of supporting institutions. The second class of benefits relate to actual benefits to the community as a result of utilizing the services of the centre (Harris, 1999; Hudson, 1999). These benefits are directed at the sustainability of individual telecentres and are contingent upon regional and institutional differences in the local milieu. For example, International Development Research Centre (IDRC) studies have revealed a significant difference in the evaluation criteria adopted for telecentre projects in Asia and Latin America. Evaluation of IDRC's Asian telecentre projects was highly focussed on individuals and people making the evalu-ation participatory and socially inclusive to gender concerns. While the IDRC analyses have placed a lot of emphasis on the local context within which the projects have been implemented, there has been little analysis of the way in which wider institutional forces have affected the take-up of the telecentre initiatives at the community level.

We suggest that use of elements from Sen's capabilities approach enables us to go beyond these other evaluation approaches in several ways, making it a useful approach for the current study. We aim to understand how the capabilities of citizens to hold government accountable, pay their bills, and generate income are constrained by wider institutional forces. Our focus is on process and our tentative findings prompt us into areas which require further investigation through fieldwork.

Our study methodology is described next. Then, in the following section, we describe three e-governance projects in India, which the author has been involved in studying over two years.

Methodology

While the present research on e-governance in India started in 2001, the author has been working in India since 1988 on projects related to the introduction of IT in Indian government. The selection of study sites for the current research was driven by the fact that most of the southern states in India have set precedence in e-governance activity. Much of the author's previous research has been focussed on Bangalore where some of the best-known experiments in participatory urban governance have taken place. From an earlier focus in Bangalore on property tax reform, our current interest is in how the city corporation uses ICT to improve accountability. The southern state of Kerala, with its unique social indices, is our second location of study. Since 2001, for 15 months, the author was involved in studying how ICT is being used to improve the delivery of routine services to citizens. Andhra Pradesh is another southern state which has been a pioneer in initiating e-governance in India, particularly in terms of bringing in transparency in governance activity. In addition to networking government offices, the state government of Andhra Pradesh has initiated various e-government projects in rural areas in order to build social and economic development within local communities.

The research approach adopted reflects the tradition of studies that can broadly be classified as 'interpretive case studies'. The author has selected a longitudinal research design in order to trace the dynamics and long-term implications of e-governance projects in India. Fieldwork is organized at roughly six-month intervals for 15 months. During the course of the research, we have conducted many interviews with government officials, politicians, NGOs, private sector employees and various citizens. For the interviews, we did not have a fixed schedule of questions, but a set of issues that we felt were important. We encouraged the respondents to comment on these issues, identify new ones and describe how they were dealing with them. Over time, as we met people for the second and third time, there already existed a shared understanding between us and we could start the conversation where we had left off at the last meeting. All the interviews were conducted in face-to-face settings in the location of the interviewee's workplace. The duration of the interviews ranged from a few minutes to an hour and we were invariably interrupted by phone calls and by the entry and exit of various other people coming in the office to meet the same person whom we were interviewing. None of the interviews were tape-recorded. Although data collection was done mainly through semi-structured interviewing, additional data was obtained through participant observation, attendance in public meetings, study of press reports, study of websites and various other sources.

Three E-governance Projects in India

Bangalore: Systems to Promote Accountability of the City Corporation

Bangalore has a long history of urban participatory movements with many international and small NGOs based in the city working to improve the conditions of citizens ranging from slum-dwellers (Madon and Sahay, 2002) to middle-class urbanites.

In 1997, the Asian Development Bank (ADB) launched a programme entitled 'Enhancing municipal service delivery capability' with the aim of helping a number of municipalities in the Asian region to enhance their capacity to deliver services (ADB, 2001). Part of ADB's municipal reform package for Bangalore includes improving accountability of the Bangalore City Corporation (BCC) in terms of financial accountability of income and expenses according to international standards, and in terms of political accountability of service provision to Bangaloreans. A representative of the ADB was appointed from within the BCC to coordinate between ADB's mission and the BCC' internal priorities and agenda.

A variety of agencies have been involved in the implementation of these reform programmes. The Centre for Development of Advanced Computing (C-DAC), a Central Government agency, was initially commissioned to develop a Customer Complaint System (CCS) as a front-end application that would record details of grievances registered at by citizens keeping a record of complainant's address, the date and nature of complaint, follow-up action taken by the concerned department and any feedback to the complainant. Attempts were made to customize the original C-DAC prototype to produce a workable system for the BCC. However, the BCC lacked both the resources to complete the enormous task of compiling and entering disaggregated data down to street-level objects for the entire city as well as inputting the backlog of citizen complaints received at the BCC by e-mail or other means since April 2002 when the incumbent Commissioner took office. The C-DAC prototype eventually reached a stalemate and the implementation of the CCS has been put out to tender by the BCC. At present, when citizens make a complaint in their local range office, they are issued with a computerized slip. In the absence of a network connection from range offices to headquarters, the information about complaints needs to be re-entered at the BCC.

The second key agency involved in promoting reform initiatives is the Bangalore Action Task Force (BATF)—a consortium spearheaded by the private sector in January 2000 with strong backing of the Chief Minister of Karnataka. Its mandate was to bring together key stakeholders in the city and to pressure them to make public pledges for qualitative and quantitative improvements in the services they provide. This was achieved through a series of bi-annual public meetings, which brought the BATF into the citizen gaze as a public-facing body committed to improving the government's image. The BATF is responsible for implementing the Fund-based Accounting System (FBAS) in the BCC, which deals with reconciliation of property tax collections according to international standards for government organizations. The BATF have appointed a consulting agency with previous experience of implementing systems within the BCC to manage implementation of the FBAS. So far, the system has been installed in 22 out of the 30 range offices in the city with each range having one or two clerks to input data. The system is updated daily through floppy disks physically brought over by range office clerks on a daily basis to a BATF unit specially set up in the BCC. In general, implementation of the FBAS has been successful in the range offices reflecting the Chief Minister and BATF's current preoccupation with revenue and back-end accountability.

Two recent initiatives aimed at promoting political accountability in Bangalore are worthy of mention. One is called the Wardworks campaign organized by the Bangalore NGO Janaagraha which the objective of enabling citizens residing in a particular ward to discuss and debate different aspects related to the implementation of works carried out by the BCC in that ward such as quality, timeliness, utility, and so on (*Janaagraha Times*, 2002). The second, spearheaded by the BATF is the recent PROOF initiative which commenced as a 10-month campaign in July 2002 by four Bangalore NGOs (PROOF, 2002). Public Record of Operations and Finance (PROOF) aims to put in place a system of government performance reporting along the lines of the private sector using US accounts and standards methodologies. A PROOF document is produced every quarter, providing citizens to become fully engaged in budget formulation and assessment through a series of bilingual training sessions organized by the NGO Janaagraha since October to help the public analyze the budget. Although it is early days for this campaign, there are concerns expressed. First, the BCC is concerned about its own ledgers not tallying with FBAS figures.

Second, it is questionable as to the extent to which PROOF quarterly meetings are 'public' and 'inclusive'—the second quarter meeting held in November 2002 held at Infosys was attended by 120 people mainly from the BATF, and other state and voluntary sector agencies.

Kerala: The FRIENDS System

The e-governance mandate for the Kerala state government has been to implement high visibility 'people-oriented' projects in parallel with more substantial governmental reform promoted by the ADB (Government of Kerala, 2001). One such people-oriented project is called FRIENDS—an acronym for Fast, Reliable, Instant, Effective Network for Disbursement of Services launched in June 2000 by the Kerala state IT Department and rolled out to all 14 districts by 2001. Each of the FRIENDS centres offers a one-stop IT-enabled payment counter where citizens can pay all their bills rather than having to personally visit individual department payment counters located in different parts of the city. The centres are open from 9 a.m. to 7 p.m. seven days a week including Sundays and offer a modern, clean environment facilitated by a computerized queue management system for citizens in their interaction with government. Each centre operates on any service, any counter basis with each counter run by one Service Officer who is a government employee from any department trained to deal with transactions of all departments. At present, seven government departments participate and help fund the scheme and around 1,000 different types of bills are paid by citizens on production of the demand notice they would receive by post from the utility company. Many new services are being planned such as the issuing of licences for electrical installations, passenger reservations for railways, long-distance road transport tickets, and other services.

Bills can be settled by cash and draft and Service Officer issues a receipt to each customer as legitimate evidence of payment. Queries generated by citizens are being dealt with by Service Officers from the concerned department liasing by phone with the department office. The current mechanism for updating departmental registers is through floppy disk being sent to each of the seven departments, printed out and then pasted onto departmental registers. Some departments are now precipitating plans to automate their records and many departments are beginning to experiment with online data transfer from FRIENDS to the department. Remittance of money from citizen payments to individual department is normally swiftly executed, via the district treasuries. Sometimes, however, the state government issues a 'treasury ban' which causes a delay in transferring money from the treasury to its final destination in individual departments.

Between April and August 2002, we conducted a study which tried to examine the FRIENDS project mainly from the perspective of key specific agents—the public, the service officers, below poverty line employees working at the centres, and employees and heads of participating department from a selection of centres in different geographic regions of the state. In general, public support for the FRIENDS centres was overwhelming among the middle class population of the state. The highest income bracket did not pay bills themselves and typically used agents while the lowest income bracket had a low number of payments to warrant making a trip to the district FRIENDS centre. The quality of service offered in FRIENDS was found to be much higher than the quality of service offered through conventional departmental counters. It was found that citizens save a monthly average of Rs 46.21 and 42.35 minutes in making payments to FRIENDS rather than at department counters. The expenditure incurred by each centre per month comes to an average of Rs 65,000—that includes maintenance charges for hardware and rent, electricity, stationery and payment of casual labourers. Although it seems that expenditure incurred by these centres could be recouped through advertising income, the Kerala government has not yet tried to utilize the centres as point of purchase advertisement spots.

One key benefit of the project has been capacity-building within government. The project has been able to demonstrate to the common citizen that with appropriate training and skill upgradation, the very same government officers who once may have been considered arrogant and customer unfriendly, could be turned to play the exact opposite role. The enthusiasm of these officers seems to have been one of the major reasons for success of the project with many centres opening side activities such as running a canteen and stalls. Out of 58 service officers in Trivandrum FRIENDS, only 10–15 have requested to return to their parent department for promotions.

The FRIENDS project is very much at the crossroads now with decentralization to subdistrict level being planned by the state government using a 'business model' involving investment from local entrepreneurs. Under the 'Akshaya' project which was officially inaugurated in October 2002, these village centres will be hubs for

IT literacy programmes among villagers, local content provision, and for the dissemination of government entitlements and utility payments. The involvement of local politicians in supporting the project and the identification and training of suitable entrepreneurs has been the key focus over the first few months since Akshaya was launched.

Kuppam, Andhra Pradesh: Community Information Centres (CICs)

World Corps was launched in 1998 as an international non-profit organization headquartered in Seattle, USA (World Corps, 2001). World Corps India (WCI) is currently working in Kuppam, which has a rural population of around 250,000. The NGO operates with a staff of five people and a Director to offer intensive training and capacity-building programmes to unemployed rural youth in order to provide them with self-employment opportunities (ibid). WCI put in place a careful selection process to identify deserving youth who would own and run CICs out of the 30,000–40,000 unemployed youth in Kuppam. Above all, WCI was interested in applicants' attitude towards community development, which was assessed by asking them to explain how they felt the CICs could realistically help the community. Based on that, WCI selected the top 10 from each *mandal* (50 in all) which was later pruned down to 27, based on how the candidates performed during an internship during which the youth were given practical assignments like cleaning out tanks or dealing with complex social issues and were observed for their skills in leadership, their concerns for the poor, their approach on discrimination based on caste, and their result-orientation. Fifteen youths were finally selected to be CIC owners in the five *mandals* in Kuppam constituency.[1]

Once the state government gave sanction for the NGO to start operations, training commenced. This involved two-and-a-half months of classroom learning on computers and the Internet, basic English and accounting skills, community outreach and entrepreneurship development including how to identify a customer base, apart from general themes of value education and personality development. Most of the training was contracted out to Aptec, a private computer company, while WCI conducted training on ethics and self-esteem. Financial assistance to set up the five CICs came from various sources including a grant from WCI, loans from the district and state administration, and contributions of 2–3 per cent of total costs of setting up the CICs from the unemployed youth themselves thereby reinforcing their commitment to the project. The Chief Minister officially inaugurated the centres in April 2002. During our fieldwork, the CICs had been running for around four months with designated roles for the three CIC owners in each *mandal*. The (Centre and administrative officer) CAO runs and manages the centre. The Government Liaison Officer (GLO) has the task of liasing with various government departments to get information on programmes and schemes, sometimes with WCI staff intervention. The Community Relations Officer (CRO) disseminates information about government schemes to around two–three villages per week promoting the products and services of the CIC. As CICs are getting busier every day, the extent of outreach possible is reduced.

ICT facilities are provided by each CIC at nominal cost to the local community and include communications facility, Internet browsing, chatting, downloading and e-mailing, computer and Internet lessons, printing, entertainment, floppy disk sales, scanning and photocopying. Non-ICT activities depend on local demand within each *mandal*. For example, Kuppam already has seven Internet cafes, therefore Internet is not a major source of income at the CIC which instead excels in other services like telephone, photocopy, fax, sale of bus passes and providing electricity meter reading services for which the GLO liases with the electricity department and recruits employees to go from house to house. The GLO also regularly visits the agriculture department to find out information about what crop to plant in which season, what medicine to use, and information about government schemes which this officer passes onto the CRO who spreads it to the 20–30 farmers from each village who attend meetings. A new service currently being planned for Kuppam CIC is being discussed with the agriculture department and farmers and involves starting a small fertilizer shop. In Guddipalle CIC, which has a 30 per cent literacy rate, the computer, telephone and xerox services, together with meter reading and the selling of bus passes, bring in some revenue but the main source of income is a shop for selling fancy items such as costume jewellery which the youth identified as a viable income-generating business. The youth purchase fancy items for sale from nearby Chittoor wholesale market and already make around Rs 600–800 per month from this activity.

Some CICs have proactively set themselves up as information centres of the government by keeping a stock of application forms required by citizens for government services, and by creating proformae for specific customers such as for schoolteachers according to demand.

For example, primary schoolteachers have requested some CICs to create forms with specific data fields to help them monitor attendance. Other CICs are proactive in generating income by obtaining data entry work from local government departments.

At the end of the first phase of the World Corps CIC project, the project appears to have established a viable business model for further expansion although there is concern as to whether the average income currently generated by the CIC owners (around Rs 1,500 per month) is sufficient. There have been some difficulties in certain CICs with owners deciding to look elsewhere to generate more income, or with owners swapping roles and responsibilities. In general, however, these problems have resolved themselves. The second phase of the World Corps CIC project commenced in January 2003 and has two specific aims. First, to establish 10 or more CICs at gram panchayat level (two gram panchayats per *mandal* CIC) using funding from Hewlett-Packard. *Mandal* volunteers have been identified and employed by World Corps to create awareness of the new centres among local villagers. Second, to offer self-employment opportunities to a further 1,000 youths, particularly women and self-help groups. Through linkages with local companies, the feasibility of various non IT-related activities such as dairy farming, poultry-rearing, agarbati-making, barber shop and running stalls has been explored.

Analysis

The previous section described three e-governance applications in India each of which are different in scope. In Bangalore, computer-based information systems have been introduced improve government accountability. The Kerala FRIENDS initiative is intended to enhance the delivery of government services to citizens. The Kuppam CIC project is intended to help people escape poverty by providing them with ICT-based income-generating possibilities. In this section, we provide examples of the type of analysis that can be conducted to study the development impact of e-governance using the framework developed earlier in the chapter.

The Bangalore Case

The language of the ADB policy document in 1997 in the first phase of the 'Enhancing municipal service delivery capability' project was striking on its emphasis on certain new public management principles such as competition, privatization, performance monitoring and computerization to create an appropriate interface with citizens (ADB, 2001). Following the publication of the 1997 ADB policy document, there have been changes in the agency's conceptualization of governance. Phase 2 of the ADB project focussed on building the capacity of participating municipalities to drive and implement change. To give an example of the changing policy focus of the ADB, in 1997 there was only one coordinator linking ADB policy with BCC priorities while since 1991, six more coordinators have been identified from within the BCC to act as a link with the ADB in its second phase of benchmarking key urban services. At this time, the ADB have also appointed an agency called the Centre for Symbiosis of Technology, Environment & Management (STEM) as domestic consultants for ADB projects. Further research is needed to trace how the ADB's policy manifesto changes over time to reflect the increasing input of domestic players such as the BATF, Janaagraha, other local NGOs, domestic consultants and BCC representatives.

The implementation of government reform projects typically takes place in municipal organizations, which do not have the institutional structure or the financial resources to support them. Initiatives to streamline the customer complaints system in Bangalore were initiated as early as the mid-1990s with the then Commissioner directing changes to make the existing system more transparent. These changes involved all range offices being provided with computers so that routine paper-based transactions would be automated. However, this automation process was stalled due to administrative bottlenecks with only the FBAS being installed because of the BCC and BATF's preoccupation with revenue. The BCC has currently decided to outsource the design and development of the CCS to private companies. A central research question for us is to investigate the extent to which the new system will enable improved accountability and diffusion of IT capabilities among these government organizations.

The Kerala Case

Implicit in the policy documents and manifestos of international agencies is a particular logic that drives the concept of e-governance from IT automation within individual departments to the notion of 'joined-up' government, to e-services applications. While earlier this logic served as a blueprint for policy statements of government agencies in developing countries, we now find on the contrary that policy discourse suggest a reversal of this logic with ICT being used for e-services applications prior to IT automation in individual government departments and to integration of departmental functions.

A reversal of this logic can be seen with Kerala's recent policy focus on establishing front-end computerization as a trigger for administrative reform even before automation of departmental records has been completed. We found that the successful implementation of FRIENDS has prompted individual government departments to automate their records but further research is clearly needed to examine the extent to which a focus on 'front-end' applications are able to trigger more substantial 'back-end' reform within government.

A most important factor in the implementation of the FRIENDS project in Kerala has been attitudinal change among Service Officers and citizens. Government functionaries from different departments, now collectively called Service Officers, are working together in the same physical location to service citizens in an integrated fashion. This activity is changing attitudes of government administrators, who earlier suffered from apathy and low morale in their work. It was extremely important for the success of the project that staff became emotionally involved with the project. The strategy was to send home the message that the individuals selected for FRIENDS have a very special responsibility on them—that of taking the state to the next millennium. The self-esteem and image of the workers went up substantially, as they could perceive themselves as computer professionals working in an excellent environment.

Citizen attitudes towards government are also changing as a result of an increased sense of trust and reciprocity developing between citizens and the state. With FRIENDS, for the first time, the government is seen as being capable of providing a reasonable level of service without corruption. A central research question is what happens over time to the social capital that has been built up as FRIENDS is decentralized to constituency level with local entrepreneurs, rather than government, owning and running the centres.

The Kuppam Case

The concept of partnership features prominently in the e-governance policy discourse of international agencies such as World Corps.

Initial World Corps policy regarding partnerships put forward by the founders of the NGO was quite different from today's vision. World Corps policy was directed towards achieving sustainable rural development that alleviates poverty, protects the environment and builds global peace through youth empowerment. This policy appeared to disguise the challenges and complexities of building partnerships between the rural poor, local industries, local government and politicians.

In the second phase of the CIC project, WCI has learnt a number of lessons about the dynamics with external agencies and partnerships. For example, where good relations have been built up between the CIC and local government departments, income-generating activities have increased significantly. In the second phase of the project, further research will be needed to study the crucial role of the *mandal* volunteers as mediators liaising with citizens, government, political representatives, and local businesses to sustain and integrate the CIC within existing structures.

In the Kuppam case, human resource development has been considered a top priority. Right from the start, WCI focussed on intensive training of young men and women aged 21–28 to establish community-based businesses in rural areas. These young people trained together in multinational teams and return to their home communities in the developing world to establish small businesses in areas such as Internet and renewable energy.

A rigorous methodology was developed by WCI to impart skills and education over a six-month period on different aspects of poverty alleviation, team-building, small business management, community outreach and development, computer and Internet training, climate change, and renewable energy sources. Early indications from the weekly reports written by the CIC owners and sent to WCI indicate that through the interaction of the CRO, the rural population are increasingly motivated to find out about government schemes. A key research question for our study is the impact of this human resource focus on the overall development of the CIC owners and the rural population.

Policy implications

This chapter has used Sen's notion of capabilities as an evaluative space for e-governance assessment. Such an emphasis demands a change in the way e-governance is assessed. The issue of what is relevant to assess has, so far, been overshadowed by the issue of what can be measured and quantified. The capabilities framework allows us to incorporate social and institutional issues, which makes it distinct from conventional measures of assessment. No analysis of the developmental impact of e-governance initiatives would be complete without understanding changes at the level of policy discourse. At the same time, no understanding of the development impact of e-governance initiatives would be complete without understanding the factors which are shaped by, but at the same time shape the implementation of policies at the grassroots level.

There are policy implications for such an approach to assessing e-governance projects. The main difficulty is to consider the relative weightage to attach to different human functionings, given scarce resources. Let us take the example of poverty among unemployed youth in Kuppam to illustrate this point. Poverty could be due to lack of schools in the area, lack of access to particular groups (say women), or due to other responsibilities which the youth have; and the difficulty at the policy level is to consider which of these constraints should be prioritized as the target for investment. Moreover, each of these different functionings may be more or less relevant to different socio-economic groupings within a locality. For example, freedom to pay bills may be more relevant for middle-income groups in Bangalore than for low-income groupings, while freedom to generate socio-economic development might be more important for unemployed youth in Kuppam than for elderly people living in the area.

Conclusion

India is one of many developing countries currently launching major e-governance projects aiming to improve government processes, connect government to citizens and build interactions within civil society. However, in a developing country like India, it remains uncertain as to what contribution, if any, e-governance initiatives can make to overall development priorities.

The principal objective of this chpater has been to explore the interpretive scope of the capability framework for analyzing the developmental impact of e-governance. We realize that our understanding of these projects needs to be built on further empirical research, which we plan to carry out over the next few years. At the same time, this framework can be seen as an opportunity to explore the relationship between policy discourse on e-governance and the way in which the capabilities of individuals and organizations shape the implementation of these projects. If key functionings such as maintaining health or holding government accountable depend on relational capabilities like the opportunity to live without being hassled by bribery, then policy must not only be directed towards the provision of e-governance infrastructure, but must also address itself to building such capability among individual and organizational end-users of the applications.

Note

1. The five *mandal* headquarters are located at Kuppam, Gudipalle, Rama Kuppam, Shanti Puram, and part of Venkatagiri Kota.

References

ADB. 2001. *Asian Municipalities Project: Enhancing Municipal Service Delivery Capability*. Project Report, Asian Development Bank. Available from www.adb.org/projects/benchmarking/implementation.asp.

Apthorne, R. 1997. 'Human Development Reporting and Social Anthropology', *Social Anthropology,* 5(1): 21–34.

Avgerou, C. 2003. 'The Link Between ICT and Economic Growth in the Discourse of Development', in M. Korpela, R. Montealegre and A. Poulymenakou (eds), *Organizational Information Systems in the Context of Globalisation*. Boston: Kluwer Academic Publishers.

Castells, M. 1998. *The End of Millennium. The Information Age: Economy, Society and Culture, Vol 3*. Oxford: Blackwell.

Castro, R. and G. Chopra. 2000. *India Country Brief 2000*. World Bank Report, September 2000.

Centre for Electronic Governance. 2001. Available from http://www.developmentgateway.org/topic/index?page_id+3647.

Chambers, R. 1995. 'Poverty and Livelihoods: Whose Reality Counts?', Institute of Development Studies, Sussex Discussion Paper, 347.

Chopra, K. and A. Duraiappah. 2001. 'Operationalising Capabilities and Freedom in a Segmented Society: The Role of Institutions', paper for conference entitled 'Examining Sen's capability approach' at Von Hugel Institute, University of Cambridge, June 2001.

Ciborra, C. 2002. *Unveiling E-government and Development: Governing at a Distance in the New War*. Public Lecture at the LSE.

Corbridge, S. 2002. 'Development as Freedom: the Spaces of Amartya Sen', *Progress in Development Studies,* 2(3): 183–217.

de Haan, A. and S. Maxwell. 1998. 'Editorial: Poverty and Social Exclusion in North and South', *IDS Bulletin*, 29(1): 1–9.

Escobar, A. 1995. *Encountering Development—The Making and Unmaking of the Third World*. Princeton: Princeton University Press.

Ferguson, J. 1990. *The Anti-Politics Machine. Development, Depoliticization and Bureaucratic Power in Lesotho*. Cambridge: Cambridge University Press.

Gardner, K. and D. Lewis. 2002. 'Dominant Paradigms Overturned or Business as Usual? Development Discourse and the White Paper on International Development', *Critique of Anthropology*, 20(7): 15–29.

Garnham, N. 2000. 'Amartya Sen's 'Capabilities' Approach to the Evaluation of Welfare and its Application to Communications', in B. Cammaerts and J.C. Burgelmans (eds), *Beyond Competition: Broadening the Scope of Telecommunications Policy* (pp. 25–37). Brussels: VUB University Press.

Gasper, D. 2001. 'Is the Capability Approach an Adequate Basis for Considering Human Development?', paper for conference entitled 'Examining Sen's capability approach' at Von Hugel Institute, University of Cambridge, June 2001.

Government of India. 1985. *Seventh Five Year Plan 1985–90*. New Delhi: Government of India, Planning Commission.

———. 2000. *Electronic Governance—A Concept Paper, Ministry of Information Technology*. New Delhi: Government of India.

Government of Kerala. 2001. *FRIENDS Project*. Department of Information Technology, Government of Kerala. Available from http://www.doitKerala.com/friends.htm.

Grillo, R.D. and R.L. Stirrat (eds). 1997. *Discourses of Development: Anthropological Perspectives*. Oxford: Berg.

Harris, R. 1999. 'Evaluating Telecentres within National Policies for ICTs in Developing Countries', in Gomez, R. and P. Hunt, (eds), *Telecentre Evaluation: A Global Perspective (Report of an International Meeting on Telecentre Evaluation)*. Quebec, Canada: Far Hills Inn.

Harris, R.W., A. Kumar and V. Balaji. 2002. 'Sustainable Telecentres? Two Cases from India', paper presented at the 7th IFIP International Conference entitled 'Information & Communication Technologies and Development: New Opportunities, Perspectives & Challenges', held in Bangalore, May 2002.

Heeks, R. 2001a. 'Understanding E-governance for Development', iGovernment Working Paper Series, Paper No. 11, Institute of Development Policy and Management, University of Manchester. Available from http://www.man.ac.uk/idpm/idpm_dp.htm#ig.

———. 2001b. 'Building E-Governance for Development', iGovernment Working Paper Series, Paper No. 11, Institute of Development Policy and Management, University of Manchester. Available from http://www.man.ac.uk/idpm/idpm_dp.htm#ig.

Hudson, H. 1999. 'Designing Research for Telecentre Evaluation', in R. Gomez, and P. Hunt, (eds), *Telecentre Evaluation: A Global perspective (Report of an International Meeting on Telecentre Evaluation)*. Quebec, Canada: Far Hills Inn.

Janaagraha Times. 'BMP PROOF QZ Analyses Workshop', *Janaagraha Times*, 15 December 2002. Volume 1, Issue, 26.

Jayai, N.G. and S. Pai. 2001. 'Governance, Poverty, and Development: An introduction', in N. Gopal Jayal and S. Pai (eds), *Democratic Governance in India*. New Delhi: Sage Publications.

Kanungo, S. 2002. 'Information Village: Bridging the Digital Divide in Rural India', paper presented at the 7th IFIP International Conference entitled 'Information & Communication Technologies and Development: New Opportunities, Perspectives & Challenges', held in Bangalore, May 2002.

Kirkman, G. (ed.) 2002. *The Global Information Technology Report 2001–02*. USA: Oxford University Press.

Madon, S. and S. Sahay. 2002. 'An Information-Based Model of NGO-Mediation for the Empowerment of Slum Dwellers in Bangalore', *The Information Society*, 18(1): 13–19.

Mansell, R. 2001. 'New Media and the Power of Networks', First Dixons Public Lecture, The London School of Economics and Political Science, 23 October 2001.

PROOF. 2002. *Public Record of Operations & Finance*. Bangalore: Public Record of Operations and Finance, a public campaign.

Rama Rao, T.P. 2003. 'Electronic Governance: Lessons from Experience', *IFIP WG9.4 Newsletter*, 13(1): 12–15.

Schech, S. 2002. 'Wired for Change: The Links Between ICTs and Development Discourse', *Journal of International Development*, 14: 13–23.

Sen, A. 1999. *Development as Freedom*. Oxford: Oxford University Press.

———. 2000. 'Social Exclusion: Concept, Application and Scrutiny', *Social Development Papers No. 1*, Asian Development Bank, Office of Environment and Social Development, June.

UNDP. 1996. *Human Development Report 1996*. United Nations Development Programme.

Unnithan, C. 2002. 'E-governance in India—Initiatives and Drivers: A Preliminary Investigation', paper presented at the 2nd European Conference on e-Government, held in Oxford, October 2002.

Wade, R.H. 2002. 'Bridging the Digital Divide: New Route to Development or New Form of Dependency?', *Global Governance*, 8(4): 53–85.

World Bank. 1999. *World Development Report 1998/99, Knowledge for Development*. Oxford: Oxford University Press.

World Corps. 2001. 'World Corps—A Global Training Institute for a New Generation of Entrepreneurs in Developing Countries'. Available from http://www.worldcorps.org/.

12

Ricocheting Gender Equations: Women Workers in the Call Centre Industry

Piush Antony
V. Gayathri

Introduction

Conventionally, lower rates of women's participation in the workforce have been understood as a reflection of unfavourable cultural factors.[1] In recent times, the emergence of new Information Technology (IT)-based industries and induction of women workers into this sector, portend great potential to influence these cultural factors in favour of women. Since the opening up of the economy in the early 1990s, India is affected by changes in IT and in the last 10 years, there has been mushrooming of what is popularly called the 'IT-enabled' sector in the urban centres. While women do find a presence in the software services sector, it is in the IT enabled services (ITES) segment that they are increasingly employed (Vijayabaskar, this volume). It has also been suggested that the IT sector offers greater ease of access to women workers than traditional sectors by way of its organizational flexibility. Hence, the new economy in general, is considered to be more favourable to women as it allows new ways of working and enables a better 'work–life' balance. The work environment is considered conducive for women, as work in this sector does not involve any physical exertion. It is also purported to be beneficial to women who opt to leave and re-enter the labour market after a period of absence for domestic reasons, mainly child bearing and child rearing, as the industry is said to be defying the traditional seniority-based mobility norms.

Further, rapid growth in this sector has prompted some segments of the Indian intelligentsia to believe that the so-called 'IT revolution' can effect fundamental transformations in gender relations. And, its potential to open up new avenues for women as a gender-neutral employer continues to enjoy popular consensus. Nevertheless, many basic questions about the sector, which form the basis of such perceptions especially about the workers, remain unanswered, thereby limiting a dispassionate assessment of its equality and empowerment potential. It is also important to understand the impact of macro-economic policies on women in relating to the panacea envisaged through this sector. In this chapter, we examine how the emergence of a new service industry—the call centre sector of the IT-enabled service industry—has impacted sexual segregation and gender equality goals. We also look at factors and processes that underlie the drafting and recruitment of women into the IT-enabled labour force.

Method

The basic premise of this chapter is that women's engagement with technological change and opportunities need to be contextualized within the accompanying changes in the larger socio-economic realm. It therefore, attempts to look beyond the logic of numbers to understand the push–pull factors that effect high participation rate of women in this sector. This framework is adopted to bring out the nature of work participation of women in the IT sector from the perspective of workers as well as from the employers. The push factors are fathomed by employing socialization theories, which take into account the aspirations of women workers, changing family norms in favour of working women/career women and the nature and organization of work in the sector. The pull factors include the perceived image of the sector and the supposed shift in gender relations at the work place. The chapter also analyzes factors related to skill requirements, 'suitability' of women to engage in such activities and emerging industry practices like perks and opportunities for career growth. Drawing from this, the chapter attempts to understand: the gender dimension of work *vis-à-vis* their access to, and control over knowledge, skills and management; and, the aspirations of women workers in the context of changing family norms and gender relations.

This chapter is based on interviews with women executives in the call centres and members of their households. Interviews were guided by an open-ended interview schedule. Seventy-five in-depth interviews were conducted with women workers employed in call centres. The sample of women workers was generated through a 'snowball' sampling technique. Through local contacts we visited four major call centres in Delhi and Hyderabad, respectively. During these visits, we identified women workers willing to participate in the study. We then conducted interviews with these workers in their homes. Since many of them were single were and staying in hostels, we could identify additional respondents for the study. The final sample of call centre agents, while not randomly generated, included women from as many as 15 different call centres in Delhi and Hyderabad.

In addition, we also conducted 30 household case studies wherein we visited the homes of women workers and interviewed family members. The selection of the household for the case studies was done keeping in mind the following criteria: type of sector that the call centre caters to, educational qualification and knowledge of computer skills/language, marital status and socio-economic background.

Women and Labour Market: Dominant Trends and Assumptions

The theorizing exercise on women workers draws heavily on their familial and socially ascribed roles in understanding their relationship with the market (Boserup, 1970). The major questions that are pertinent in an analysis of women workers, therefore, posit women *vis-à-vis* their social roles. First, given the multiple roles that women perform and primarily, motherhood as an over-riding responsibility stretched over a major part of their life cycle, what determines the choice of women to enter the labour market? Second, given the differential access to resources and services, predisposed by the patriarchal notion of family which disadvantages women to be skilled workers, how are women workers positioned in the labour market? As a corollary to this, subjected to a gendered socialization, do women aspire to be workers? Third, being a gendered space reflecting the discriminations in the society, how does the labour market treat women as workers, in relation to opportunities, remuneration and organization of work?

An overview of analyses addressing these questions throws up certain dominant assumptions and trends. Historically, the entry of women into the labour market signified the need for extra hands in a phase of expanded accumulation. Women as 'care givers' in the family and 'support providers' of the family enterprises were absorbed into workforce to supplement the labour of male skilled workers. This led women to have a supplementary status in the labour market due to the 'male breadwinner' concept; wages were lower, as it was meant to supplement the family income. This trend, however, codified the status of women workers in the labour market with regard to lower wages, nature of employment and work. In the larger context of a welfare state, the dominant assumptions that guided such practices were primarily the social construct of 'motherly women' and 'bread-winning men'. Further, the industrial revolution entailed not only a separation of home and work, but also a spatial reorganization of gender relations that created home as a female sphere and the public world of work and politics as a male one. There have been ample criticisms of this 'myth of separate worlds', highlighting the multiple ways in which the home remained a workplace even after the industrial revolution, and the ways in which the family processes have always intertwined with both women's and men's income earnings (Prugal, 1999). The exclusion of work from home has been a key factor in the articulation of a new gender order that emerged with the industrial revolution. This new gender order needs to be understood against the prevailing system in the pre-industrial agriculture and artisan households where wives and children contributed economically. Though male honour and power were linked to 'property and skill', legitimizing men's status as heads of households, the insecurity of income precluded reliance on one breadwinner. The industrial revolution fundamentally changed this family structure, creating the role of a single male income earner, moving men into factories. The notion of the male breadwinner was employed to re-create male dominance under different rules and situations (ibid.). In the new form of industrial patriarchy, the full-time housewife became a symbol of working-class respectability (Armstrong and Armstrong, 1990). In other words, the binary opposite of male breadwinner constituted the domestic woman. What legitimized this domestic woman was her childbearing capacity. All women, regardless of whether they had or wanted children, were potential mothers and this notion dominated the policies of the welfare state. These policies reinforced the notion that motherhood requires women to stay at home to ensure the well-being of their offspring, for which they hold sole responsibility.

While the transition to more knowledge-intensive and service dominant economic growth was uneventful with no significant changes in these notions, the development agenda during this period affected women's relationship with the market. Though women's movements that gained strength during this period articulated discrimination in work, it is argued that it had only supplementary impacts. But what contributed to a significant difference was the perceived need for technocratic intervention to enable economic development, on the one hand and population planning, on the other. In this modernization agenda of development, traditional women were routine child-bearers, and household industries maintained them in that position. Insofar as work kept women close to the traditional family and made child-rearing easy, it slowed down the demographic transition and retarded modernization. Therefore, the modernization agenda suggested that women who had hitherto worked as family labour in 'traditional' agriculture are to be drawn into 'modern' industry as wageworkers (ILO, 1927). This discovery of women as the preferred labour force in the market contributing to development and propagation of paid employment as the most effective route to empowerment and demographic transition gained additional support by Women and Development (WAD) agenda networks. Nevertheless, beyond this development agenda, many could read the market agenda of the new industries springing up in the third world. These industries relied largely on unskilled labour, but their intensive use of such labour meant that wage costs formed the major portion of their overall costs. Furthermore, the seasonality of demand in industries like clothing introduced an inherent instability into their production cycle. They needed a cheap labour force to cut down on their costs and they also needed a 'flexible' labour force, which could be drawn into and expelled from employment in response to changing levels of demand (Joekes, 1995; Kabeer, 2000; Standing, 1989).

Many studies also brought out the deliberate promotion of various gender stereotyping by employers. In their attempts to extract maximum profit from their female workforce, this often formed the criterion in the work place (Elson, 1996; Kabeer, 2000). Young, unmarried women in the Third World constituted an ideal workforce on grounds, which reflected an intersection of the 'economics' of demand and the 'culture' of supply (ibid.). Many studies that probed the organization of work in modern factories highlighted that these factories actively promoted cultural norms of femininity that legitimized employers' exploitation of the female workforce (Elson, 1996).

By the close of the 20th century, the ways in which traditional norms and practices employed in the labour processes of modern factories were strategized to suit their production needs was well established (Kalpagam, 2001). It was also pointed out that women's entry into paid labour force had done little to challenge the structures of patriarchy within the family and community, and instead was re-enforcing (Kapadia, 1999).

Thus, in comparison to the detailed attention paid to employers' motivation in hiring a young and female workforce, there was deafening silence in these early studies on who these women were, why they had sought factory employment and what their jobs meant to them. They were treated as 'un-differentiated, homogeneous, faceless and voiceless' members of a puppet-like reserve army of labour whose behavioural strings were pulled by capital, not social actors who thought about, struggled against and acted upon their own conditions. Such approaches attributed far more attention to capital than to female labour (Kabeer, 2000).

Based on the discussion on the historical evolution of women workforce,[2] certain major trends over time can be delineated. These are summarized next to extend the same to the specific context of the chapter.

- Skills and roles in families and societies are assigned in the overarching framework of patriarchy; thereby women are disadvantaged in skills and choices and tend to contribute mostly to low-end jobs.
- Women's large-scale entry into a particular activity or female intensiveness reduces the wages, mostly due to the surplus supply of labour and also due to the interplay of factors constituting biological and cultural determinism.
- Jobs that involve less physical exertion and are more time-intensive and dexterous are more likely to be feminized. Feminized activities carry biological determinism as an underlying principle, but cultural determinism is strategically used to create it as a rational choice for the workers as well.
- Lower wages or discriminatory wages of feminized activities is not only restricted to unskilled work but also in certain specific skills

and professions. These are culturally sanctioned as suitable jobs for women, for which certain feminine attributes are strategically incorporated, such as emotional traits and biological roles. This explains female intensiveness of certain professions such as hospitality and care services, teaching and other customer care services.

- Women's engagement in the family as care providers leaves them with lesser time at their disposal to engage in the formal labour market, which in turn forces them, among other reasons, to involve in seasonal/flexible/home-based activities.
- Women's status as supplementary workers in the family and subsidiary workers in the market allow them to resort to the labour market only to tide over the contingencies of the household. The ideal type of a woman in a patriarchal-welfare state model is that of a care provider in the family; and the status of worker can be, at best, an aberration that can be conveniently engaged in and moved out.

Against these generalized trends ranging over two centuries of transformations in women's relationship with the labour market, we try to juxtapose the shifts and trends in the Indian labour market *vis-à-vis* women workers.

Women in the Indian Labour Market

In this section, as a prelude to our discussion on women workers in the call centre industry, we provide an overview of skilled employment, to illustrate women's position in the labour market. We then present a snapshot view of women's employment patterns in the ITES.

In India, women comprise only a small proportion of the total workforce. The proportion of females among marginal workers is about 90 per cent, while that of males is barely 10 per cent. While over 90 per cent of the total workforce is in the unorganized sector, over 95 per cent of the total women workers belong to this sector. Sector-wise, women are more concentrated in the primary sector, compared to their male counterparts. As a whole, 70 per cent of all women workers are employed in the primary sector as compared to 47 per cent males and correspondingly, the share of female workers in secondary and tertiary sectors is much lower than those of male workers.

It is noteworthy that the percentage of women who participate in economic activities in urban areas is almost half that of rural areas. Further, skilled employment, both in the organized and unorganized sectors, has been and continues to be excessively male-dominated. According to the statistics published by the Ministry of Labour in 1999, women comprised only 17 per cent of the total workforce in this sector. A sector-wise profile of employment shows that the proportion of women is highest in the most backward and low-paying segment of organized industry, such as agriculture, hunting, forestry and fishing and lowest in electricity, gas and water. In terms of absolute numbers, the largest concentration of women workers are found in community, social and personal services.

Two broad distinctions can be made in the organized sector employment—that of the public and the private sector. While the public sector is the largest employer of both men and women in the organized sector in India, it employs a larger percentage of men as compared to women. Many studies have explored the factors responsible for this (Duraisamy and Duraisamy, 1995; Srivastava, 1999); and the arguments can be briefly summarized as follows. Higher wages and more job security characterize public sector employment. In a job-scarce economy, where societal attitudes favour men and women's employment is considered supplementary, women naturally find themselves at a disadvantageous position while looking for jobs. Further, gender bias and patriarchal values deny girls and women access to appropriate education and training, which are pre-requisites to finding a job in the organized public sector. For example, public sector is almost monopolistic in areas such as electricity generation and transmission, petroleum and gas. As can be seen from the sector-wise distribution of jobs, men outnumber women by a large margin in these sectors and this is attributed to the low proportion of qualified women available in these fields.

Women's presence in skilled employment also needs to be contextualized in the existing politico-economic situation of the country. Since 1981, with some minor exceptions, organized sector employment, in general, has been declining along with employment in the public sector. However, it is also true that the private sector has seen a marginal rise in employment rate, particularly in the financial service sector and in recent years, in the IT sector. These sectors are seen as sunrise sectors and even the Planning Commission, in its plan document on Employment Generation in 2002, emphasizes the need to promote these sectors, as they are seen as employment-friendly and gender-neutral sectors.

Much has been written about the growth of IT sector in India and about the phenomenal expansion in the software development and IT-enabled sectors. Although the business is large and expanding, especially in the IT-enabled sector, the Indian business is primarily export-driven. It is reported that through the 1990s, the IT sector growth exceeded 50 per cent annually (NASSCOM, 2000). Even with the global downturn in this sector in 2000–01, exports from this sector has been substantial. The figures for the IT enabled sector are even more promising. According to NASSCOM, professional employment in ITES reached 4,15,000 in 2005–06. A major survey conducted by NASSCOM–McKinsey estimates that 1.1 million jobs would be generated in the IT-enabled sector by 2008. Further, it also forecasts that this will be a major employment absorber for women in the years to come.

Although IT is a new field, gendered division of labour patterns are already emerging. Studies have pointed out that women tend to be concentrated in end-user, lower-skilled IT jobs, related to word processing or data entry and make up small percentages of managerial, maintenance and design personnel in networks, operating systems or software. Jobs have also been created for women in particular divisions of the call centre industries, such as information processing, banking, insurance, finance, printing and publishing, where skill requirements are relatively lower than in software development. Further, within these service sectors, the major employment for women is in information processing jobs, particularly involving data entry (Rothboeck, Vijaybaskar and Gayathri, 2002). It is also highlighted that many of the new jobs are in call centres. India[3] and Malaysia have cornered the bulk of these jobs, but they are expanding to other areas as well. The main attraction for these jobs seems to be salaries that are generally better than in other locally available jobs requiring comparable skill levels.

Women Workers: Typologies of Participation

In the past two decades, a number of studies have been carried out in the area of feminization of work and higher participation of women in certain occupations, particularly due to technological changes. These studies have problematized issues ranging from gender-specific socialization patterns resulting in gender-segregated labour markets to lower levels of women's career aspirations. Most scholars have explored the barriers hindering women's equal participation and those unusual situations in which women have been a majority in the workforce. It has been established by many studies that women in India derive their primary identity from their family. Studies pertaining to women in the labour market also corroborate this point. It has also been established that the compromises which women are conditioned to make in terms of time, nature of work and wages basically derive from the institutional primacy of their familial identity. This implicit identification of women in the familial and domestic sphere has been perpetuated by the state as well.

Given these trends, what can be an analytical framework in understanding the dynamics of women's relationship with labour market? Certain typologies are identified from these generalized trends to allow for more focussed discussion on the push–pull factors that affect individual and household-level decisions in women's work participation. These typologies are, however, illustrative and do not form mutually exclusive categories and need to be understood in the context of larger societal norms disadvantaging women. While gender differential access to resources pre-determined by cultural, religious and social factors make women less equipped to become workers, gendered socialization guarantees their acceptance of these disadvantaged positions. Further, the labour market continues to be a gendered space, functioning as a microstructure of the society with inherent societal discriminations effecting segmentation of the market. At the same time, the lower status of women and dominant patriarchal notions of body contribute to the making of public spheres insecure and unsafe for women, restricting their freedom of physical mobility. Such an understanding allows us to identify the following typologies of women workers largely drawn from the motivational aspects for work. They are:

- Survival/subsistence workers
- Supplementary/subsidiary workers
- Transient workers
- Professional career aspirants
- Self-promotion workers

Most of the women workers from poor income families constitute the first category by participating directly in the market, by selling their labour power and time, if not skills. A large majority of women workers in the unorganized sector constitute survival/subsistence workers. These women cannot afford to be unemployed or full-time mothers due to their low income levels. Agricultural, seasonal, contractual, casual and home-based workers are examples of this. The second category of workers are also from poor income families as well as from those families in which adult male earning is not sufficient, and sustenance of family enterprise needs women's labour. Transient workers are those pushed into the labour market as skilled and unskilled workers for a particular period of time for economic, social and personal reasons. Seasonal/circular migrant workers constitute the bulk of this category from the lowest classes. Though not to the same extent, in the upper economic classes too, especially in urban areas, women opt to become transient workers. Marriage is a significant marker in a woman's life, as it initiates them into a different social role. Given the accompanying values and socialization associated with marriage as an institution that largely disfavours women's direct involvement with the labour market, it is not surprising that many unmarried women opt to be transient workers till marriage. It has also been noted that many married women who have taken a break from work to cater to the demands of early childhood care, get back to work as transient workers. They tend to continue in work till the next period in their familial life cycle wherein their gender responsibilities at home preclude any involvement in productive work. Most export processing zones tend to have transient workers comprising unmarried women who will retire at the time of marriage.

Of more recent origin is the section of women workers who enter into the labour market for reasons of self-promotion. And, the presence of these workers is clearly an indication of the changing gender norms in the society as well as in personal relationships. Their understanding of labour market engagement may not be based on their rights, but a deliberate choice that gives them more leverage in bargaining for personal space and freedom within the family and within the marital relationship. These workers attribute increased self-esteem to their work status and they range from skilled, semi-skilled to professional workers.

A closely related category is that of professional career aspirant workers. In absolute numbers, they constitute the minority among women workers as they signify a different gender socialization that the family/community adopted over the years or re-socialization that individuals opted for, during the course of their education. However, women workers in the self-promotion and professional career aspirant categories need to be understood as products of modernity and within the larger changes that happened in the development agenda. Given this understanding, we proceed to an analysis of call centres and women workers.

Call Centre: Characteristics

The world of call centres is a heterogeneous one. Most of the call centres express their organizational size in terms of the seating capacity of employees, whether the business is sales or customer service. To provide an understanding of the call centre business, we have, therefore, tried to classify them along the lines of the nature of the business and capital. In fact, call centres vary according to the activities carried out, size and place within the production cycle and market. Therefore, it is possible to identify a variety of typologies of call centres, with different models for work organization and flexibility use which affect working conditions. Four types of call centres can be identified in India:

- In-house
- Large call centres in outsourcing
- Small call centres in outsourcing
- Automated call centres

The first category is largely that of multinational companies which work as either in-house maintenance or technical support units or as maintenance of clientele/customer service divisions of the parent company, which are popularly referred to as routine customer care service. The next category is that of large call centres in outsourcing, which are primarily involved in campaigning for new service products or financial products. Although some form a part of the multinational units, most of these large-seater call centres are stand-alone operations. Automated call centres too are often part of large in-house operations, especially connected to the banking and insurance sectors. This forms a very small proportion in India. The last category comprises small call centres in outsourcing, which are very flexible and are often in common parlance referred to as 'fly-by-night' operations. Although a large majority of them perform outsourced sales and marketing functions, a sizable proportion is also involved in data entry, medical transcription and back office support. In general, most call centres perform functions such as inquiries, booking, debt-collection, sales and telemarketing, which seem to require fairly simple tasks. The larger call centres in outsourcing and in-house customer care centres service sectors such as credit and insurance, social health services, technical assistance and so-called creative centres, require a relatively complex set of skills. In call centres that work as *knowledge-based organizations*—although not so common in this universe—traditional skills are combined with information and communication technology (ICT) skills.

Delhi is purported to have the largest number of call centres in the country. Although a large majority of them are located in either Gurgaon or NOIDA, many of the smaller operations are located in various commercial centres spread across the city. Only very large outsourced and in-house centres are located within techno-parks and industrial zones. In the case of Hyderabad too, call centres are as old as the Hitech city, implying the proactive role of the state towards IT sector investment and signifies the infrastructure created especially for this sector. However, like in Delhi, many of the new call centres are not a part of this infrastructure and have their offices in the commercial centres of the city. Though we do not have the exact number of call centres functioning in either of the cities, estimations of per-sonnel at the higher executive level place this number at over 100 for Delhi and at 20–25 for Hyderabad. While Delhi boasts of a number of large in-house and outsourced call centres involved in activities ranging from knowledge-based organizations to simple customer care and service centres, having an average of 300 agents in large outsourced centres to over 2,000 in in-house operations, the call centres in Hyderabad are smaller in seating capacity, with the average larger call centres seating not more than 350 agents. Further, most centres in Hyderabad perform the function of tele-sales/marketing or customer servicing. A very large majority of call centres in both cities do not have an independent identity of their own, but function either as a satellite outfit of the parent company or as a unit within the company, depending on the size of the parent company and the nature of business. They are very often located in commercial areas and thus form the informal sector of the IT industry.

Call centre workers are essentially young, often graduates, but primarily women with relatively high level of education. The over-representation of women manifests as a new form of female segregation. The quantum and nature of supply of workers are ensured through large-scale print media advertisements, which highlight the aspects of English speaking, urban young girls heading their way towards a successful career. The recent years have also seen many training centres for such jobs, which undertake placements as well. These schools provide training in English fluency, pronunciation, diction correction and appropriate usages in spoken English suitable to the clientele. However, a major part of the training consists of exercises to neutralize the accent. The in-house training soon after the induction also involves accent neutralization and familiarity with script or campaign. It is interesting to note that most of the call centre executives are given pseudo-names for easy recognition by the client. These names are chosen from the popular names in the client country and these names remain as their business identity.

There is a hierarchy that can be observed within the call centre in-dustry. Financial services such as insurance, credit cards and stock market enjoy a better status. Within each segment also, there is social rating. Customer care is rated high, compared to marketing and sales. One of the main reasons for this can be attributed to the fact that sales exerts more pressure and is more stressful, compared to customer care which is very quality-oriented, as against being output-oriented.

In the earlier years, call centres are said to have offered better salaries in the range of Rs 10,000–12,000, but it is reported that, over the years, there has been a steady decline of salaries. Even among the 'MNC-pioneers' of this industry, this trend has set in. During the year 2002, it was reported that the recruitments of the first half of the year were better salaried than those in the later half of the year. The management unanimously attributes it to the reduced contract rates from their clients outside India. They also attribute it to the increased competition that they face in international bidding from other companies in the Asian region. However, a more realistic and convincing answer would be the argument of reserve army of labour affecting the labour market trends.

The contractual nature of the job precludes many of the essential security mechanisms. During the interviews, it was noticed that many of the respondents were not informed about benefits and in some cases were unaware of provisions such as Provident Fund, Gratuity, Medical Insurance, and so on.

Profile of Women Call Centre Agents

The collective profile of the workers in the sample of call centre executives is that of a group of young women and mostly unmarried.

The nascent stage of the industry is evident in its skewed pattern of age distribution. While 20–30 is the predominant age cohort, women in their early thirties also constitute the workforce. The age profile of the sample indicates the perceived lack of need for any specific training and experience. In all cases, the call centre operators train their 'agents' or 'executives'.

Further, as most firms are located in metropolitan centres, professionals with urban backgrounds find easy access to these firms. This urban bias appears to have a class bias too. Given that English-speaking skill is a pre-requisite, exclusion of persons from rural backgrounds and lower class groups, even in urban centres is but natural. This exclusiveness becomes even more evident in the case of women professionals wherein a majority of them belong to metropolitan cities and none come from the rural districts. On an average, they have had graduate education, although some have had engineering or computer application education. The presence of large numbers of graduates may be a reflection of the perception of work and the gross shortage of employment opportunities. This may also imply strong entry-level barriers for non-engineering graduates to enter bigger or medium to high-end IT firms and forces many women to enter the call centre industry.

Advertisements are a major source of entry-level recruitment, followed by walk-ins and employee referrals. While the use of recruiters is less important, the larger call centres approach recruiters for filling higher-level positions. Selection is done normally through a written test and then an interview. The nature of the written test, however, varies. In larger call centres, 'repartee' skills are tested essentially rather than knowledge of language. However, speaking English with a neutral accent is a necessary condition. As the head of Human Resources department of one respondent centre puts it 'We are bothered about the "learnability" of the person recruited. So, what we look for in people is a pleasant, soothing voice, an ability to learn quickly and a knack for handling difficult situations'.

There is a wide disparity in wage levels across firm types. This is also true for firm sizes. The striking feature is the low wages prevailing in the smaller outsourced firms. In fact, these rates are comparable to salary levels in secretarial professions and hospitality industries. The salary levels are the highest in in-house MNCs as well as in a few large call centres in outsourcing. This wage differential would obviously act as an important incentive for employees to move from less-paying to high-paying firms.

All the sampled women workers acknowledge that they work under conditions of high stress levels, since most call centres have systems of monthly performance appraisals or ratings. Though on an average, employees report to work in regular fixed shifts of eight hours a day, work intensity varies between small periods of lull, when there are no new calls to high stress hours to complete deadlines and/or meet sales figures. In-house call centres do not face this problem, as their work is not dependent on the marketable skills of the agents. High turnover rate affects all call centres, although differences can be found in the sector and typology of call centres. Work repetitiveness, stress and lack of career prospects are the most frequent motives, which lead workers to quit.

Motivations and Aspirations: Through the Lens of Socialization

A major part of the interviews were devoted to an in-depth understanding of the motivational aspects of workers. In each category of enquiry, such as educational qualification, earlier work experience, work in call centres and family support, detailed discussions were held to bring out changes and continuity in motivations. These were supplemented by questions related to their career aspirations. Three types of explanations are apparent in women's narratives of why they entered into call centre work.

In what follows, we analyze these explanations—the household factors and processes that underlie these constructions. Most of the women have taken the decision to enter into the workforce in the context of their position as a member of a household and of the immediate community in which they live. Hence, in this analysis, we would like to recall the typologies of women's work participation discussed earlier (section on 'Women Workers: Typologies of Participation'), namely, survival/subsistence workers, supplementary/subsidiary workers, transient workers, professional career aspirants and self-promotion workers. However, applying the same kind of typologies to explain the participation of women workers in call centres needs considerable substantiation as it goes against the popular notions/image of industry, of being a gender-neutral employer and having the potential of altering the gender equations in work place as well as within families.

For a simplistic understanding of a complex process of decision-making that involves social norms, family values and individual aspirations, we follow narratives of individuals whom we interviewed to dispel many of the myths about the industry. These myths, to reiterate, include better remuneration, stress-free, women-friendly/suited, environment with gender equity. We have taken up each of these issues as queries posited in the interview schedule.

Why Have You Joined the Call Centre?[4]

Though the specified basic qualification is graduation in any discipline, it is found that many professional diploma holders from hospitality services also form a sizeable proportion of the call centre workforce. Equally significant is the presence of many highly qualified workers with technical qualification. With varying educational backgrounds and work experience, motivation to join this sector reflects the constraints of choice that these women workers face *vis-à-vis* family and labour market. Given here are excerpts from the narrations of respondents which illustrate these constraints.

> I did my MCA (Master of Computer Applications) from Pondicherry University. But by the time I passed out, the software market had collapsed. If one doesn't have hands-on experience with projects, nobody will take an MCA person. My first job through campus placement was as a web-designer. But then I got married and came to Hyderabad. Here, this was the only opportunity I got.
>
> (Manju, 27, married, senior telemarketing executive, medium-sized outsourced centre, marketing of client-specific credit cards)

> I have never been a good student. I always wanted to be a good housewife and mother. But circumstances forced me to take up this job. My husband lost his job in the US and though my in-laws are looking after my needs, I thought this will be a good diversion and will also provide me with some financial autonomy.
>
> (Rekha, 27, a graduate in Arts, telemarketing executive in a large outsourced centre involved in Dental Insurance Campaign in the US)

> Where else will I get such a job and a salary? I am just a graduate in English literature. Picking up a job in Calcutta with just a degree is impossible. Possibly, I can get a job as a receptionist, but salaries are low and I do not think my parents will allow me to work as a receptionist. Moreover, as you know, travelling in Calcutta is very difficult. Here in Hyderabad, I stay in a hostel close to the office with other colleagues from my state, and office provides transport for us.... Anyway, till marriage, I wanted to enjoy and did not want to stay with my parents doing nothing. After marriage, how can I be in a job where I am out for the whole night? Otherwise also, after

marriage, continuing in a job depends on family responsibilities, husband's attitude and the nature of his work.

(Mahua, 22, unmarried, telemarketing executive for an MNC call centre involved in customer service of a Petrocard)

I have been a good student and wanted to do post-graduation. My parents did not want me to leave the city to do a Master's degree in Biochemistry. I always wanted to be on my own and this call centre job came in very handy. My father is comfortable with the security arrangements of my office. After marriage, I do not know, but I will try my best to negotiate. Even with my father, I negotiated to look for alliances from progressive families.

(Shamim, 23, unmarried, telemarketing executive for an MNC call centre involved in marketing consumer goods)

After my MSW (Master of Social Work), I worked for two years in a development organization working in Hyderabad slums. My work was stressful to the extent that I had long hours of work and every day I used to walk around for a minimum of 5–6 km. My salary was only Rs 3500. Now, I am much more comfortable as I have better salary, better work environment and I think the status of this job is better than that of a social worker in the slums.

(Vanita Rachel, 26, unmarried, MNC call centre, customer care service)

After hotel management, I joined the hotel industry with great enthusiasm, but the hours were long and the nature of the job was physically tiring. The salary was very low and my promotion chances looked very bleak. When my family moved from Mumbai to Vishakapatnam, I had to rethink my career plan. A girl working in a hotel industry was also not considered in good taste here. My father was worried that it would affect good marriage alliances. Now I am doing very well in this industry.

(Shobha, 24, unmarried, large in-house customer care service of a financial institution)

After BE (Bachelor of Engineering), I wanted to write GRE and go abroad. My parents dissuaded me by saying that it will be difficult to get a match for me after doing an MS in the US. They are looking for alliances of software engineers for me. When I go to the US after marriage, I do not know whether I will be able to study or work. Right now, I am in this job because after doing BE, I do not feel like sitting at home and this is the only time I may work in my life.

(Haritha, 22, unmarried, marketing executive in MNC call centre, marketing of office maintenance services)

It is evident from the narrations that there is a complex interplay of factors that determine induction of women into this sector. While graduates find this job the only decent opportunity of being employed for their 'limited' qualification, professionals and other technically qualified women perceive this as the last opportunity to remain in the IT sector. Women from other diverse educational backgrounds such as hotel management, social work, teaching, and so on, have personal and familial reasons rather than market-related reasons. However, most of them enjoy a favourable backing from their families as compared to their previous professions. This is largely because of the media hype surrounding this industry and consequent eulogized social status associated with being a part of the 'most happening industry'.

Do You Feel Call Centre Jobs are Better Suited to Women Professionals?

All the respondents felt that call centre jobs are most suitable for them. However, justification of suitability varied between biological and social traits of women to socio-cultural factors favouring women's employment in this sector. Though they acknowledge the presence of male colleagues in their offices, women executives were unanimous in reporting that employers prefer women.

Though there are male executives in call centres, one can comfortably say that this is a female-intensive industry. The employers use certain feminine emotional traits such as patience, perseverance, and softer and sweeter voice to justify the preponderance of women over men. It is interesting that many women executives also believe that they are better suited for this and offer similar explanations to justify their presence in large numbers on the shop floor. Nevertheless, an analysis of the industry in terms of nature of employment, business and organization of work reveals that the call centre industry thrives on the availability of cheap, flexible labour. Semi-skilled urban women workers naturally form a homogeneous group that can be recruited from across the cities. To reiterate, women workers with no long-term career plans are better suited for contractual employment wherein the nature of job does not impart any other marketable skills, requires longer working hours, lower remuneration, lack of career prospects and lack of job security. Many Human Resource professionals feel that women workers are passive and can be better moulded into a 'good employee'. It was also reported that attrition rate of male workers tend to be higher as they have less constraints in shifting jobs across the industry or even in changing the geographical locations. See the following narrations that favoured women's suitability.

> It's not that men are not making sales. But, I think women have certain advantages in dealing with clients abroad. The client abroad, especially when it is a stranger, would feel less intimidated if there is a sweet female voice on the other side. Not only this, with some tough customers, lot of patience is needed. Women tend to be more patient than men, even in normal arguments at home and with friends.
>
> (Manju)[5]

> Most importantly, there is complete physical security for women. There is free transport provided to women from home to office. If the woman is alone, the driver is accompanied by a woman.
>
> (Rekha)

How Does Your Family Relate to Your Work?

Decision to take up the job in call centres, especially with night shifts, seems to have familial and social factors influencing it than individual aspirations. Many mothers of unmarried workers whom we interviewed during our household interaction reiterated this stance. There was an unanimous opinion that though *these girls may not be able to work after marriage or may work till the last stages of pregnancy, their working status at present is good for them*. Some of these mothers justified this opinion by a very compassionate understanding of women's lives from their personal experience that *this is the only time she can enjoy herself*.

At the same time, there were mothers who have a very pragmatic approach to their daughters' present status. From their understanding of the marriage market and the demands of it, they coherently articulated certain trends, which they may not favour but accept as a norm. Many of them pointed out that a working girl is considered smart and especially one who is working in the computer industry is accorded 'social skills and manners' which are essential to become the wife of a *successful, well-settled, urban, cosmopolitan husband*.

The mothers-in-law of the married women workers also articulated a similar argument. Compassionate mothers-in-law felt that their daughters-in-law can enjoy life as long as they can provide the present support and till children are young. It is observed that as children grow up, grandmothers are less effective as substitute to mothers. Many mothers-in-law were of the opinion that when it comes to studies, their grand children will not listen to them and that is the time when mothers need to pay more attention to children. 'After all, we want our children to grow up as successful individuals. If not for this, what else is family life for. If you concentrate on your work, your children do not turn up well, can you pardon yourself?' And, this forms the most common reinforcing statement about a woman's primary identity and responsibility in life. The following narratives vividly show these reinforcements.

> My father-in-law is very proud that I am a call centre executive. My mother-in-law is very happy because the timings are convenient for a young mother as I work from evening to late night or late evening to morning and I can finish my family responsibilities before I leave for work. She is also happy that there is a free and safe transport system.
>
> (Rekha)

> In my earlier job, I was a secretary to the Managing Director of a small social service organization. Now I am part of the computer industry. So, although my salary is not very different, my family feels that this is a better job.
>
> (Lakhsmi, 22, unmarried, small outsourcing call centre, customer care)

> Soon after I did Bachelor of Education (B.Ed.), I was teaching in an English medium school. Since I live in a joint family, I found it difficult to spend time on student's workbooks and test papers at home, which is inevitable in a teaching job. My mother-in-law's remarks often made me conscious of this. When I got this opportunity to join as a Telemarketing executive, I felt that it suits me better than the earlier job because there is no carry-over work for home.
>
> (Anita, 26, married, large outsourced call centre, financial customer services)

Why do You Think There are Not Many Women in Managerial Jobs in This Industry?

It is rather difficult to be conclusive about gender equality at all levels in the industry, as the very nature of the job 'blamed on geography' precludes women who have to perform a larger responsibility in their families. To explain, the industry recruits younger women who inevitably retire after a few years due to marriage as call centre jobs, by definition, are performed in the night. Therefore, women need to, on their own 'free-will', forego the promotion chances to accommodate their life cycle demands. These compromises are inevitable as larger familial and social pressures persuade these career-aspiring women to make a choice in favour of domestic responsibilities. The prescriptions of motherhood and other marital responsibilities are so internalized that many women who even attempted to begin a career often spoke with guilt and were apprehensive about the sustenance of their interest in career. This trend is often quoted as one of the reasons for lower representation of women in the higher cadres of the industry. Therefore, the issue of gender equity seems very precarious. The concept of 'glass ceiling', which gains validity in terms of disproportion in absolute numbers of men and women in higher positions, gets diluted because of the informed choice involved in the compromises that women are expected to make. The narratives given here are indicative of this.

> Men have the advantage in that they can take independent decisions. They can postpone marriage, they can change the location of their work. So, they will have a better career planning.
>
> (Vandana, 21, unmarried, small outsourced call centre, data entry)

> There is so much uncertainty in my life. I guess, it is true for most of my female colleagues. Now my son is only two years old. As of now, my mother-in-law manages him with the help of a maid. But I do not know how long she can do this and I do not know what decisions to take when my son is in school.
>
> (Deepa, 29, married, large in-house call centre)

> I do not know how many women would like to take up managerial positions. In my case, I am an aspirant for that, but I have a good support system now. If your husband is in an equally demanding job and you are living as a nuclear family, how can you opt for it? It is strange that all the female team leaders that I know in other call centres in the city are either single women or separated or have a strong family support.
>
> Promotions come after a period of time. But by then, my family responsibilities, especially my duty towards my children, will increase. So I am not sure if I will be working long enough to be in a managerial position.
>
> (Radhika, 29, married, B.Tech., in-house call centre)

> I do not have any personal experience of gender discrimination, but my friends often narrate cases wherein women declined from accepting a promotion that involves on-board training and extensive travelling and cases where companies show reluctance in training women.
>
> (Madhumitha, 25, married, in-house call centre)

Where would You Like to See Yourself, Five Years from Now?

Very few women in the sample perceive career growth in this industry or even relate to this present job in such terms. All of the call centre executives from Calcutta working in an MNC outsourced call centre in the city of Hyderabad exemplify this. These graduates see this job as an interim arrangement in their lives as they/their families do not want to invest further on studies and want to utilize the waiting period until marriage doing something which is remunerative and respectable. A large majority also feel that working in this industry particularly enhances their chances of getting a better marriage alliance. While internalized gendered socialization filters the career aspirations of most of the women workers, those who aspire for a career tend to be apprehensive about the practicality of such an eventuality *vis-à-vis* their household responsibilities. Articulated by the respondents, the illustrations given next substantiate the earlier-mentioned observations—aspirations that are filtered by socialization; angst-ridden due to family responsibilities; and dispassionate due to lack of decision-making capacity.

> Five years from now I will be married and will have children by then. Work of this nature is ruled out. In my community women don't work when they have small children to attend to. If it is a nuclear family, the question does not arise at all.
>
> (Mahua)
>
> I would like to be a team leader because I am qualified, Anyway, if we decide to have a child, my priorities may change depending upon the support I can get from my in laws or from my parents. Ideally I would not like to take more than an year's break.
>
> (Manju)
>
> I do not know how long I will continue in this job. Once my husband comes back from the US...it's meaningless to plan. If he is not finding a job and I am working, my mother-in-law will be upset that her son is sitting at home and I am going to work. Therefore, career planning is meaningless in my case.
>
> (Rekha)
>
> My husband was very unhappy when I got a promotion from the telemarketing unit to the quality unit. Here I have to prepare reports on quality of calls and sales, which he thinks reduces the time I spend at home. I am not sure how long I can continue in this.
>
> (Parvati, 24, large outsourced call centre)

Describe Your Work Environment—Disadvantages and Advantages?

All the women workers that we interviewed were unanimous in reporting their individual and families, satisfaction with the present job, which they articulated mostly in terms of a secure physical environment and free transportation that the industry provides. However, their stress levels were reported to be high, especially for those involved in sales. Few of them even complained about ear pain as an occupational hazard of their work. However, disadvantages were not entirely articulated in relation to health factors as revealed in the following narrations.

> After we went online, the pressure for sales has been slowly building up. The management has increased the cut off speed. It is too stressful at times. You get difficult customers who would not entertain a conversation, however polite it will be. And some times the lines are so bad that the customer hangs the phone down when all that is left is to take the credit card details for the billing. It is really frustrating.
>
> (Reshmi, 21, unmarried, graduate, MNC outsourced call centre)
>
> In my office so many of my colleagues were asked to resign. Sometimes, it is difficult to understand the criteria followed by the HR department. There is a cut off speed for sales that one needs to maintain, which varies from office to office. In my office average sales per day is fixed at three. If one is not maintaining the same sales consistently for two months he or she is supposed to resign and as you know sales is a matter of luck too.
>
> The promotion chances are very minimal for a person like me. Next level is senior executive, which means only a marginal increase in salary and work profile remains the same. But to become a team leader seems difficult for me, as many technically qualified people are aiming for it.
>
> I feel sad at times. Many of my colleagues are MCA, M.Tech. and B.Tech., but when it comes to sales, our performance levels however do not vary.
>
> (Rekha)

One of the main advantages of this sector that is favourable to women among other factors is its broad-based qualification requirements for the job. However, it can be seen that this option is also used by highly/technically-qualified women to be in a job with little career growth, which is suited for the gender roles that they play outside the work sphere. It is interesting to see that very few could relate to the 'hire and fire' policy of these firms as a disadvantage. One possible reason was that many were relatively new. Similarly, most of them did not consider lack of provisions/benefits such as Provident Fund, gratuity, group insurance, maternity leave as disadvantages of this sector.

If You Decide to Leave the Firm, What could be the Main Reason?

Though many workers hold the opinion that call centre jobs are suited for women, we have seen that very few workers aspire for career growth in this sector. Moreover, most of them were not sure of continuing in work. Probing the reasons that would lead them to leave this work, it was revealed that in the long run, night shift inconveniences many workers due to factors affecting their general well-being including health and family responsibilities. From the responses that articulated inconveniences related to night shifts and other stress factors related to this job, selected ones are given next.

> I want to change this job, so that I can avoid night shifts and be with my son. I also have ear pain due to constantly using the head set. I tried to look for opportunities in customer care services in financial institutions. But as you have seen in the advertisements, the upper age limit is 25 years and I am 27 years old.
>
> (Rekha)

> With this experience, I would like to look for other customer care call centres which cater to domestic clients. Due to night shifts, I spend time with my husband and children only during the weed end. My idea is to gain sufficient experience and then shift to a day time job.
>
> (Latha, 29, married, Telemarketing executive in a medium-sized firm)

> There are times I get tired of the night shifts but where else will I get a job. So I will continue in this as long as I can, physically and psychologically. And, I need this money.
>
> (Sasikala, 26 years, graduate, customer care unit of a telephone enquiry service)

> My husband works in two dental clinics in the city. He is really busy and comes home very late. With a business administration degree from a local institution, this is the only job that does not involve any travel. But when we decide to have a child or when he can set up a clinic of his own, I will leave this job. Who wants to work for a night shift then?
>
> (Lathika, M.B.A., 27 years, quality control officer of customer care unit of a financial institution)

> I feel that nobody can continue in this job for a long time. I get headache quite often as I can't sleep well during daytime. I have gained lot of weight and feel unhealthy due to increased intake of tea and coffee during working hours. I need it to keep myself awake.
>
> (Merlyn, Graduate, 30 years, telemarketing executive of a large-sized MNC call centre)

> I think it will be less strenuous to work in a customer care centre than a telemarketing centre. Pressure for sales is very stressful. Some days you only get clients who are very rude and it is really depressing, especially when you have to meet the sales target. I would like to leave this job as soon as I get another one with the same salary.
>
> (Kalyani, 23 years, Diploma in Apparel designing, telemarketing executive of a medium-sized call centre)

> Can anyone get a sound sleep of six hours during daytime, when everybody else is moving around in the house? I would like to shift to another job. Wish I had another option where my life remains normal.
>
> (Varalakshmi, 24 years, customer care unit of a financial institution)

By picking out selected queries from the interviews with the respondents, the attempt was to understand the workers and through them this new sector with promises for women workers. Selection of queries were to facilitate an analysis of this sector with regard to its work culture favouring women workers and also to bring out the dynamics of socialization factors of these workers moulding their motivations and aspirations for work while engaging in this sector.

Between History and Rhetoric

To conclude from the earlier discussion, it can be discerned that all that has been promised and seems attractive about this industry *vis-à-vis* women workers is not unprecedented in history. A cursory glance of the major developments that pushed women into the labour market at various historical junctures exhibits striking similarities to what is happening in the call centre industry. Possibly, the only dissimilarity is that the target of beneficiaries of this sector is the urban middle class. The well-documented experiences of industrial revolution and emergence of export processing zones, especially the garment sector boom, provides ample illustrations to this. In all these cases, women formed the pliable, flexible, cheap labour and their participation did not alter the structure of gender relations within the families or in the larger society. Here, we would also like to recall the typologies we have identified as 'ideal types' of women's labour market participation in the context of India.

As already mentioned, the parallels that can be found in history allow us to situate these typologies in understanding women workers in the call centre industry. Except for the first category of survival workers, our sample represents the other four typologies. The exemption of the first category is evident from the fact that the industry pre-empts the involvement of the income poor women. There is a considerable proportion of workers that can come under the typology of subsidiary/supplementary workers. These workers consider their earnings essential to maintain their existing levels of middle class consumerist demands. These demands vary from very 'feminine indulgence' of an unmarried woman to a middle class mother's aspiration to put her child in a better play school. From the family point of view, these women workers are a reflection of the modern values of the family and their work status add esteem of the family—being supportive of a working woman. However, the concept of work is restricted to a means to create a monthly supplementary income and therefore, participation in work does not include aspirations for a career. In other words, the changed family values are supportive of a working women but not of a career women. And, it is feared that the sacrifices that a career women have to make in terms of reduced time with family members is detrimental to the welfare of the family and children.

The typology of transient workers probably forms the largest category in absolute numbers. For instance, in the city of Hyderabad, a hostel was provided exclusively for young, unmarried girls from other cities such as Calcutta, Bangalore and Vishakapatnam. All these unmarried women are transient workers and most of them exhibit an insouciant attitude towards work and career. This insouciance is not incidental among them. They all share a similar kind of socialization, which conclusively equate womanhood with motherhood. This internalized value attached to marriage gives them 'no shock' when they are forced to quit or 'no shakes' when they are in work. In a way, these transient workers are well prepared by their gendered socialization, and their expectations at different points of their life cycle fall very much within the limited parameters of accepted norms of autonomy.

Women workers who engage in work for self-promotion constitute the binary opposite category to transient workers. In most cases, earnings form an insignificant amount, which they largely spend on themselves. A part of their income is also spent on participating in 'gift-economy' within their natal and marital homes to gain leverage for their autonomy. For them, 'to be different' in terms of their preferences in life and autonomy that they need is intrinsically linked to their status as a working woman. Some justify it in terms of their need to work because of their qualification, time at their disposal, personal capabilities and less involvement in managing their responsibilities towards extended families.

The smallest category of workers that we found in our sample was that of professional/career aspirant workers. It is important to notice here that they even hold a different attitude towards womanhood. They consider work as important as their education. While being aware of the compromises that women make in favour of their personal lives, these women aspire for a balance by attempting to negotiate with both worlds. They believe in renegotiating gender roles within the family to accommodate their career aspirations.

What can be gathered by this analysis of women's work participation in terms of motivation and aspirations of individuals is that families and societal values are yet to undergo structural changes that can favour women to become professional workers. The immutability attributed to gender roles and hence, the unaltered ways in which women are positioned in 'care economy' as care providers to the male breadwinner, children and older family members, impede them to shift from the typologies of subsidiary or transient worker status to women workers who can work with an agenda of self-promotion and professionalism. Therefore, the issue of whether call centre industry is a gender-neutral employer or ensures gender equity assumes lesser importance in the context of women's partial engagement with the labour market. This relationship tends to remain the same as long as the larger socialization processes equates womanhood to motherhood and distinguishes between work and career.

Therefore, it is more of rhetoric than any actuality implied in attributing to a new industry the potential to alter gender equations, as any work environment tends to mirror or embed dominant ideologies of the larger society. This is particularly so, in the case of call centres which actually utilizes the 'comparative disadvantage' of women to their advantage and indulges in gender-stereotyping.

Conclusion

Women's participation in labour market in the specific context of call centres, is indeed, an individual preference of a homogeneous category of urban, middle class, English-speaking graduates. However, these individual preferences are to be perceived as a matter of choice, which in turn, reflects the wider social context of asymmetries and inequalities that are intrinsic to their immediate world. These inequalities not only act as impeding factors on women in exercising their choices, but also are inherently built into the construction of choice. Therefore, the preferences that these women workers have, choices they make, the way they relate to, and conceive their identities and responsibilities, are indicative of their individual life experiences and subjectivities, which are guided by the norms and values of their societies.

In the evolution of a choice, these norms and values represent collective understandings of appropriate ways of behaviour for men and women. Since this collective belief and values in society emanate from the structural inequalities of the wider society, it is argued that the choices that individuals adopt are unlikely to be neutral in their content or implications (Kabeer, 2000). Therefore, even when we talk of the choice that women workers make in joining the call centre industry, we need to go beyond a rational choice to a more social choice frame in which less of economics and more of gender, cultural and class constructs of work play a significant role in determining women's choice in utilizing disposable time for socially sanctioned forms of work. We would like to highlight this as our study reveals that the call centre industry thrives on transient workers more than professional and career aspirant workers. When individual women seek to negotiate gender equations by engaging in industries with presumed emancipatory potentials, they only confront stereotypes of *sweeter voice* and *better mothers*.

Acknowledgement

We would like to acknowledge our gratitude to Jhunu Roy and B. Srinivasa Naik for their capable research assistance. We would also like to place on record our sincere thanks to M. Rekha Rao for providing us with contacts in the industry.

Notes

1. For a review and critique of how culture is emphasized as an explanatory factor, see Kabeer (1994).
2. However, less attention has been paid to the shifts that have happened in the understanding of work and discourses on private/public dichotomies that accrued status of work to some and not to certain other activities. This omission is not incidental but to keep the focus unaltered on the relationship women have had with the labour market.

3. While NASSCOM lists 211 ITES firms, popular media estimates indicate that there may be as many as 500 call centres in India alone.
4. For obvious reasons, the names have been changed.
5. Details of age, qualification, marital status and nature of firms are not repeated for respondents whose details are given earlier.

References

Armstrong, P. and H. Armstrong. 1990. *Theorizing Women's Work*. Toronto: Garamond Press.

Boserup, E. 1970. *Women's Role in Economic Development*. New York: St Martin's Press.

Duraisamy, M. and P. Duraisamy. 1995. 'Implications of Structural Reforms for Public and Private Sector Wage Differences in India', *The Indian Journal of Labour Economics,* 38(Conference Issue, October–December).

Elson, D. 1996. 'Appraising Recent Developments in the World Market for Nimble Fingers', in A. Chhachhi and R. Putin (eds), *Confronting State, Captial, and Patriarchy: Women Organizing in the Process of Industrialisation* (pp. 35–55). Bassingstoke: MacMillan in Association with ISS, The Hague.

ILO. 1927. *International Labour Conference, 10th Session: Proceedings, Vol. 1*. Geneva: International Labour Organization.

Joekes, S. 1995. 'Trade Related Employment for Women in Industry and Services in Developing Countries', Occasional Paper No. 5. UNRISD, Geneva.

Kabeer, N. 1994. *Reversed Realities: Gender Hierarchies in Development Though*. London: Verso.

———. 2000. *Bangladeshi Women Workers and Labour Market Decisions: The Power to Choose*. New Delhi: Vistaar Publications.

Kalpagam, U. 2001. 'Globalisation, Liberalisation and Women Workers in the Informal Sector', in A. Kundu and A.N. Sharma (eds), *Informal Sector in India: Perspectives and Policies*. New Delhi: Institute of Human Development and Manohar Publications.

Kapadia, K. 1999. 'Gender Ideologies and the Formation of Rural Industrial Classes in South India Today', *Contributions to Indian Sociology,* 33(1 and 2): 329–52.

Kundu, Amitabh and Alakh N. Sharma (eds). 2001. *Informal Sector in India: Perspectives and Policies*. New Delhi: Manohar.

Prugal, E. 1999. *The Global Construction of Gender: Home-Based Work in the Political Economy of the 20th Century*. New York: Columbia University Press.

Rothboeck, S., M. Vijayabaskar and V. Gayathri. 2002. *Labour in the New Economy: The Case of the Indian Software Labour Market*. New Delhi: International Labour Organization.

Srivastava, N. 1999. 'Striving for a Toe-hold: Women in the Organised Sector', in T.S. Papola and A.N. Sharma (eds), *Gender and Employment in India* (pp. 181–206). Delhi: Indian Society of Labour Economics and Institute for Economic Growth in association with Vikas Publishing House Pvt. Ltd.

Standing, G. 1989. 'Global Feminization through Flexible Labour', *World Development,* 17(7):1077–95.

13

Constructing Work and Identity in the Indian Outsourced ITES Sector[1]

M. Vijayabaskar

Introduction

Studies on labour market implications of informatization and servitization in the advanced capitalist economies have not been matched by similar studies in low-income economies that are increasingly in-corporated into these sectors through new global divisions of labour. Important questions like the quality of employment, work and labour markets generated by new kinds of services, often seen as a growth pole in low-income economies, are yet to be adequately addressed. Such questions, when posed in the context of economies characterized by different factor market proportions and patterns of institutional regulation, may yield different, but important answers. This chapter is a step in this direction. It addresses these questions by examining the construction of work and employment in the Indian Information Technology (IT) enabled services (ITES) sector, a sector primarily driven by global capital in search of low-cost locations. Through this exercise, the chapter examines some of the propositions that are generally made with regard to labour market changes in the 'new' economy.

ITES as Growth Poles and Employment Generators in India

ICTs enable a number of services that were earlier tied to specific locations to be dispersed globally. The call centre is an important part of such fast growing services sector enabled by information and communication technologies (ICTs). It acts as an interface between the producer and the consumer, either through attending to service-related queries or act as remote selling agents. The growth spurt for this sector stems from two features of late capitalist accumulation; one, the growing servitization of economies, and two, competition between firms through provision of quality services, both by producers and retailers.

With advances in ICTs, agents located in different parts of the globe can attend to calls from customers located elsewhere. This trend makes India a relatively favourable destination for firms in the advanced capitalist economies to locate their call centres. Low-cost, educated labour, capable of being trained to converse in English—American and British—proves to be an attractive factor. While cheap labour may be found elsewhere, it is the knowledge of English language that has been a prime factor. In India, its share in total exports of IT services has increased from 14 per cent in 1999–2000 to 24 per cent in 2002–03. That this trend is going to be a long-term phenomenon, is stressed by many industry experts (http://www.nasscom.org/artdisplay.asp?cat_id=325; Chengappa and Goyal, 2002).

Global capital, it is understood, does not produce homogeneous labour and employment outcomes across spaces. Outcomes are constituted through a configuration of global and local relations. In other words, labour market outcomes are determined both horizontally and vertically (Fine, 1998). In this chapter, we address the process of labour force formation in a specific peripheral region shaped by these relations and in the process seek to identify the specifities of the global workforce in a peripheral region. Service sector work involves exchange of intangibles like knowledge and information, which requires a relationship to be established between the service provider and the consumer. The rise of the services sectors is generally associated with 'feminization', in terms of both higher women's work participation rate and rise of jobs that require skills/attributes typically seen as being vested with women (McDowell, 2003: 27). To be sure, there are also jobs like that in software development which require problem-solving skills; but in recent times, it is the set of jobs that require lower skills and exercise of emotional labour which have risen (McDowell, 2003: 31). In the Indian context too, the share of ITES has slowly increased in the share of IT services exports, which requires exercise of emotional labour rather than intellectual labour. Interestingly, this sector is seen in policy circles to be critical to employment generation, given the stagnation in other sectors (Taskforce, Ministry of HRD; ITES Policy, 2002, Tamil Nadu).

Simultaneously, the growing use of ICTs in the workplace and the rise of ICT-enabled sectors also pose questions about the relationship between new technologies and the quality of work and employment. There are debates about whether ICTs enable enskilled or deskilled work, whether the workplace is differently gendered and whether it promotes autonomous and creative work. Against this background, it is pertinent to pose the following questions:

- What are the sourcing patterns and specificities of the workforce in Indian ITES?
- What are the skills and attributes privileged in this sector?
- Are organizational structures flatter and decentralized? What are the prospects of a life-long career in the Indian ITES sector?
- To what extent does the presence of male employees in the ITES sector alter the gendered nature of work?
- How do managerial and technical innovations alter the quality of work?
- What are the constraints and opportunities for collective organization and resistance to managerial control and work discipline?

Towards this, we relate the workforce, work patterns and regimes of work control in the Indian workforce with that observed elsewhere. This exercise also contributes to a growing body of literature that ethnographically maps the formation of a global labour force by looking at specific regions and/or sectors (Burawoy, 2000).

Though our primary focus in this chapter is on call centres, we also cursorily examine the work and employment conditions in two other major ITES segments in India—the medical transcription (MT) segment and back office processing (BOP) segment. Our case firms consisted of one medical transcription firm, four call centres and one back office processing firm. Except for one call centre that catered to a domestic telecom firm, the rest were catering to global clients. This distribution does not correspond to their distribution in terms of numbers or revenues generated, but do represent the spectrum of activities in the Indian ITES sector. Fieldwork proved to be extremely difficult. In many cases, managers in firms stated that allowing of interviews would undermine the secrecy clause that they enter into with client firms. In fact, one firm allowed us to talk to employees only during their lunch break, or as they were leaving work. Another factor appears to be a lack of interest in disclosing hiring and employing practices of firms or information about employees, given the high labour turnover in the industry.

Apart from interviews with employees, we conducted in-depth interviews with key informants associated with the industry. They included HR heads, recruitment consultants, trainers and those involved in starting-up of ITES firms. In total, 60 employees were interviewed, with roughly 10 from each firm. Among them, we chose 20 employees to do in-depth case studies of their work and life histories and their perceptions of how work in such firms are constructed. Answers to questions were taken as leads to elicit more information about their understanding of work and its meanings, and the role that this work plays in the overall context of their perceptions of a successful career and life. Importantly, we sought to understand the way in which the workforce perceives itself in relation to the workforce in other sectors like traditional manufacturing or other services and how such perceptions condition scope for mobilization and resistance. Given the highly 'gendered' nature of work in ITES firms, we also sought to understand the way men and women employees code their work.

Of the 60 employees, while 50 were chosen in random, the remaining 10 were chosen purposively to ensure that all levels in the organization are represented. The small size of sample employees is partly compensated by interviews with key informants. Secondary literature pertaining to ITES industry in India is confined mostly to newspaper reports and articles in business and trade journals.

It must be remembered that we have confined our study mostly to the low-end segment of the ITES sector. In high-end services, conditions of work, especially remuneration and skill requirements may be quite different from that depicted here. However, it is the low-end segment that is the most dominant and seen to have the potential to generate diffused employment on a larger scale and hence, important from the point of view of employment generation. It is this segment that can employ the 'less skilled', a trait that is said to characterize most of the Indian workforce, and hence seen to generate employment for the 'common people'.

Demographic Profile

Fifty-five per cent of the 50 workers chosen at random are women with only minor variations in the proportion across firms. In fact, a study by Ramesh comes up with a figure of 40 per cent which is even less (Ramesh, 2005). Interestingly, this ratio is less when compared to that found elsewhere. Studies undertaken in Europe and Canada reveal that women account for a larger proportion of the workforce (70–72 per cent in Canada according to Buchanan and Koch-Schulte, 2000: v; 70 per cent in Europe according to Belt et al., 2000b: 4). Thus, even as this sector em-ploys a larger share of women as compared to other sectors in India, their share when compared to other countries appears to be less.

Another distinct demographic feature of the workforce is its age. The average age of the sample workforce is only 22 years. The relatively nascent stage of the sector is probably the major factor behind this dominance of young employees apart from the fact that it requires few skills nurtured in other sectors, especially manufacturing. However, an important difference that needs to be noted is that call centres in the West also employ a number of senior persons on temporary contracts. The oldest firm studied being not more than 4–5 years old, most employees are inexperienced, with a maximum of three years of work experience. Even employees at higher levels in the organization have work experience outside this sector and have entered only recently. Among the workforce studied, there were eight employees who are in the 28–35 age group, whereas 15 of them are in the 17–21 age group. Primarily, employees in the medical transcription firm studied constituted the former segment. We find that there are quite a few employees in this firm with a few years of experience either in pharmaceuticals or in healthcare sectors. In the call centre case firms, however, there were only five employees in that age group out of the total 40 interviewed. No sex-wise differences in the age profile can be discerned.

The younger age profile also implies that most of them are unmarried. In fact, out of the 50 interviewed at random, only four women and two men were married. Interestingly, time and again throughout our interviews, a recurrent note on marriage hindering work in this sector came up among the women employees.[2] To men, however, marriage does not seem to pose such restrictions. Even among the 10 employees employed in senior positions, seven are unmarried.

The next important characteristic of the workforce is the income and social groups they come from. The workforce is highly urban-centric. All the employees hail from urban areas, metropolitan, in fact, in most cases. Only in the domestic call centre, we found a few employees hailing from non-metropolitan but big towns. All the employees interviewed but for a couple of them in the domestic call centre, hail from higher income groups. Invariably, parents are employed in professional occupations either in the government or private corporate sector, a sector that accounts for only seven per cent of all employment in the economy. This is quite unlike the profile of employees elsewhere. In the call centres in Canada, for instance, there are a lot of poor immigrants employed and in the Caribbean, women employees in back office processing centres opt for work either in garment factories or in these firms and hail from poorer and less elite backgrounds (Freeman, 2000). In India, the workforce that enters garment factories and those that enter ITES firms belong to entirely different socio-economic backgrounds.

Weaving these aspects together, a few observations stand out. One, visions of creating more diffused and broad-based employment as compared to software services, for a less-skilled workforce is not supported by facts. Urban elites dominate the workforce. This is true especially for the women employees. This exclusivity implies that the sector demands a set of attributes/skills that are not distributed across all segments in society. Even within such a narrow segment from which the sector draws its workforce, there are further barriers to entry like age. Even though the younger age profile can partly be explained by the nascent stage of the sector, barriers to entry into the workforce for older segments of the population seem to exist. The age-related barrier, it appears, is also tied to the marital status of employees, especially, women. That such entry barriers exist poses important questions about the constitution of 'skill requirement' and the factors behind such a skewed employment pattern.

Recruitment Processes, and What Do Employers Look for?

Studies of work in a servitized economy in general, and in call centres in particular, indicate the need for what are referred to as 'soft' skills and use of 'emotional labour' (Belt et al., 2000b). Soft skills essentially refer to inter-personal skills like ability to communicate and convince clients. Emotional labour involves the capacity to 'publicly display an emotion that they may not necessarily feel' (Wharton, 1993: 28, cited in Belt et al., 2000a: 14). An ability to 'smile down the telephone' is seen to vest more with women. Given the 'genderedness' of such skills, women are expected to find better employment and growth prospects in the new economy.

Belt et al. (2000b: 42) present a hierarchy of skill requirements in a declining order; communication skills, customer service, product knowledge and IT skills. However, there are differences in levels of product knowledge. Customer support services for hardware producers, for example, warrant a higher level of technical knowledge than one that caters to banking customers. Communication skills rank high in all call centre segments. In India, this need for communication skills translate into differing requirements depending upon the clientele that the firm caters to. Since catering to foreign clients, especially in the USA and the UK, is a major growth segment, understanding of English language in addition to an ability to follow the respective accents and respond in the same becomes very important. In the call centre that provides support for hardware, recruitment was done solely from a pool of engineering graduates.

Selection to the lower levels, that is, agents or customer care representatives, is done through a written test followed by an interview. While the tests seek to examine the knowledge of English language along with their IQ and an ability to react to situations quickly, interviews seek to primarily evaluate the communicative capabilities of the candidate. Though such skills need not be confined to only the formally trained/educated, firms do have restrictions on the minimum formal educational qualifications. While most firms insist on a minimum of a Bachelor's degree, a few firms insist only on completion of school. HR managers too do not hint at any preference for women employees or those with 'sweet' voices as cited by Buchanan and Koch-Schulte in the context of Canadian call centres. Employee respondents concede this sex-neutral approach to recruitment.

Among the 50 employees interviewed at random, 47 are graduates, one a diploma holder and the remaining two have completed only schooling. There were not many differences between call centres and other ITES firms like medical transcription and back office processing. Only in the MT case firm studied, there was a qualified medical professional and a couple of employees with a pharmacology degree. We also found that among the 46 employees studied, seven have engineering degrees, and all but one are employed in the hardware support centre.

Employment Contracts and Systems of Remuneration

Unlike call centres in the West where many types of contracts co-exist even within a firm, here, in all the firms studied, employees are employed on a permanent basis. They are put through a year's probation after which they become permanent employees. This makes them eligible for gratuity and provident fund benefits. However, since the export-oriented call centres are located in export parks, they need not comply with exit procedures that are supposed to render labour markets in India rigid.

Remuneration is more in firms catering to global clients and also varies according to the stress and skill levels required. Starting salaries hover in the range of Rs 6,000 to Rs 10,000 per month. Lack of adequate employment opportunities in other segments of the economy due to decline in public sector and inadequate labour absorption in other sectors seem to make employment in call centres attractive. While the quantum of overall income is high, basic pay, which determines other social security benefits like gratuity and provident fund is quite less. This is compensated through performance-based incentives.

Elaborate measurement standards are in place, which enable workers to gauge their performance and strive for additional income through monetary incentives. By the end of first year, most employees earn around Rs 12,000–15,000, the additional income accruing by way of performance incentives. In addition to incentives for individual performance, incentive systems also allow for group performance enhancement. Agents are placed in teams of 8 to 12, each team headed by a team leader. While at times, the clients they cater to differentiate teams, or the kind of work they handle, quite often they perform similar work (Ramesh, 2005). Such a system also allows for work assistance from senior well-trained employees for the younger inexperienced workers. Personnel in higher levels in the call centre firm earn higher amounts ranging from Rs 15,000–20,000 for team leaders to Rs 20,000–25,000 for supervisors and more for managers. The salary levels are relatively lower in both the MT firm and the back office processing firm studied.

Training and Skill Acquisition

The most important characteristic as far as skill requirements are concerned is the distinction made by respondents between what are formally acquired skills and what are seen as personality traits like 'ability to listen', 'patience' and the wherewithal required to be pleasant when confronted by angry or abusive clients. Such attributes, typically labelled 'feminine', render call centre work highly gendered. While to some, this implies a better deal for women as it draws them into paid employment and into the public sphere, this movement also reinforces gendered work roles. The extent to which such participation is likely to undermine existing patterns of gendering in the workplace is something that needs more examination. Apart from such attributes that help perform 'emotional labour', the skill sets of the workforce recruited need to be supplemented by further training within the firm. Training involves primarily familiarity and an ability to communicate in addition to product-specific knowledge.

Quality of Work in Call Centres

Much has been written about the 'sweatshop' reminiscent conditions of work in call centres (Bain and Taylor, 2002). Work consists primarily of two types, inbound and outbound. Employees in inbound work provide after-sales service over the telephone, a substantial portion of which does not require face-to-face interaction or physical presence on the customer's site. Outbound work involves telemarketing, that is, making efforts to sell a product or service to potential customers. Work schedules in call centres are normally organized into three eight-hour shifts.[3] Since most client firms are located in either the USA or the UK, the night shift is the more important one as most calls are made during daytime in client countries. There are generally no gender differences in working in the various shifts.

In big, established call centres,[4] firms provide their own means of transport to the employees. This mode is resorted to primarily because of the prevalence of night shifts and the apprehensions that parents in India have about sending their daughters to work at night.

As soon as a caller dials, she is taken through a series of automated steps that not only identifies the caller but also answers a few basic queries. Only when the caller needs further information is the call transferred to an agent. The agent then listens to the caller's query and responds accordingly. In the case of inbound work, calls are routed through a system that identifies agents who are free and directs the calls to them. Once the call reaches the agent, with the help of computer-telephony integration (CTI) systems, information about the caller pops up on the screen as soon as the caller identifies herself.

The responses are increasingly routinized. Agents are provided with 'scripts' that gives the agent appropriate responses to be made to a set of questions. Only in three to four per cent of the total calls received do queries go beyond the script, which warrants exercise of the agent's discretion to respond. Many a time, such occasions call for intervention of the team leader or supervisor. In sum, work for agents in call centres involves very little use of one's discretion or ingenuity. Rather, it requires them to put on a demeanour that may often conflict with what they actually feel at the moment. As an employee who has completed a year's work in the domestic call centre responded poignantly, 'Work here has taught me to be patient. I mean I have become more patient even outside the call centre. In the initial days, I used to be quite aggressive with clients because I am normally that kind of a person. I used to get angry soon and argue with my friends and at home. But within two weeks after joining here, I was warned by the supervisor that I can't be rude to customers! Since then, I have changed a lot. I try to empathize with the customers and realize that had I been in the customer's position, I too would be angry like him.' Interestingly, even firms highlight this ability to fashion attitudes in the workplace! 'I see a definite positive change in Varun since he has started work here. He's so much more responsible and refined as a person... Varun has done us proud,' says the uncle of a customer support executive in a recruitment advertisement released by a leading ITES firm (*Times of India*, Bangalore, 23 November 2005: IX).

High-end work like technical support service providers for, say, hardware or software definitely requires more utilization of technical skills acquired, though routine solutions to most problems are available here as well. Work experience here definitely brings in a degree of learning, not matched by work in other kinds of call centre firms. Also, work here is less stressful, as more time is allowed for each call, and a respondent even cited instances of 'making friends' with some of her clients.

Employees in back office processing or in medical transcription firms do not exercise any kind of 'emotional labour'. Nevertheless, work here too does not require use of one's judgement or intelligence. Scripts are, once again, provided to back office processing workers. For instance, work that involves examining whether a particular applicant is eligible for a credit card or whether a claimant is eligible for insurance claims, and if eligible, the amount he/she is eligible for, are all standardized. Employees are expected to follow a series of standard checks to ensure that the claim is valid.

In call centres, customers often tend to be upset over the quality of product or service delivered, and agents find themselves at the receiving end of their ire. Or else, as some respondents point out, when customers realize that the call is being attended to from India, they are reluctant to divulge their personal information required for identification. Their reactions are often tinged with racism, refusing to talk to an Indian and insist on being put on to an American agent. Even under extreme provocation, employees can ill-afford to react angrily. They can only explain politely that the customer is not being helpful by being angry or rude. Such bottling-up of emotions can prove to be a frustrating experience, especially in the initial months, after which, respondents say they learn to cope with it. This is also a period that witnesses a lot of turnover.

A frustrating aspect of such moments is their total lack of responsibility for the mistakes made and their inability to redress the caller's grievance as at best they can only inform others that things have to be rectified. A crucial component of their work, it appears, includes acting as an intermediary to absorb negative reactions from customers. Stress at work is further compounded by the lack of any time gap between two calls. During busy hours, agents just move from one call to the other without any let-up, forcing one respondent to say 'At times, I feel like cutting off my ears and run. Even when I get back home, I can't stand listening to others talk!'.

In the case of outbound call centre work, employees make 'cold calls', that is, random calls to potential customers to sell them a product or a service. To that extent, they do have control over the time taken over and between calls. At the same time, here, since the agents make the calls and not the customer, often, the customers react angrily for being disturbed in between important work. So, the chances of encountering irate customers are much more than in inbound calls. It is for this reason that more respondents preferred to work in inbound calls than in outbound calls as when customers call, they feel that they are easier to 'handle'. In outbound work, employees are also expected to meet targets like the number of successful calls made in proportion to the total calls made, amount of time spent per call, and so on.

Medical transcription firms are given only 24 hours time period for the entire audio data on tapes to be transcripted and sent. Apart from delays due to quality and problems associated with specific accents, there are frequent transmission problems that render sticking to the deadline a tough task. At times, the data received contains too much of noise or may be delayed due to loss of transmission links. On these occasions, employees are expected to stay on to complete the work.

Technological innovations only serve to routinize and intensify work. In call centres, the major trend has been towards routinization of work on the one hand and increasing efficiency through automation on the other. While such automation definitely implies liberation of workers from routine work, other innovations have only reinforced the routinization trajectory. Development of elaborate scripts restricts exercise of one's intelligence or creativity. This trend is likely to increase in future as more and more dimensions of service delivery are routinized.

In medical transcription firms, there has been a movement from hand-operated mechanism for playing back or pausing the audio player to one that enables employees to use their feet for this purpose. The latter technique almost doubles the speed of a transcriptionist's work. In the case of back office processing too, scripts are available which take an employee through successive steps to the final evaluation. We need to note that all these innovations have to be complemented by innovations in communications infrastructure that enables firms to outsource such work. In sum, it does appear that work in ITES firms is subject to deskilling, an observation that undermines claims to a new, enriching work culture in the informational economy. Or else, automation may further reduce its labour absorption potential.

Disciplinary Practices in the Workplace

While the assembly line epitomises the dominant mode of worker control in the Fordist model of capitalist work organization, the electronic eye can easily be the new metaphor for worker control in the informational economy. ICTs provide management with powerful systems of monitoring and control of both pace and quality of work. It may be true that under different social conditions, similar technologies can empower workers at lower levels in the hierarchy or for that matter, even reduce the need for such hierarchized production structures. In the Indian ITES sector however, ICTs have been used largely to reduce the autonomy of workers and enhance the ability of management to monitor the workforce.

We could not obtain enough information on these aspects in the MT and BPO firms studied. In the call centre industry, various parameters of work performance are continuously measured and available on a real-time basis to management. The time taken per call, the rate of success, the time spent between calls, and so on, are not only available but also used by management to monitor performance and introduce incentive systems as well. Such systems also enable management to reduce the need for supervision and middle management, a factor that may be responsible for low number of levels in the organization. Managers can and do resort to random monitoring of calls. The fact that such a possibility exists also forces workers to put in their best effort all the time. As one respondent said, 'The floor manager can listen to our calls any time he wants. So, we need to be careful all the time about how we talk. It keeps you on your toes all the time'.

Also, such systems allow management to develop comprehensive performance evaluation systems that give a long-term view of an individual's performance. Apart from such technology-supported systems, there are also other managerial practices to enhance performance like putting up periodically in common bulletin boards, the best and worst performing teams, which exert considerable pressure on the various teams to improve their performance. Such practices seriously question the claims to greater worker autonomy and decentralization of power in an informatized economy.

As of now, resistance to such high work intensity seems to be primarily by way of exit. The high turnover is testimony to this phenomenon. Such individualized responses reduce the prospects for collective mobilization. In fact, none of the respondents felt the need for such collective resistance. This, again contrasts with the fair degree of unionization of call centre workers in Europe. In the UK, for instance, 40 per cent of workers are unionized. Unions play a role in ensuring a degree of social security and in the Netherlands, work councils intervene to reduce the intensity of work. They have been able to ensure breaks for employees between each call, to reduce stress and also have limits to the number of calls that can be attended by an employee (Belt, Richardson and Webster, 2000a).

Next, we examine the prospects for a career in the ITES segment through mobility, intra-firm or inter-firm.

Scope for Mobility and Career Prospects

Recruitment calls to this sector highlight its identification with modern careers.[5] However, in the course of our fieldwork, we find that few employees perceive a career in this sector.

The flat organizational structure with few hierarchical levels, however, do not imply higher mobility or decentralized authority. On the contrary, it is a reflection of the ability to exercise greater control over the workforce without recourse to an elaborate administrative and managerial structure. Mobility, intra-firm or inter-firm, is crucial to ensure that jobs in ITES firms can be a part of or lead to a life-long career and not remain as just jobs. Traditional organizational structures offer the prospects of intra-firm vertical mobility with its multiple levels of hierarchy. New organizational forms are purported to be smaller, and hence offer less prospects for intra-firm mobility but increase scope for inter-firm mobility. Even within firms, it is claimed that since levels are lesser and decision-making powers more decentralized, mobility becomes easier. Evidence from the Indian ITES sector strongly contests such positive claims to mobility and career development.

Given the small number of levels in the organization, prospects for intra-firm mobility are highly limited. The lower number of levels, however, does not imply flatter structures. In fact, in the call centres studied, there are a large number of agents and very few numbers employed in the managerial levels. The first move up the organizational ladder is to become a team leader. As there is only one team leader for 10 agents, only one of the 10 agents stands a chance of getting promoted. Given the rapid growth of the sector, promotions do not take a long time, and we find team leaders with even one year of work experience as an agent. Normally, it takes approximately two years for agents to become team leaders. Team leaders then go on to become supervisors after, say, another two years or so. Also, quite often, we observe that supervisors and managers are recruited from other sectors due to the sector's origins in the recent past and consequent lack of experienced human power. When supervisors are recruited from other sectors, they are trained to help them understand the work processes in call centres. Nevertheless, given the few positions available at such levels, the industry offers little scope for vertical mobility within the firm. This scenario is true of both MT and BOP firms as well.

Interestingly, even HR heads are of the opinion that employees cannot work in call centres as agents for too long as they are likely to experience burnout within 3–4 years of working as an agent. While some may continue due to movement into higher positions, working as an agent for too long is extremely difficult. This is even true of employees working in the hardware technical support centre, where, even though mobility prospects are better due to existence of more levels in the organization, the number of such openings continues to remain small. Employment in the ITES sector in India does not constitute a life-long career unlike work in traditional sectors.

Importantly, the sector has little to offer even by way of inter-firm mobility. With the growing number of call centres and a relative scarcity of experienced agents, there is a demand for such agents from new call centres, who lure them with higher salaries. Employees do have the option of moving from one firm to the other, but only within the call centre industry. But more importantly, they are confined to only being agents.[6] Inter-firm mobility is therefore not accompanied by vertical mobility in most cases. Very few (only five) say that work here helps them to build a career outside the call centre industry. The training they receive in communication skills, they feel, does help them to find jobs in public relations field in other sectors. Eleven others saw their jobs as leading to a career within the ITES sector.

To most respondents in call centres, work essentially meant a job that does not hold much for their future. Sixty per cent of the respondents said that they are working only to save some money to join some educational course later. Except two, none of the female respondents felt that they could continue in this sector after their wedding. Night shifts were cited as the major hindering factor. This inability is also evident from the fact that there are very few married women working in the firms studied.

Careers may be a possibility even in the absence of vertical mobility if there is scope for horizontal mobility and salary hikes linked to experience. In the bigger call centres, employees are provided with an option of moving from one client to the other or from inbound work to outbound work. Definitely, such movement offers relatively better quality of work than being stuck to one type of client or one kind of work in a smaller call centre. Respondents do look forward to such changes. There are, once again, limits to the creative content of such horizontal mobility. Earlier, when call centre or service operations were integral activities of the vertically integrated larger firm, employees in the service division did have the possibility of moving to other departments and hence enjoyed better prospects for non-monotonous work. The enhanced inter-firm division of labour enabled by developments in ICTs restricts such opportunities for employees in call centres and other such ITES firms.

Given the strenuous work conditions, deskilled work and limited prospects for a long-term career, it is indeed surprising that the industry continues to draw a large number of workers. As stated earlier, in high-income countries, such work is largely taken up by members occupying lower levels of the social hierarchy with few other employment prospects. Here, however, we do observe that the workforce is drawn from urban elite or middle-income families. To them, such work may not offer sufficient incentive to join the labour force. Our fieldwork, surprisingly, proves to the contrary. Few employees thought that work was strenuous or stressful and hence degrading. In fact, 60 per cent of respondents opined that they liked their work and some even went to the extent of saying that it is challenging and creative. As for the rest, to some, it was stressful work that makes them 'want to throw the headset and run', 'cut off my ears', and so on. To others, it is work and you cannot complain about it. As one of them puts it, 'You cannot crib about work, that is life. You need to cope with it'. Such fatalistic perceptions stand in contrast to the enthusiasm evinced by the rest that includes male employees as well. It is therefore important to understand the way in which perceptions of such work is being constructed in India and the possible factors influencing the process.

Meanings of Work and Production of Consent

In the course of our interviews, we could identify two sources of interest in call centre work claimed by the respondents. One stems from the nature of work content itself and another from its use outside the workplace. To some of the respondents, work in call centres is not strenuous or for that matter, even monotonous. Each call is a new experience of interacting with a new customer and hence not repetitive at all. Further, in the case of outbound work, making a successful call is seen as a challenge by agents, especially male respondents. To quote a respondent, 'I love talking in general, so working in a call centre is fun for me. It is also challenging to sell a product to a customer. Each customer is different. We need to find out the best way to talk to them to convince them. That motivates you all the time.' Seeing work as a challenge is interestingly confined to male respondents, while to women employees who claim to enjoy work, it is fun talking to strangers, helping them sort out problems.

Men are not only well represented, but male respondents do not perceive this work as women's work. Though this dimension needs further examination that is beyond the scope of this study, pointers towards possible explanation can be put forth. One, male employees, confronted with few employment opportunities in other sectors, are forced into work in the ITES sector. Once employed, they may seek to re-encode their work so as to render it into work that men too can be comfortable with. Viewing outbound calls as a challenge may be one indicator of such recoding processes at work. As indicated, only male employees hinted at this dimension of their work. Such perceptions, coming as they do from male agents, imply that gender codes need not be rigid and can be fluid enough to accommodate changes to ensure reproduction of the required labour force. It also shows that work can be 'recoded' to facilitate entry of men into a domain that is historically constructed as women's work. Such recoding has been also recorded by Hensen and Rogers (2001) and McDowell (2003), among others.

The extent to which training programmes aid such recasting of perceptions is not clear. They may well play an important role. Also, since the call centre is part of a new segment providing 'fashionable' work environments, unhindered by the burden of tradition, male employees may find it a lot easier to invoke such notions about work. As a society moves towards a 'post-industrial' phase where work that is 'physical' and 'strenuous', and hence 'masculine' is declining, employment opportunities are available only in the service sectors that demand both intellectual and emotional labour. By attributing intellectual content to jobs that are traditionally seen as involving only emotional labour, men may seek to recode work so as to facilitate their participation in the 'new economy'. What we witness in call centres may be an indication of such new gendering processes under new capitalist work conditions. Also, such recoding questions some feminist positions that envisage a liberatory potential in new technologies as they privilege skills that are seen to be vested more with women.

Consent for such stressful and monotonous work is also secured by the ways in which such work facilitates accumulation of cultural capital outside the workplace. In India, knowledge of spoken English is confined to a small, elite segment of the populace, though aspired by large segments. The knowledge brings with it therefore, a sense of elite belonging and hence, high social status. Further, the growing globalization of some segments of the economy places a large premium on such language communication skills. Trained to communicate fluently in British or American English, call centre employees find themselves in an enviable position, equipped with a skill that enhances their social position outside the workplace and helps them habitate exclusive spaces.

Firms, through the organization of employees' leisure reinforce such identity formation. Regular parties are organized apart from picnics for employees. In one call centre, housed in an international technology park along with other software firms, dance parties are held on the last Friday of every month. Liberal dress codes also impart a sense of freedom to employees, evident in the oft-cited remarks by respondents, 'It is like a college campus'. Or, 'I love working here, it is fun.' One respondent who had moved out of a call centre and works as a secretary to a research team elsewhere said that she misses all the fun. 'Here, I don't have anyone to talk to. People are not in my age group or my wavelength. But there, we used to be eight of us hanging out together. Every weekend, we used to go out for movies, shopping'.... Only because my mother didn't want me to work in night shifts, I had to quit.'

Also, none of the employees whom we interviewed felt the need for formation of trade unions in this sector. Trade unions would spoil the growth of the industry, was one common refrain. Another reason was that a cordial work atmosphere exists between managers and the agents. A report by a newspaper too confirms this (*Business Line*, 7 November 2005: 1).

The re-creation of work in call centres as one filled with fun appears to play an important role in drawing or retaining labour. Relying on a young workforce from fairly well-off families too aids this process. They see this job as only a temporary one that can help them earn a bit of money to meet their immediate consumption needs as they continue to be dependent upon their parents and family for other reproductive needs. Disposable incomes are also higher as their transport requirements are taken care of by firms. It however, needs reiteration that the latter and even the formation of new consumption identities are true only for bigger, export-oriented call centres and not in other kinds of ITES firms.

Conclusion

There is a strong deskilling trend in the call centre industry. In a nascent sector, drawing labour from new sources and hence, not yet collectively organized, capital has been able to introduce technological innovations with a clear deskilling trajectory. Even here, there has been automation of routine tasks to a certain extent though the routinization of work does overarch such tendencies. Given the fact that the industry is only a few years old in India, it is difficult to discern any tendencies towards skill polarization. Nevertheless, the creation of such sectors with low-paying, low-skilled and dead-end jobs by itself may reflect such polarization processes at work at a macro-level.

Call centre firms are characterized by low levels of middle management. Unburdened by any previous organizational baggage, capital has been able to fashion organizations that use advances in ICTs to almost eliminate the need for any middle management. Interestingly, such organizations are not flatter than conventional bureaucratic firms, but more hierarchical with power less diffused in the organization. Thus, the scope for disintermediation and worker autonomy stands thwarted by virtue of capital's ability to wield greater power in the workplace.

Efficient monitoring and controlling ICT-based systems are built, which leave little scope for worker autonomy. Lack of any resistance on the part of workers to such intense surveillance is a major factor in strengthening the trend.

The growth of the ITES segment, at least some segments of it, by itself is conditioned by reliance on external labour to fulfil in-house requirements. The rise of 'third party' call centres which provide services to clients in the US or the UK is representative of ICTs' ability to enable firms to rely on distant labour pools for their service provision. However, within such 'third party' call centres, we do not find firms relying on external labour. This is due to two factors. One, and the most important, is the need for quality control of service delivery. Second, by having access to a large set of internal employees, firms have the advantage of deploying them in various segments or across different clients as per demand.

ICTs do undermine some of the conditions that are seen to have hitherto excluded women from entering into the labour market, like reduced need for physical labour. However, such possibilities do not automatically translate into greater avenues for women's employment. There are more important institutional barriers like social construction of work. The ITES sector, however, employs a significant share of women, largely because of the fact that attributes required for such work like 'emotional labour' are seen to vest with women. It must also be mentioned that gendered construction of work is quite fluid and can transform over time. To this extent, ICTs do have the potential to enhance the employment prospects of women. Simultaneously, we also observe that even work in call centres is being recoded as 'masculine' work by the male employees in the workforce. Efforts to use ICTs as tools to promote women's employment need to take into account the social construction of gendering processes in the workplace.

Notes

1. This paper is largely based on fieldwork undertaken for a study on 'Labour Market Implications of Diffusion of ICTs in Developing Countries' funded by APDIP and IDRC. The study, completed in 2003, was also supported by a research grant from the Foundation for Urban and Regional Studies, University of Esssex, UK.
2. The reasons for this phenomenon will be discussed in a later section.
3. 24/7, interestingly, is the name of a successful call centre in Bangalore.
4. By 'big', we refer to call centres that have 100 or more seats.
5 'Choose the New Age Profession', 'Start a great career with X...Discover a new life', are examples of typical recruitment calls by ITES firms in India.
6. Of late, HR heads hint at getting into agreements with other heads to prevent poaching agents from each other.

References

Bain, P. and P. Taylor. 2002. 'Ringing the Changes? Union Recognition and Organisation in Call Centres in the UK Financial Sector', *Industrial Relations Journal*, 33(3): 246–61.

Belt, V., R. Richardson and J. Webster. 2000a. *Work Opportunities for Women in the Information Society: Call Centre Teleworking*. Centre for Urban and Regional Development Studies, University of Newcastle upon Tyne, UK.

———. 2000b. 'Women's Work in the Information Economy: The Case of Telephone Call Centres', *Information, Communication and Society*, 3(3): 366–85.

Buchanan, R. and S. Koch-Schulte. 2000. *Gender on the Line: Technology, Restructuring and the Reorganisation of Work in the Call Centre Industry*. Ottawa: Status of Women Canada.

Burawoy, M. 2000. *Global Ethnography: Forces, Connections, and Imaginations in a Postmodern World*. Berkeley: University of California Press.

Chengappa, R. and M. Goyal. 2002. 'Call Centres: Housekeepers to the World', *India Today*, 27(46): 46–58.

Fine, B. 1998. *Labour Market Theory: A Constructive Reassessment*. London: Routledge.

Freeman, C.F. 2000. *High Tech and High Heels in the Global Economy: Women, Work and Pink-Collar Identities in the Caribbean*. Durham NC: Duke University Press.

Hensen, K.D. and J.K. Rogers. 2001. '"Why Marcia You've Changed!"—Male Clerical Temporary Workers Doing Masculinity in a Feminized Occupation', *Gender and Society*, 15(2): 218–38.

McDowell, L. 2003. *Redundant Masculinities: Employment Change and Working Class Youth*. Oxford: Blackwell Publishing Ltd.

Ramesh, B.P. 2005. 'Dynamics of Service Work in the Information Society: Work Organisation, Controls and 'Empowerment' in Call Centres', paper presented at the International Conference on New Global Workforces and Virtual Workplaces: Connections, Culture and Control, 12–13 August 2005, National Institute of Advanced Studies, Bangalore.

14

Ethnic Transnational Middle Classes in Formation: A Case Study of Indian Information Technology Professionals

Xiang Biao

Introduction

The spectacular growth of India's Information Technology (IT) industry has attracted much attention worldwide. The then Indian Prime Minister Atal Bihari Vajpayee (1998) remarked that '*India's Tomorrow is IT*'. The former US President Bill Clinton suggested that if he were the Indian prime minister, he would put promoting the high technology industry and education before the India–Pakistan tension and Hindu–Muslim conflict as his priorities (Joshi, 2003). In discussing IT's impacts on India's development, existing literature either treats IT as a technology, concerning its contribution to increasing productivity and efficiency; or treats it as an industry, assessing its implications for employment and revenue generation. This chapter moves away from these perspectives and focuses on another aspect, namely the emergence of IT professionals as a special social group.

Though it has been widely recognized that IT and the 'New Economy' are human capital-based, little has been discussed on how the human capital is produced and appropriated globally and what social consequences it may have. In the Indian context, IT as a technology has so far not been applied to such an extent that would enhance India's overall development capability. As an industry, IT has very limited forward and backward linkages. What deserves more attention may be the fact that the astonishing amount of wealth created by IT is confined to a very small population. While the gross revenues from the IT sector accounted for more than 12 per cent of GDP generated from the services sector which itself accounted for 50 per cent of the total GDP, its share in the total workforce was only .02 per cent (Chandrasekhar, 2005). Examining how the IT professionals have emerged as a new social group may offer a fresh angle for understanding IT's impacts on India.

This chapter demonstrates that India's position in the global IT industry has relied on its function of producing a large number of IT labour; and Indian IT professionals become a distinct group both in India and globally by establishing a special association with global capital, particularly through the 'body shopping' scheme. The body shopping scheme is critical for the establishment of a flexible IT labour force on a global scale, for the development of the IT industry in India and also serves as an arena where Indian IT workers move up in their career. Ethnic networks play an important role in this process. Ethnic networks here are not only a cultural matter, but reflect the position of India as a nation in the international economic system. Through the body shopping scheme and ethnic networks, Indian IT professionals on one hand pump local surplus value to global system and, on the other hand, they accumulate their own resources on transnational circuits and bring new development dynamisms to India. On how to balance these two sides should become a concern for the Indian government and development agencies.

I conceptualize the Indian IT professionals as part of 'ethnic transnational middle classes'. 'Transnational middle classes' (TMCs) cover a wide range of professionals such as employees of multinational corporations, international business solicitors, accountants, 'head hunting' agents, petty bourgeois involved in international businesses, language interpreters and so on. Their collective interest lies in serving the internationalization of production, capital circulation and trade. They are the key players to translate economic globalization into factual social relations, yet they have no direct possessions over global capital. 'Ethnic transnational middle classes' (ETMCs) are those TMC members who have originated from developing countries. I call them 'ethnic' because they are connected to global capital in ways distinct from those of their 'mainstream', particularly Anglo-Saxon, counterparts. Furthermore, this fact is consciously recognized and reinforced by both the Indian IT professionals themselves and others, which process I refer to as 'ethnicization'. The ETMCs are vital for global capital not only as cheap skilled labour, but also as agents facilitating the penetration of global capital to the developing world. They are not only in the 'middle' between capital and labour, but also between global and local.

In the remaining part of this chapter, I will first establish the fact that IT professionals have become a special social category in India. They are special because they are 'transnational', that is, they are employed by global capital and global mobility is part of their working life. Then I point out that it is due to the dynamisms of global capital that this transnational labour force has emerged. This is followed by a description of how the Indian IT professionals are managed transnationally through the body shopping scheme to meet the demand of global capital. The section after this is devoted to the process of 'ethnicization', exploring the question why body shopping has been largely a uniquely Indian phenomenon. Finally, based on the empirical evidence, I elaborate the concept of ETMCs and propose related action agenda.

This chapter is based on my 22 months' fieldwork in Sydney (January 2000–June 2001), Andhra Pradesh (AP) from June to September 2001, mainly in the capital city, Hyderabad, and New Delhi (September to October 2001). In-depth interviews and participant observation were the main investigation methods. I had open-ended interviews with 180 IT professionals from relevant institutes and lived with about 20 Indian IT workers in Sydney and Hyderabad. I also spent 10 days in three villages in West Godavari (a district in AP) from which a number of IT professionals originated. Besides first-hand data, I accumulated considerable information through documentary study, particularly by collecting media reports. In order to maintain privacy and confidentiality, most of the individual interviewees' names mentioned in this chapter are pseudonyms.

'Computer People'

In the daily life of urban South India, 'computer people', a colloquial reference for IT professionals, have become a special social category. The common image of 'computer people' is that they are rich, competitive, and most importantly, can go overseas anytime. An IT professional in South India earns, on an average, two to five times more than a non-IT engineer. In 1999, the entry salary for IT professionals was around Rs 10,000, which is almost the highest payment that a senior non-IT engineer in big companies can expect. Stories of how IT professionals went to the US and dramatically changed their whole families' lives are widely transmitted. The Non-Resident Indian (NRI) account, a special account set up by the Indian government in the early 1970s to attract foreign currency deposits from overseas Indians, is an important indicator of an NRI's social status. A middle-aged informant talked about young IT professionals working in the US: 'A boy, 22 or 23 years old, can deposit Rs 40,000 to 50,000 every month in his NRI account for his family! In dollars! For rich families, it used to be (that) rupees make rupees, now rupees make dollars. Is there any business better than this?' An IT groom of the Kamma or Reddy caste working in a big company is expected to receive a dowry of Rs 10–50 lakhs[1] in Hyderabad, more than double that of a non-IT engineer. If the boy is US-based, the figure could go up to US$ 120,000 (Rs 80 lakhs).

It is no surprise that an occupation offering opportunities of going overseas became something special in India. What intrigued me is that 'computer people' also became a special category in the Indian community in Australia. Laxman, a 39-year-old software engineer working in Nortel, described to me the life of Indian professionals in Sydney. Revealing his Brahman background, he chose his words carefully in the beginning:

> There are two...kinds of Indians here. IT people and other conventional technicians are different. People in Liverpool[2] are stuck. They can't go back (to India). If we go back, the life is similar. But the Liverpool people can't even maintain a two-wheeler there. From India to Australia is a big jump for them [...] We don't mix too much. They ('conventional technicians') stick together, who found a new job, who took an IT course, (they) gossip all the time. They can't see beyond the corner where they live. Real IT people have their professional line. We are really looking at the world [...] You can say these are two classes!

Ganga, a young Telugu doctor in Saudi Arabia whom I got to know through an online chat room, asked me to hunt for brides in Australia for him so that he could migrate there. Ken, a Tamil civil engineer in Sydney, was busy searching for grooms for his daughter. When I told Ken about Ganga, he turned it down immediately: 'No doctors.

We are looking for computer people only'. The reason is simple. While an IT professional's Permanent Residency and job in Australia are guaranteed, doctors are subject to years' re-examination before settling down. Besides, with three to five years' experience in Australia, an IT professional's income can easily be double that of other professionals. I did not pass this on to Ganga. Several weeks later, Ganga told me that he was thinking of taking an IT course in order 'to get out of the present situation and move on'.

Thus, what matters is not the distinction between migrants and those staying in India, but also the distinction between IT and non-IT. IT professionals with secure positions, be they in the US, Australia or India, are viewed in largely the same way. The IT professionals are regarded special because they are 'transnational' due to their connections with global capital and their global mobility.

First, the IT professionals' economic status is based on the fact that they are hired by global, rather than domestic, capital. Sky-rocketing wages in IT, rising 25 per cent a year in the late 1990s, have mainly been driven by Multinational Corporations (MNCs). The multinationals typically offer salaries 20 per cent higher than the average to attract a new workforce when setting up their Indian operations (interview, also see Lateef, 1997). In Sydney, an estimated 75–80 per cent of the Indian IT temporary migrants (457 visa holders) and at least 60 per cent of all the Indian IT professionals (except students and intra-corporate transferees) work in multinational companies or Australian companies with stable overseas operations (such as Telstra). The Australian government encourages temporary immigration of IT professionals because this migration stream is seen as a crucial means of attracting foreign investment. In highlighting the importance of the long-term work (457) visa, the Immigration Minister made it clear: 'If Bill Gates were to arrive in Australia tomorrow to establish a new enterprise, he would be able to get as many IT staff as he wanted on long-term temporary visas' (Ruddock, 1999: 10). Thus, wherever they are, the IT professionals are working in the same global market and are paid by the same global capital.

Second, global mobility is an important part of the IT professionals' working life. Two months before moving to Canada, Panika, an IT worker in his thirties, explained to me why he decided to do so:

> It is my fundamental idea to go further [...] You have to do survey often (by browsing the Internet) to see where you are. The way we think is: what's the position you will have in two years? You have to follow the

latest technologies, and know what salary they are getting. We have to think it in a big way. In three years' time, I may go to America. IT people are never happy with one place. This is IT professionals' train. The plan is always open. The train can go anywhere.

According to a report from Canada (Badrinath Rao, 2000: 4), 29 out of the 47 Asian IT workers surveyed planned to migrate to the US, compared to two out of eight Canadian-born workers who expressed their desire to do so. A common estimate by my informants was that as many as half of Indian IT professionals in Sydney plan to move to the US and about 20 per cent had done it. An informant in New Delhi likened IT professionals to e-mails because both of them 'can go anywhere in the world'. Multinational work experiences are critical for an IT professionals' promotion in career, particularly when one wants to move up to senior positions. One reason that Indian IT professionals are keen to move to the US is that the US is thought to have a high 'returning value', that is, an IT professional with US work experience is better valued than others when returning to India. In AP, the dowry rate for male IT professionals is closely associated with the country where the groom is based. Most families that I visited had good knowledge of the incomes and expenditures in the US and some also know about the UK and Australia well.

Global Capital and Transnational Labour

IT and Global Capital

In explaining the emergence of the transnational labour, particularly the temporary skilled migrants whose numbers increased dramatically worldwide throughout the 1990s, research and public media commonly point to technical reasons such as the demand generated by the new (IT) industry, universal computer language and internationalized work standards. However, a closer look suggests that the new dynamism of capital accumulation at this stage was at least as important as technological change. Though the global IT industry came into being in the 1960s, it was only in the late 1990s, when IT was widely applied in various businesses in the form of software and particularly when the Internet was commercialized, that the movement of IT professionals became a global issue. This is because at this stage, IT became a crucial tool for the 'flexible accumulation' (Harvey, 1989) of global capital.

First, business-integrated IT makes 'deterritorialized' production and management technologically practical. E-commerce is arguably leading to a 'tariff-free' economy and enables capital to pursue profits globally through free trade regardless of its origin and size. A White House document states: '[f]or over 50 years, nations have negotiated tariff reductions [...] [T]he Internet is truly a global medium, it makes little sense to introduce tariffs on goods and services delivered over the Internet' (White House, 1997). E-commerce software solutions exported from India were worth US$ 500 million in 1999–2000 and they were expected to reach US$ 10 billion by 2008 (NASSCOM, 2000).

Second, IT became the major arena where global capital concentrated, particularly through the stock market Nasdaq. Starting trading in 1971, Nasdaq surpassed the New York Stock Exchange in the mid-1990s to become the largest stock exchange in the US. Dominated by high-technology firms, Nasdaq is also the first fully 'electronicalized' stock exchange and literally absorbs money from all over the world. Nasdaq transmits real-time quote and trade data to more than 1.3 million users in 83 countries (Nasdaq, 2001). In terms of the source of the capital, the most globalized fraction of capital—finance capital—was the first major commercial investor in IT and remains to be a critical consumer. According to a McKinsey report (cited in Jain, 2002), banks in the US invested more than US$ 5,000 per employee in IT, or around 12 times what the rest of the private sector did. Software for the banking sector was the focus of most Indian software developers in the early 1990s (*Financial Times*, 1995, 1). Other established global companies such as airlines are also major clients of the IT industry.

Third, the IT sector has been a major generator of global capital throughout the second half of the 1990s. California has overtaken Italy and France to become the world's fifth largest economy, mainly due to its IT industry (Computer Today, 2001: 25). Bill Gates' personal wealth in 1998 was more than the GNPs of 85 nations and more than the combined GNP of the 29 poorest nations. Using a mere 40 per cent of his wealth, he could wipe out India's fiscal deficit (*Magnet*, 15 November 1998). The rosy expectations about IT attract further investments. In sum, IT is expected to fully liberate the spirit of capital, namely to maximize itself by circulating freely.

These dynamisms of global capital have led to a peculiar 'virtual shortage of labour' in the IT sector. 'Virtual shortage' refers to the situation where employers experience skill shortages but cannot prove the shortages to be 'real'. Industrial bodies in all major developed countries have produced astonishing estimates of IT-skill shortages. Australia reported a shortage of 31,500 for 1999; 89,300 for the period 1999–2002; and over 180,200 by 2004 (Australian IT&T Skills Task Force, 1999). Even in mid-2001, when recruitment froze and redundancy had become a common practice, the Information Technology Association of America (ITAA) still projected that 425,000 IT jobs would be unfilled in the US during 2001–02 (ITAA, 2001). Major software companies have formed powerful lobbying groups to push the liberalization of migration regulations for IT professionals, particularly the schemes of the H-1B visa in the US and 457 visa in Australia. The US introduced H-1B visa in the early 1990s, specifically for highly skilled temporary migrants, with an annual quota of 65,000. The quota was raised to 115,000 for fiscal years 1999 and 2000, and in 2000 the quota was further expanded to 195,000 for the following three years. More than half of the H-1B visa holders are IT professionals.

On the other hand, there has been a vociferous argument that the 'shortage' is a myth created by the industry in order to import cheap labour from overseas. In the US, H-1B workers are reported to be paid 25 per cent lower than the average workers and are not paid any compensation for overwork (Bacon, 2000). A Congress-commissioned study in the US (Committee on Workforce Needs in Information Technology et al., 2001) found no evidence of an overall shortage of IT workers in the last half of the 1990s. Wages in IT occupations grew only a little more rapidly than wages of other highly educated workers (Appelbaum and Rouse, 2001). In the IT labour market, only about 2 per cent of software job applicants were accepted by employers in the US even in the peak time (Matloff, cited in Manfred, 2001). Big placement agents (see the following) in Sydney admitted to me that their rejection rate is higher than 90 per cent in processing job applications, something that one would not expect for a really tight labour market. There have also been cases wherein H-1B workers displaced permanent US employees or US employees were even required to train H-1B workers to replace themselves. It is also important to note that most reports on shortages are based on surveys conducted by large placement agents and industry bodies.

This debate on the IT labour shortage actually reveals a basic fact: the IT industry relies on not only labour, but on cheap, internationally mobile, and highly qualified labour who can be hired and fired anytime. John Ridge, President of the Australian Computer Society, argued that the skills shortage should not only be addressed, but should be over-addressed. According to him, an increase in supply would result in more demand as more business opportunities and projects become possible locally. Multinationals will base their operations in Australia when an oversupply is created (cited in Woods, 2001: 25). In response to a proposed bill to reduce H-1B visas, Bill Gates remarked: 'If you want to prevent companies like ours from doing work in the US, this [bill] is a masterpiece' (Mills, 1995, cited in Stalker, 2000: 135). A cheap and flexible labour force is indispensable for the global IT industry for the following reasons:

First, as opposed to what is usually assumed, both software package development and software services are highly labour-intensive. The work normally includes four stages: designing, programming, implementing and testing. Designing, particularly developing a large package, is a process full of uncertainties. The most successful software development models over the last 30 years have essentially dealt with this by an extensive—rather than intensive—growth, that is, through large increases in the number of software labour rather than through increases in process productivity or quality through rationalization (Eischen, 2000: 31). 'Even the most celebrated success stories [...] rely on large amounts of unpaid labour to overcome the fundamental inefficiencies and bottlenecks of the software process itself' (ibid. 33). Programming (coding) consists in translating a design into codes that are understandable to computers. Testing implies finding out and removing errors in the programme designed (the error is usually called 'bug' and testing is accordingly called 'debugging'). These two stages are even more labour consuming. The work is so monotonous that it is sometimes called 'donkey work'. These two job areas are where the majority of Indian IT professionals have concentrated in so far.

Second, software is essentially an 'enabling' industry, that is, software creates value only by enabling other operations. In order to enable other practices, detailed understanding of the client's needs and the specific environment where the software programme is to be installed are critical. Implementation and testing are, by definition, to be done on-shore.

Recently, an increasing number of companies in the US and India have adopted the Business Process Outsourcing (BPO) model. Outsourcing may contribute to the stabilization of IT professionals' mobility in terms of absolute volume, but the offshore model is unlikely to substitute for the mobility altogether. A key reason for this is that software development is more an 'informational' rather than a 'technological' process. A technological process means that the process is embodied in coded knowledge, while an 'informational' process comprises continuous communications and constant mutual adjustments. With the growth of BPO came the significant increase of the usage of the L-1 visa, a visa for intra-corporate transferees, by Indian companies to send their workers to the US. Between 1998 and 2002, the number of L-1 visas issued jumped by 50 per cent and climbed an additional 10 per cent in the first five months of fiscal 2003 (US State Department, cited in Grow, 2003). In early 2003, TCS, Infosys and Wipro, the three largest Indian software companies, had respectively 2,500, 1,000 and 500 employees working in the US on L-1 visas (Grow, 2003).

Third, the volatility of the IT industry, which is inevitable due to its close link to global capital, forces companies to hire cheap labour in a highly flexible way. In March 2001, gloomy profit reports from Cisco System and Microsoft spurred global panic in the IT area. Nasdaq share prices fell nearly by half, compared to early 2000. The major casualty has been of the 'dot-com' companies, which had been at the forefront of the IT boom. Three hundred and thirty dot-com companies closed down in the US during the first half of 2001 (Nag, 2001). Dot-com, once dot-bomb, became dot-tomb. As a result, dot-com companies in the US announced 98,522 layoffs, particularly of those with Java skills, through November 2001 alone (*The Times of India*, 30 December 2001). Many of them were migrant workers and some had to move on to new countries to search for new jobs.

The following two sections detail how the Indian IT transnational labour is managed to meet the needs of global capital, particularly through the scheme of 'body shopping', and how the body shopping scheme is 'ethnicized'. These two sections reveal that the Indian IT professionals become transnational and 'ethnic' labour not only because the IT industry is a global industry and Indian culture is powerful, but also because the global IT sector is based on the established unequal world economic system at the macro-level and transnational mobility and ethnic connections are taken on by the workers as a way to cope with the high uncertainty in the global labour market at the micro-level.

The Body Shopping Scheme

Inexpensive yet qualified labour is so important for the global IT industry that to some extent, a new international division of labour has been developed under which some societies specialize in producing the labour as a special commodity. India is so far the most typical country that fills in this position. During the period from May 1998 to July 1999, nearly half of all the US H-1B visas were issued to Indians [US Immigration and Naturalization Service (INS), 2000]. Of all the computer systems analysts and programmers on H-1B visas in the US, 74 per cent were born in India (INS, 2000). In the UK, Indians made up 63 per cent of all foreign IT professionals entering in 2000 (Khadira, 2001: 51). IT training in India has become a booming industry, with the revenue increasing from Rs 33 crore in 1994–95 to 125 crore in 1998–99 (NASSCOM, 2000). According to my field data, about Rs 420–680 crore was channelled to IT training in AP through private channels over 2000–01, a large part being surplus value from the rural society (Xiang, 2002).

The newly-developed Indian IT industry is more a result than a cause of the large number of IT professionals. India's IT industry is based on and driven by exports of IT services, which made up more than 70 per cent of the total revenue of the IT industry in the end of the 1990s (10.5 per cent of India's total exports) (NASSCOM, 2000). The exports are essentially the exports of labour. According to one estimate, the Indian IT industry would need to send about 200,000 professionals to the US alone to sustain its 40–50 per cent growth rate over the year 2000–05 (Ramesh, 1999). The H-1B programme is so vital to Wipro's expansion plans that it listed the possibility that the visa programme might be curtailed as a major risk factor in disclosures to the Securities and Exchange Commission when the company floated an initial public offering on the New York Stock Exchange in 2000 (Thompson, 2001). The expected cutback in the number of H1-B visas was thought to 'send shock waves' through the IT sector in India (Malhotra, 2003). In turn, Indian IT professionals establish wide overseas business connections through migration, which is a crucial asset when they set up their own IT businesses back in India later on (Arora et al., 1999).

The most typical way by which Indian IT professionals enter the global market is through a special scheme known as 'body shopping'. Body shopping is the practice whereby a firm ('body shop') recruits IT workers in India and then farms them out to clients for a particular project, though the firm itself is not involved in the project. Unlike the conventional recruitment agents who *introduce* employees to employers, body shops *manage* workers for the employers. Workers managed by body shops do not form formal employment relationships with the employers and can be laid off anytime. In the migration documents, the workers managed by the body shops appear to be inter-firm transferees, that is, those who are sent by the body shops to clients to carry out certain tasks. Body shops in different countries are associated with each other. They send workers from one place to another when required. By doing so, a flexible labour management system on a global scale is achieved. There were a total of 35–60 Indian body shops in Sydney by late 2000 managing 1,500–2,000 contract workers. In the US in 2000, there were estimated to be hundreds of agents specializing in temporary Indian IT workers in Northern California and perhaps over a thousand across the country (Lubman, 2000).

Body shops do not act alone. They are associated with a series of recruitment agents of different sizes, including big agents such as Ernst and Young, Anderson, IT Manpower and ICON and Candle in Australia, and form 'agent chains'. In the chain, body shops focus on recruiting workers from India and managing the workers in the destination countries, particularly when they are unemployed. The key characteristic of body shops is their 'benching' practice, which the bigger agents normally do not adopt. Doing 'bench' means sponsoring a worker to move to a new country without any job opportunity in existence, though the body shop claims so when applying for the sponsorship. Therefore, the workers have to wait 'on the bench' after they arrive. During the bench period, workers are not paid or are only given a nominal 'bench salary'. At the same time, body shops rely on bigger agents to find jobs for the workers. The IT industry has been so dependent on the placement agents, as a result of its pursuit of labour flexibility, that IT companies need agents who fully understand their overall business to manage their comprehensive labour tasks. Only big agents have access to major clients. In Sydney, the biggest 5–7 agents in the IT area are estimated to take 40–60 per cent of market share. Big agents are willing to work with body shops because it is highly cost effective compared to recruiting workers from overseas directly in short notices from their clients, and this also frees big agents from managing the workers.

Thus, workers who are brought in by body shops find jobs typically through three or four agents. In this situation, the worker is still sponsored by the body shop. But this body shop does not contact or even know anything about the real employer. The real employer signs a contract with the last agent only and sends the worker's salary to the last agent. Each agent along the placement chain takes away a part of the worker's wage as its profit and sometimes the total amount taken away is as much as half of the original pay. Indian IT workers who are managed by body shops have an average annual income of about A$ 30,000 in Sydney, after considering the parts taken away and the unpaid benching period. This is far below the prevailing level.

The agent chains partly explain a puzzling phenomenon, namely the co-existence of the reported high demand for labour and the high rate of short-term unemployment and low payment of the migrant IT workers. As an Indian H-1B worker in the US put it, the body shopping scheme is similar to the way airlines overbook flights, to have workers available in case there is need (cited in Kirby, 2000). One of my informants put it this way: 'The companies need a big reserve of workers. But it is very expensive to keep all the workers for one big agent. Then these body shops came in. Each consultancy keeps a couple of workers. Every one can afford it.'

Ethnicization

The body shopping scheme is largely a uniquely Indian phenomenon. There are many Chinese, Filipino and Brazilian IT professionals, but the practice of body shopping has not been found among them. The scope of the body shopping scheme is so large that it is partly responsible for an 'ethnicization' of the entire global IT labour force. Twenty-three per cent of Microsoft's workforce has ties to India, whereas Intel, IBM and Oracle have 21, 19 and 18 per cent, respectively (cited in Bentsen, 2000). Lucent Technologies in the US hired so many Indians that it even has an Indian Cultural Association within the company. An informant told me that when one of his friends applied for a job as an accountant in the US, she was given a job as computer programmer. The employer said to her: 'You are Indian, you can do this'. H-1B visa is also sometimes referred to as 'Indian visa'.

This process of 'ethnicization', that is, an objective process whereby disproportionately large numbers of Indians cluster in the IT sector, as well as a discursive process which emphasizes the special connections between India and IT, takes place from both 'without' and 'within'. Generally speaking, ethnicization from without refers to effects of other groups, particularly the mainstream society, of making an ethnic group, with a typical example being 'labelling' (for example, Mitchell, 1969; Williams, 1989). By comparison, ethnicization from within means that the 'ethnic' actors voluntarily mobilize their cultural symbols and intensify networks to create social closure (for examples, see Cohen, 1969; Watson, 1958). In what follows, I will describe these two aspects of ethnicization of the international IT labour system respectively. The key argument is that the fundamental reason for the ethnicization is the position of India as a nation in the international economic system and global capital's demand for cheap and flexible labour.

Ethnicization from without Paradoxically enough, intensified global mobility and connections have made ethnicity more, rather than less, important in the IT labour market. For example, a common stereotype is that Indian IT workers are good at general skills and particularly the Mainframe technology, but they are not easy to manage, Filipinos are good in the C language and Chinese are good in engineer-related programmes. Placement agents, the key player in the IT labour market, deal with a large labour pool worldwide[3] and have to rely on certain stereotypes to screen applicants quickly. A placement manager told me that the applicant's nationality is one of the first things that an agent checks when processing an application, and he advised applicants to conform with the stereotype in order to have a better chance to get a job. It is significant to note that these ethnic stereotypes are largely nationality-based and have little to do with everyday cultural practice (such as language). Tamils, Telugus, Punjabis and Gujarati Muslims, sometimes even Pakistanis and Bangladeshis, are all lumped together as 'Indians', and are thought to be similar in terms of their skill capability and work style.

My own survey suggests an association between ethnicity and occupational rank in the IT sector. A disproportionately high percentage (20–30 per cent) of the AP-origin IT professionals with permanent residence in Sydney worked in the public sector, where jobs were easier to obtain, more secure, and involve less managerial tasks. The overwhelming majority of my interviewees, in both the public and private sectors, were at the middle level, such as Project Leader and Senior Programmer, and only 1–5 per cent of the Indian IT professionals were estimated to be among the highest 20 per cent positions. By contrast, 85 per cent of Indians are at the middle levels, compared to the average proportion of 53.8 per cent (see Table 14.1). Earlier surveys by *Dataquest* (March, 1989, cited in Lakha, 1991: 32) and Lakha (1991) reached similar conclusions.

Apart from covert discrimination (for example, placement agents commonly turn to the US and the UK for 'white' candidates when

TABLE 14.1 **Structure of IT Work-force in Australia, by Responsibility Level (percentage)**

	Average	*Indians*
Level 1	6.1	10–15
Level 2	17.1	—
Level 3	28.8	85
Level 4	27.0	—
Level 5	12.0	1–5
Above Level 5	9.0	—
Total	100	100

Source: Australian Computer Society, Remuneration Survey Report 1999. The Association of Professional Engineers, Scientists and Managers, Australia. Interviews by the author.

they recruit senior managers because workers in Australia are said to be uncomfortable with 'brown' bosses), the new organizational pattern in the high-technology industry, characterized by horizontal network instead of clear hierarchical structure and the constant mobility of staff, makes tacit knowledge of communication more important for developing social relations and promotions at workplaces. Indians often lack the skills to communicate well with British, American or Australian supervisors. Indians are also thought to be 'nerds' without marketing and management skills that are crucial for senior positions in the IT industry. After the September 11 attacks, ethnicity became a particularly salient issue. Many US firms ordered criminal background checks on ethnic employees, both on the rolls and prospective, particularly those working in the areas of Internet security and network-system support (Vidyasagar, 2001). I-flex Solutions and TCS explicitly asked their staff working in the US to stick to a Western dress code (Karkhanis, 2001).

In the recent heated debates on outsourcing and immigrant IT workers, particularly in the US, India has been a standard target. When the German government planned to allow 20,000 non-European Union IT workers in the country in May 2000, anti-immigration groups in Germany shouted *Kinder statt Inder* (children instead of Indians) (Wallace, 2001). A quick browse of the websites of the most active organizations against importing foreign skilled workers in the US, such as the Institute of Electrical and Electronics Engineers Inc., USA, Communications Workers of America, International Federation for Professional and Technical Engineers, Washington Alliance of Technology Workers, and the Programmers' Guild, India is repeatedly singled out as the country responsible for US workers' job loss. In November 2002, after being replaced by IT professional sent by TCS, a US employee set up a site called www.hannatroup.com with one section devoted to revealing the names and contact addresses of 'Our Indian Replacements from Tata Consulting India'. Anti-H1-B movement also entered universities recently. A group of Stanford students who call themselves 'H1B non-believers' handed out fliers where Indians and Asians were identified as the main problem, and this triggered considerable controversy in the campus (Kim, 2003). Mainly in response to this growing sentiment, NASSCOM, the national representative of the Indian IT industry, has engaged the public relations firm Hill and Knowlton to work with businesses and politicians in the US to promote the positive image of the Indian software industry (Rai, 2003).

Ethnicization from within Far more important than ethnicization from without is ethnicization from within, namely Indian migrants, industry bodies and the government voluntarily, and increasingly in an orchestrated way, highlight the linkage between 'India' and 'IT'. Indian business organizations across the world (such as the NASSCOM, The IndUS Entrepreneurs, originated in Silicon Valley with 40 locations around the world, and the Indian IT Forum in Australia) have been working hard to promote the image of 'Indian IT talents', or, as the Indians call it, the 'brand'. A body shop-cum-software development house in Sydney was seeking to have long-term working relationships with journalists. The owner complained to me that the Indian IT Forum in Sydney has not done enough in 'propaganda' (that is, publicity) and he suggested the lack of an 'Indian IT profile' in Australia was the reason why body shopping there was not as successful as in the US. The government in India, both at the federal and state levels, also join this promotion. A typical case is Mr Chandrababu Naidu, the Chief Minister of AP. He called himself the CEO of AP and bid to turn Hyderabad into a world centre of software development, 'Cyberabad'. Using IT as the key leverage, Naidu has been particularly successful in promoting a public image of AP's IT sector on the international stage. In 1996, he secured a talk with Bill Gates[4] and eventually per-suaded Microsoft to set up a research centre in Hyderabad, Microsoft's first one outside the US.

Ethnicization of the IT labour system is also pushed by the larger migrant politics circle. In 1998, a group of Indian IT workers in the

US set up the Immigrants Support Network (ISN). By July 2001, it had about 22,000 members and had raised US$ 74,000, mostly for the lobbying efforts for a more liberal skilled migration regime (Kirby, 2001). Norman Matloff, a computer science professor at the University of California at Davis, commented that ISN's leader 'has bought into the feeling, currently prevalent among many India-immigrant programmers, that Indians are the saviours of Silicon Valley' (ibid.). In 2001, several Indian Americans ran for state legislative seats in New Jersey for the first time. With 'emulating the Jewish American community' as a declared goal, political activist Vinit Dhruva pointed out that Indians need their own people in Congress to protect the H-1B visa programme (Ginsberg, 2001).

More fundamental than the efforts on publicity, ethnicization from within is driven by the body shopping scheme's reliance on ethnic networks. Ethnic networks play three roles here. First, ethnic networks mobilize potential IT labour force in India to join the global market. For instance, connections with overseas Indians are instrumental in running emigration-oriented IT training courses in India. These connections send home the information of the latest global market trends and technologies. Ethnicity-specific discourses and perception, such as those related to dowry, caste and social status, also play an important role.

Second, the co-ethnic relations between body shop owners and workers enable body shops to circumvent certain state regulations in recruitment (see the following section). Body shops never bring in non-Indian workers unless there are genuine job openings ready for them. When these non-Indians (mostly Filipinos) are laid off, body shops still pay them part of the salaries and the bench time is recorded as a vacation or medical leave. One body shop in Sydney once sent a few local Australian IT workers to the US, but this practice stopped immediately because the white workers refused to share bedrooms with Indians. It is far more costly to manage non-Indians than Indians, particularly given the highly uncertain labour market situation.

Third, ethnic networks enable body shops to form associations among themselves across the world, which is essential for managing IT workers' global mobility. Interestingly, the most important way for body shops in Australia to establish connections in a new country is through India. Body shops and IT firms in India often put Australia-based body shops in contact with body shops in other countries. For workers, before moving on to a new country from Australia, they typically go back to India to attend IT courses to upgrade their skills and consult body shops there for more information about the world labour market. India is not only a source country of IT labour, but also a monitoring centre for its IT professionals' global mobility.

These three functions of ethnic networks, namely labour mobilization, internal management and external expansion of the body shopping scheme, are also crucial for the reproduction of the body shopping business, making the scheme sustainable. A fundamental institutional background of the body shopping scheme is the friction between the State and corporate interest regarding the immigration of skilled labour in developed countries. According to Australian Commonwealth migration regulations, a firm should not sponsor an overseas worker until a job offer is finalized, and the firm has a series of legal obligations to the worker throughout the sponsorship period. This is meant to protect local labour and minimize the possible burden arising from the international mobility of labour to the state welfare system. But this clashes with corporate interest which requires a completely smooth flow of labour. As shown earlier, the agent chains and body shopping scheme circumvent these State regulations and create a labour pool where corporations can select and dispose of workers anytime, yet without producing costs to the state. The frictions between the State and the corporate interest created body shops' business niche, but also imposed serious difficulties on body shops. First of all, body shops must prevent workers from leaving them soon after their arrival in order to make more money out of them. However, it is certainly unpleasant for a worker to be on the bench or receive a salary only half of the original pay (after various agents cutting their margins), particularly when there is an acute demand for IT professionals in the market. Second, body shops must prevent workers from raising complaints to the government that will bring about severe punishment. Workers have several grounds to do so. For example, the government regulations obligate body shops to pay the workers once they are brought in regardless they are working or on the bench.

The body shopping scheme can be sustainable in this circumstance because of three inter-related reasons. First, workers expect that the time in body shops is only a transition and they will move up to a better position, such as being managed by a large placement agent or obtaining a permanent residency in Australia, through this. Workers rarely stay with a body shop more than two years. An interesting phenomenon in the body shopping practice is the role of 'resume forwarders'. Resume forwarders are the middlemen between body shops and workers. They help workers who want to migrate to Australia, or to change jobs or sponsorships in Australia by forwarding their resumes to body shops. All the body shops are connected to resumes forwarder and take on this as an important way to recruit reliable workers. When conflicts occur between the workers and the body shops, the resume forwarders may intervene in the body shops' favour, though the workers are often connected to forwarders through personal networks. Resume forwarders are mostly workers who were previously brought to Australia by body shops and some of them are still managed by the body shops. They make small money by charging the workers or the body shop, or both. But more importantly, they plan to set up their own IT businesses in the future, typically by starting with body shopping, and take on resume forwarding as preparation for this. The role of resume forwarders vividly reflects a reproduction mechanism of the body shopping scheme, namely former employees of the body shops move up to help the body shops and then further move up to set up their own body shops.

The second condition for the sustainability for the body shopping scheme lies in the large numbers of new Indian IT professionals whom body shops can recruit, manage and make profit out of constantly while the old IT workers move up. The huge IT labour pool in India who are eager to emigrate is the most important basis for body shops' business. Apart from cutting margins from the workers' salaries as profit, body shops often charge a large amount of money from the workers when they offer sponsorships for going abroad. In 2000, some body shops in Hyderabad charged Rs 1–3 lakhs for sending a worker to the US; Rs 1–2 lakhs for sending to the UK, Germany and Australia; and Rs 80,000–Rs 1.2 lakhs for Singapore.

Third, the body shopping scheme can be sustainable because a particular body shop can move up into more stable businesses, such as software development. Almost all the body shop owners that I interviewed wanted to have genuine IT technology firms eventually and they saw body shopping as the first step. Body shopping is an easy way to accumulate the initial capital since it does not require large amount of investment and have low financial risks. But given the difficulties in 'dealing with people' (the workers) and the dangers resulted from their often unlawful practices, the owners are very keen to diversify their businesses. Body shops achieve this typically through an 'overlapping' model that once again relies on ethnic networks.

Body shops overlap their businesses in two senses. First, different businesses overlap within one body shop. A body shop may do placement, project development, and particularly, IT skill training business, simultaneously. When a worker is on the bench, the firm may use the worker as trainers in its training courses. Invariably, these training courses in Sydney target Indian immigrants who arrived in Australia after the 1970s, for example, as mechanical engineers, and suffer from unemployment or underemployment now due to the shrinking of the manufacturing sector. The body shops advertise their courses through Indian newspapers, posters in Indian groceries, or by setting up stalls in ethnic functions. Interestingly, all the training institutes stress that they use Indian IT curricula. Curricula from India were thought to be better than Australia's because the Indian version is of the US standard. Students trained with Indian curricula are not only more competitive in Australia, but can also move on to any other country.

Second, a group of companies in different countries overlap through a series of joint projects. Suppose a body shop in Sydney has associates in Singapore, India and the US. Its associates in India and Singapore help the firm recruit workers for the placement business and search for clients for its project development. The Sydney body shop does the same for its associates. All the associates are Indian companies of the same nature. Therefore, based on ethnic networks, a group of associates works together like a single MNC. When body shops move into the businesses of software services or development, they again rely on low-cost Indian labour as a main source of their competitiveness. In the US, India-based firms or US tech firms run by Indians made up 10 of the top 25 companies who had the largest number of H-1B workers in 2000. Most of these H-1B visa holders were from India (Thompson, 2001).

Discussion: the Concept and Action Agenda

This chapter has attempted to establish, both empirically and analytically, that Indian 'computer people' have become a special social category. Empirically, the chapter demonstrates that IT professionals are regarded special in the everyday life in India, Indian IT professionals are special because they are 'transnational labour' and ethnicity plays an important role in the body shopping scheme and even in the entire international IT labour system. Analytically, the chapter clarifies that the emergence of the transnational labour is due to global capital's demand for a cheap and flexible labour force, and that the 'ethnicization' of the international IT labour system reflects India's position in the global IT industry as a major producer of IT labour. This also distinguishes migrant Indian IT professionals from the earlier South–North migrants. While other migrants' transnational connections are derived from migration processes, Indian IT professionals' transnational connections are based on the dynamism of global capital and their physical mobility is only a reflection of their transnationalism.

I conceptualize the Indian IT professionals as part of 'ethnic transnational middle classes'. There has been some literature on transnational class formation in the context of globalization, but it has mainly confined to the capitalist class. For example, dependency theorists such as Barnet and Mueller (1974) and Hymer (1979) proposed the concept of 'international capitalist class', and more recently, Sklair (1996a,b, 1998, 2001) and Robinson and Harris (2000) called attention to 'transnational capitalist class' (TCC). According to Sklair, TCC is a ruling bloc consisting of a range of interlocked positions and they control the accumulation of capital on a global scale. TCC comprises four groups: Multinational Corporations' executives and their local affiliates; globalizing bureaucratics; capitalist-inspired politicians and professionals; and consumerist elites. Superficially, Indian IT professionals fall into the category of 'capitalist-inspired professionals'. But clearly, these IT workers are not part of the global ruling power bloc. Indeed, Sklair's definition of TCC may be too broad. First, economic globalization is increasingly a technology-intensive process. A large group of technical workers, such as employees of multinational corporations at the middle and junior levels, are indispensable to globalization, yet they have no direct influence on the process. Second, globalization is not only about flows and links, but also about penetration, rooting and mobilization. The classical way by which MNCs expand business is to set up factories and sale outlets in developing countries which needs to mobilize both labour and consumers there. Agents in these countries, such as local representatives of MNC and small- and medium-sized enterprises involved in international businesses, are needed for these tasks. Body shops and IT-enabled services (ITES) providers (such as call centres) in India, contractors of US companies in China, and managers of Japanese-invested factories in Malaysia, all belong to this group and they are clearly not part of the TCC.

I hypothesize that economic globalization is bringing about 'transnational middle classes' (TMCs). This category covers IT professionals and other groups as mentioned earlier. Similar to national 'middle classes', TMCs are internally highly heterogeneous. But they are in the same position and assume the same function in the process of globalization. TMCs' collective interest lies in serving the global capital, rather than bounded local or national economy, but they have no active possession[5] over global capital. They tend to advocate liberal economic ideology and are cynical about governments' regulations in general.[6] TMCs are not only in the middle between the capital and labour, but also between global and local. They facilitate global capital's penetration to the local society, supervise local labour process on behalf of global capital, and mobilize local resources to serve global capital.

ETMCs are the TMC members from developing countries. Calling people from developing countries 'ethnic' is clearly problematic since in theory anyone, be it an Indian or a white American, has his/her ethnicity. However, in reality, 'white' is 'no colour', and 'ethnicity' is an issue only for those who are out of the mainstream. Institutions such as Commission of Ethnic Affairs are set up for the minorities. This chapter deliberately adopts this problematic but conventional usage to reflect this reality. The term ETMC indicates that the TMC members from the periphery countries are in a different position in the global labour market compared to their counterparts from developed countries. Apart from my own field data as mentioned earlier, Table 14.2 illustrates the obvious income disparity between the ETMCs and Western TMCs in Singapore.

TABLE 14.2 **Household Income of Expatriates in Singapore by Nationality (percentage)**

Nationality	*Less than S$ 5,000*	*S$ 5,000–10,000*	*S$ 10,001–15,000*	*More than S$ 15,000*
'Asian category' (including Japanese)	38.5	46.2	15.4	0
'Western' category	12.5	37.5	18.8	31.3

Source: Yoeh and Knoo (1998: 184, Table 2).

The concept of ETMCs also attempts to highlight the importance of their home countries. As this paper has demonstrated, the international IT labour system is ethnicized because ethnic networks consistently channel new labour from India to the global market and ethnic networks enable those who have entered the global market to upgrade themselves into a better position. This leads to an important question, namely, why India has produced so many software engineers. In explaining this, existing literature often points to the factors such as the large English-speaking population and the government's long-standing emphasis on higher education. However, a crucial factor being widely ignored is the highly skewed class and gender relations in India that enable surplus value to be effectively siphoned off from the low end of the society to produce IT professionals. For example, mothers' and wives' free labour at household, which is essential for producing any labour, is never counted as part of the costs of producing IT professionals. All the middle- and small-sized IT companies that I visited in Hyderabad hired office boys to serve tea, buy lunch and mop the floor. They are usually 8–15 years old and are paid Rs 500–800 per month. It is these women, children and many other men's invisible or undervalued work which make the costs of producing Indian IT labour so low in monetary terms. Transnational and ethnic networks also help the Indian IT professionals cope with market uncertainty: they can stay on in the destination country through body shops' document when they lose jobs,[7] share rooms with co-Indians to minimize living expenditure and obtain support from their extended families back in India. It is therefore hard to assess how—an 'untouchable' female maid, a Kamma H-1B man, or a white venture capitalist—contributes the most to Silicon Valley's glory. But the return from the IT sector is not evenly distributed at all. It is in this sense that I suggest that the ETMCs pump local resources up to serve the global system.[8]

Despite this fact, Indians of low classes are already declared to be irrelevant in the process of ethnicization in the global context. Mr Kanwal Rekhi, former chief technology officer at Novell Inc. and widely regarded as a role model for Indian IT professionals, recently became an advisor of the ISN because he held that skilled migration programmes must be liberalized at the expense of the family reunion programme. He said:

> When immigrants were first allowed in the 1960s, they were engineers and highly skilled people. Then, there was family reunification [...] primary immigration of professionals became secondary immigration of taxi-drivers and non-professionals. That secondary immigration was of very poor quality, and that caused a backlash. For one engineer, you would get 10 others. (Din, 2001)

As mentioned earlier, even ordinary engineers are regarded as a different 'class' from the IT professionals. Indeed, an important current agenda of many ethnic Indian associations outside of India is to promote the image of Indian IT professionals and distance themselves from low-skilled migrants. This reflects an intriguing intertwining between ethnicity and class formation in the time of globalization. Therefore, not only economically but also socially and ideologically, the emergence of the IT professionals may contribute to the development of a new 'dual structure' in India: a global sector where resources are accumulated in US dollars and a local sector which provides surplus value, and is yet excluded from economic globalization.

Having said this, the action agenda that I would like to propose for policy makers and development agencies is not to stop the emergence of ETMCs, but rather to balance this particular group with the larger society. A central agenda advocated by the leftists in the face of globalization has been to launch a labour- and people-centred 'globalization from below' to counter the 'globalization from above' of capital and the corporate power (for example, Tilly, 1995; Breitenfellner, 1997; Waterman, 1993), or to 'stop' economic globalization altogether by way of 'global resistance'. These agendas have gained wide popularity, as evidenced by the repeated 'anti-globalization' movements in various world cities in the last few years. However, the feasibility of this agenda is doubtful. As I showed earlier, even IT workers, who are well connected and are most capable to form global alliance, are highly fragmented along the line of nationality (say, native IT workers in the US versus India). It is hard to imagine that the really poor and disadvantaged can form a powerful 'global resistance'. Furthermore, the experiences of the Newly Industrialized Economies (NIEs) show that close connections with the global market have been critical for their development. These successful NIEs' global links do not lead to a global–local 'dual structure' mainly because they have been able to diffuse the benefits from the global links to a larger society, thanks to the rapidly developed infrastructure, less skewed class relation and a much higher level of primary education. Therefore, what the Indian government and development agencies should do is, perhaps, to focus on 'local articulation' rather than join 'global resistance'. 'Local articulation' means to enhance the linkage between the ETMCs and the other social groups by way of, among others, investing in primary education, developing small-sized technology firms and promoting ITES that have great potential for employment generation. These may be more important than helping the IT industry to move up in the value chain. One thing that should and can be done immediately is to inform the public, and particularly the IT professionals themselves, of the fact that India's IT success is essentially based on the labour of the seemingly irrelevant masses, and therefore to fight the elitist developmental ideology before it goes too far.

Notes

1. In line with most documents in contemporary India, this chapter uses lakh and crore to measure rupee value and million and billion for dollar value.
2. Liverpool is a large suburb in South-western Sydney. Most Indians in Liverpool are mechanical technicians.
3. A middle-sized agent in Australia, Ixp3 Ltd, has resumed for upto 17 million people worldwide and about 500,000 in Australia (Mulcaster, 2001).
4. According to Naidu, the appointment with Bill Gates was finalized through the then US Ambassador to India (Naidu, 2000: 7–9). But a few informants told me that some Telugu IT professionals working with Bill Gates helped. There is no evidence to support this speculation. But the perception of the ubiquitous presence of Telugu people in the world IT industry and of its influence is interesting in itself.
5. Active possession means that a person has both ownership and managerial power over a certain asset. An individual who holds shares of a multinational corporation through stock market, though is an 'owner' in theory, cannot exert any significant influence on the corporation's management of capital in practice, and therefore is not a member of TCC.
6. It is a strong consensus among the IT professionals that India's IT industry performs well precisely because it is free from government's interference. A manager of an IT firm in Hyderabad argued that the industry should be tax-free forever, since the industry 'does not take anything from this country'. When asked about income tax policy towards IT professionals, a high-level official who is an IT technician himself, asked: 'Why should they (pay income tax)? Muslims can have four wives and have ten children. Hindus have one wife and four children. The more children you have, the more subsidies the government gives you. Gives you free food, free houses. Why should they (IT people) pay tax to feed these people?
7. In both Australia and the US, workers cannot stay on employment-based visas once they lose the jobs. But body shops falsely claim that they are still employing the workers and this is an important reason why workers are willing to cooperate with body shops.
8. Paying a huge amount of money to attend Western business schools is another key means by which members from local ruling and middle-upper classes in periphery countries transform themselves into transnational middle classes. This has drained away a significant amount of foreign reserve from these countries.

References

Appelbaum, E. and C. Rouse. 2001. 'Easing Restrictions on Visas Doesn't Help Any High-Tech Workers.' Available at Economic Policy Institute http://www.epinet.org/webfeatures/viewpoints/H-1Bvisas.html.

Arora, A., V.S. Arunachalam, J. Asundi and R. Fernandes. 1999. 'The Indian Software Services Industry: Structure and Prospects', working paper of Carnegie Mellon Software Industry Center, November.

Australian IT&T Skills Task Force. 1999. *Future Demand for IT&T Skills in Australia, 1999–2004*.

Bacon, D. 2000. 'Contract Labor Program by Any Name Hurts All Workers', *AsianWeek*, 4–10 August.

Badrinath Rao. 2000. 'Economic Migrants in a Global Labour Market: A Report on the Recruitment and Retention of Asian Computer Professionals by Canadian High Tech Firms', report for Strategic Policy, Planning and Research, Citizenship and Immigration Canada and Labour Market Policy, Human Resource Development Canada, CPRN Working Paper.

Barnet, R. and R. Mueller. 1974. *Global Reach: The Power of the Multinational Corporation*. New York: Simon and Schuster.

Bentsen, C. 2000. 'Mehta in the Middle', *CIO Magazine*, 1 December.

Breitenfeller, A. 1997. 'Global Unionism: A Potential Player', *International Labor Review*, 136(4): 531–55.

Chandrasekhar C.P. 2005. 'IT-services as locomotive', *Frontline*, 22(13) Jun 18–Jul 01.

Cohen, A. 1969. *Custom and Politics in Africa*. London: Routledge and Kegan Paul.

Committee on Workforce Needs in Information Technology, Board on Testing and Assessment, Board on Science, Technology, and Economic Policy, Office of Scientific and Engineering Personnel, National Research Council. 2001. *Building a Workforce for the Information Economy*. Washington D.C: National Academy Press.

Computer Today. 2001. 'Study Reinforces Recession Fears in US', *Computer Today*, 16–23 July: 25.

Din, S. 2001. 'Kanwal Rekhi Takes up Immigration Issue'. *Rediff News* available at www.rediff.com/news/2001/apr/28us1.htm.

Eischen, K. 2000. 'Information Technology: History, Practice and Implications for Development', working paper series, Center for Global, International & Regional Studies. Santa Cruz: University of California.

Financial Times. 1995. 'Review of Information Technology: India's Software Industry', *Financial Times*, 6 December: I–VI.

Ginsberg, T. 2001. 'Asian Indians Seek Clout with Census Gains', *The Philadelphia Inquirer*, 6 June.

Grow, B. 2003. 'A Visa Loophole as Big as a Mainframe', *BusinessWeek*, 10 March.

Harvey, D. 1989. *The Condition of Postmodernity*. Cambridge MA: Basil Blackwell.

Hymer, S. 1979. *The Multinational Corporation: A Radical Approach*. Cambridge: Cambridge University Press.

INS (US Immigration and Naturalization Service). February 2000. 'Characteristics of Specialty Occupation Workers (H-1B): May 1998 to July 1999'.

ITAA (Information Technology Association of America). 2 April 2001. 'Major New Study Documents Demand for IT Workers Continues'. Press Release. Available from http://www.itaa.org/news/pr/PressRelease.cfm?ReleaseID=986219529.

Jain, S. 2002. 'Not All IT's Cracked Up To Be', The *Indian Express*, 26 January.
Joshi, M. 2003. 'World Can't Afford India to be a Pygmy: Clinton', *The Times of India*, 2 March.
Karkhanis, T.C. 2001. 'Training Firms Eye Overseas Market', *The Times of India, Hyderabad*, 16 July.
Khadria, B. 2001. 'Shifting Paradigms of Globalization: The 21st Century Transition Towards Generics in Skilled Migration from India', *International Migration* 39(5) SI 1/2001: 45–71.
Kim, H. 2003. 'Fliers Claim Int'l Grad Students Take U.S. Jobs', *Stanford Daily*, 12 March.
Kirby, C. 2000. 'New H-1B Visa Law a Life Changer', *The San Francisco Chronicle*, 5 October.
———. 2001. 'Documenting an American Dream', *The San Francisco Chronicle*, 1 July.
Lakha, S. 1991. 'Indian Computer Professionals in Australia', *BIR Bulletin*, 5: 32–33.
Lateef, A. 1997. 'Linking Up with the Global Economy: A Case Study of the Bangalore Software Industry', ILO working paper.
Lubman, S. 2000. 'Labor Contractor of All Stripes Abound in Bay Area and the Nation', *Mercury News*, 19 November.
Magent. 1998. 'Astonishing Facts About Bill Gates!', *Magent*, 15 November.
Malhotra, T.C. 2003. 'US Visa Cutbacks a Blow to Indian IT Industry'. *CNSNews.com*, 19 February. Available at http://www.cnsnews.com/ForeignBureaus/archive/200302/FOR20030219g.html.
Manfred, E. 2001. 'IT Labour Shortage: A Myth?', *The NextBestThing* (tnbt.com), 31 January.
Mills, M. 1995. 'Gates Assails Bill to Curb Immigration', *Washington Post*, 29 November.
Mitchell, C. (ed.) 1969. *Social Networks in Urban Situation: Analysis of Personal Relationships in Central African Towns*, Manchester: Manchester University Press for the Institute of Social Research, University of Zambia.
Mulcaster, G. 2001. 'Headhunters Join Forces', *IT Jobs* (Australia), 6 November.
Nag, K. 2001. 'IT Jobs Are Here Again!', *Times of India, Hyderabad*, 9 July: 16.
Naidu, N.C. (with Sevanti Ninan). 2000. *Plain Speaking*. New Delhi: Penguin.
Nasdaq. 2001. 'About Nasdaq'. Available at http://www.nasdaq.com/about/about_nasdaq.stm
NASSCOM. 2000. 'IT Software & Services Industry in India Grows by 53 per cent in 1999–2000'. Available at http://www.nasscom.org/template/itinindia.htm.
Rai, S. 2003. 'Software Success Has India Worried', *The New York Times*, 13 February.
Ramesh, J. 1999. 'Yankee Go Home, But Take Me With You: Yet Another Perspective on Indo-American Relations', *Economic and Political Weekly*, Special Articles, 11–17 December.
Robison, W.I. and J. Harris. 2000. 'Towards a Global Ruling Class? Globalisation and The Transnational Capitalist Class', *Science & Society*, 64(1): 11–54.
Ruddock, P. (Minister of DIMA, Australian) 1999. 'The Coalition Government's Position on Immigration and Population Policy', *People and Place*, 7(4): 6–16.
Sklair, L. 1996a. 'Conceptualising and Researching the Transnational Capitalist Class in Australia', *The Australia and New Zealand Journal of Sociology*, 32(August): 1–19.
———. 1996b. 'Who are the Globalisers? A Study of Key Globalisers in Australia', *Australian Political Economy*, 38(December): 1–30.

Sklair, L. 1998. 'Transnational Practices and the Analysis of the Global System', paper presented at the Transnational Communities Programme seminar series, 22 May, Oxford University.

———. 2001. *The Transnational Capitalist Class*. Oxford: Blackwell.

Stalker, P. 2000. *Workers Without Frontiers: The Impact of Globalization on International Migration*. ILO/Lynne Rienner.

The Times of India. 2001. 'Dotcoms See Hope Despite Gloom', *Times of India*, 30 December.

Thompson, M. 2001. 'More Bangalore for their Bucks', *eBusiness*, 9 April.

Tilly, C. 1995. 'Globalization Threatens Labor's Rights', *International Labor and Working Class History*, 47.

Vajpayee, A.B. 1998. Speech at Inauguration of Satyamonline, 24 November, Hyderabad.

Vidyasagar, N. 2001. 'Destination US May Not be Easy Any More', *The Times of India*, New Delhi, 5 October.

Wallace, C.P. 2001. 'Germany's New Recruits', *Time Europe*, 25 June.

Waterman, P. 1993. 'Social-Movement Unionism. A New Union Model for a New World Order?', *Review*, XVI (3): 245–278.

Watson, W. 1958. *Tribal Cohesion in a Money Economy*. Manchester: Manchester University press for Rhodes-Livingstone Institute.

White House. 1997. 'A Framework for Global Electronic Commerce', 1 July.

Williams, B.F. 1989. 'A Class Act: Anthropology and the Race to Nation Across Ethnic Terrain', *Annual Review of Anthropology*, 18: 401–44.

Woods, R. 2001. 'Addressing the Skills Gap', *Information Age*, February/March: 24–28.

Xiang, B. 2002. 'Global "Body Shopping": A New International Labour System in the Information Technology Industry', PhD dissertation, Oxford University.

15

Virtual India: Indian IT Labour and the Nation–State

Peter van der Veer

Introduction

The Indian IT sector has shown a phenomenal growth over the last decade. From 1990 to 2005, its output has doubled roughly every two years. The IT output has grown from 0.38 per cent of GDP in 1990 to over 4 per cent in 2005–06. This growth has been based on expansion of exports of services and labour for the global market. Such growth has occurred in a policy environment of liberalization, privatization and globalization. Culturally and politically, IT is seen as the success-story of individual entrepreneurship, set free from the shackles of State intervention, of the 'licence raj'. More than being just a successful industry, the Indian IT sector has become an icon of a new form of development, of bringing India into the 21st century. This is brought out in the extraordinary attention to IT in the Indian media. Especially the gurus (mentors) of the industry, such as N.R. Narayana Murthy of Infosys and Sabeer Bhatia of Hotmail, are seen as the heroes of today's India. It is not only their quick financial successes as self-made millionaires, but their attitudes, beliefs, and the image of the IT industry itself, which make them exemplars in middle-class dreams of endless global opportunities. Globally, the development of modern IT has been welcomed as a revolution, resembling the Industrial Revolution. That revolutionary image has led to visions of a New Economy, guiding huge investments in dot.com start-ups and bringing citizens from all over the world to invest their wealth and dreams in the new industry. The simultaneous production of these hopes and hypes all over the world points to a global modern culture that is integral to the nexus of capital and labour and to that of production and consumption.

The IT industry, the new economy, and the information revolution depend on hypes of various kinds. In Auguste Comte's foundational perspective, the function of the social sciences is to get rid of society's myths. In the field of IT there are many myths, such as the idea that Internet promotes democratization, that the new economy is based on entirely different principles than the old economy, that the IT industry will allow developing economies, such as the Indian, to leapfrog into the 21st century without going through all kinds of reforms that were necessary in other societies. Each of these ideas needs reflection, discussion, and often rebuttal, but at the same time it is perhaps good to remind ourselves that hype and optimism are not only cultural reflections on 'progress', but are also instrumental in furthering this field of activity. The extent to which Indians of all sections of the population 'believe' in the IT industry and in its possibilities for individual and collective advancement is incredible. It appears that religious belief and belief in IT are the two dominant topics in newspapers and general discussion, just as there are software training centres and temples in every small town. The government—national and regional—fully participates in this culture of magical belief and even the big slump in the new economy in 2001 has not affected it greatly. It may be that India is particularly in need of hope and hype. India's development is quite dismal. In terms of size of population and geographical location, it makes sense to compare India and China. In 1950, India was better off than China, but today, the average Chinese citizen earns US$ 860 a year, compared with US$ 460 a year for the average Indian. The Chinese economy has been growing at 8–10 per cent a year for the last two decades, while the Indian economy has been growing at 8 per cent, only over the last decade, and the Indian population growth is twice that of China. To rationalize this differential development out of the way by pointing at 'the success of India's democracy' (as the title of a recent book by Atul Kohli has it) may convince political scientists but no one else, and certainly not in India. The success story of the IT sector is a welcome antidote to cynicism about India's development and the corruption of State institutions.

For many young Indians, the IT industry is seen as a means of social and spatial mobility, just as the Internet is seen as a means of personal emancipation. They think that careers in this industry are based on individual skills and professional achievement rather than on ascription, based on caste and class. The internal organization of the workplace is viewed as demanding, but equal and un-hierarchical. The American ideal of the self-made man, the pioneering hero, is one that is constantly invoked in the IT industry globally, but in India it is specifically contrasted with the Homo Hierarchicus of traditional Indian industry. There are two other related aspects to working in this industry which are particularly appealing. The comparatively high salaries as well as the opportunity to go abroad are attractive. This is related to the global organization of IT work. In the IT industry, Indians get a chance to go abroad, earn an income that is in Indian rupee terms phenomenal, and get an opportunity to stay abroad as NRIs. To be an NRI in especially the US is a dream for many young people in India. This dream is immediately connected to one's position in that other market of desire, marriage. Green card holders are the most eligible of marriage candidates and IT professionals in possession of this coveted document surpass even IAS officers and medical doctors in the amount of dowry that they are able to obtain. Dowry prices have gone up all over India, and are especially steep in the context of such candidates from the IT sector.

In this chapter, I want to address the transnational organization of IT labour. Second, I want to draw attention to the role of the nation–state and especially to some of the ideological or cultural aspects of globalization that are essential to the understanding of India's IT industry.

Body Shopping

The IT industry in India is not fully integrated in the information revolution, but follows an old pattern of providing cheap Indian labour and services to a global market. Much of the media attention goes to high-skilled, IIT-trained engineers who have made it big in the US. The stories about their success are also retailed in business magazines like Silicon India. While they are not untrue, they are only a small part of a complex reality. It has been observed that Graduate Schools of Engineering in the US in the 1970s would have had to close because of too little interest in engineering careers in the US population, if not for the influx of Chinese and Indian students. Between 1985 and 1996, 62 per cent of all the doctorates in science and engineering granted to foreigners in the US were given to Chinese and Indians. This coincided with the Immigration Act of 1965 in the US, which allowed for a much larger quota for immigration for highly trained and educated professionals. The Immigration and Nationality Act of 1990 tripled the number of visas granted on the basis of occupational skills and thus further enhanced the immigration of engineers. These new laws made it also possible for Indian students to stay on in the US. The earlier preference for going to the UK was replaced with a craze for 'green cards'. Nehru's vision of an Indian modernity, which would be created by Indian engineers and scientists and would allow India to be fully self-sufficient, ironically laid the basis for the emergence of a transnational cadre of Indian engineers who would be instrumental in creating wealth in the West (van der Veer, 2003). The very top of the education pyramid in India is formed by the IITs which have a competitive selection that can only be compared to the top echelon of American institutions such as MIT and CalTech. Below that, one finds a whole range of engineering colleges of varying quality. This system of higher education depends on a huge system of pre-college education in which maths and science are emphasized. It is this elite that has become entrepreneurial in the sense that they are crossing boundaries between engineering and entrepreneurship, between old economy and new economy, and of course by being transnational. Combined with the extraordinary emphasis in this field on innovation, breaking away from traditional hierarchies, on youth and on labour flexibility, this kind of person is the exact dream of the 'new man' (homo novus), celebrated in European thought.

An important feature of the high-tech industry is that more than any other industry, it is based on rapid innovation. Engineers have been able to gain so much wealth through their innovations that they also could become entrepreneurs. The information revolution in the US is largely created by engineers who also took control over capital investments. And a large number of these engineers are born in China or India. There is a recent study of Chinese and Indian entrepreneurs in Silicon Valley by Saxenian (1999) which shows that between 1975 and 1990, the foreign-born population of the region doubled to almost 350,000. According to the 1990 census, more than half of Asian-born engineers in the region were of Chinese or Indian descent. By the end of the 1990s, half of the H-1B visas issued to foreign employees in the high-tech industry were to professionals from India. In 1998, 24 per cent of the high-tech firms started in Silicon Valley were run by a Chinese or Indian CEO. Together, they employed more than 58,000 workers. A difference between Chinese-run and Indian-run firms is that the former are more concentrated in computer and electronic hardware manufacturing and trade, while the latter are in software and business services. Saxenian relates that to the superior English language skills of Indian immigrants.

The transnational migration of what sometimes are called 'knowledge workers' has economic effects on both the country of immigration and on the country of origin. Saxenian's study demonstrates clearly that the politically motivated fears in California about Asian immigrants who take jobs away from native workers are false, because these Asians have massively created new jobs. There is a clear awareness in the American IT business community that the arrival of Asian engineers and scientists has been a great blessing for the American economy. The effects of the departure of these professionals on the Indian economy have generally been captured under the negative term 'brain-drain'. Clearly, many more Indian professionals stay in the US after having benefited from Indian education than return. In short, India's brain-drain is the USA's brain-gain. The effects of this can be partly calculated in terms of investments in education versus remittances.

Binod Khadria (1999) distinguishes financial resource flows, technological resource flows and human resource flows back to India and comes to a very sceptical assessment of the benefits for India thus far. However, what he does not seem to realize fully is that transnational connections are fundamental to India's position in the IT industry. Some members of this cadre invest 'back home' in Bangalore, Mumbai or Hyderabad and it is in the transnational nature of their business activities that they become the cutting edge. Those who lead a life- and business-style of shuttling between country of origin and the US are called 'astronauts'. They are nodes in technological–industrial–business networks that facilitate flows of capital, labour and products. Saxenian (2002) interprets this as at least a partial reversal of the earlier brain-drain and calls it 'brain-circulation'. She reports that the majority of Silicon Valley's Indian (76 per cent) and Chinese immigrants

(73 per cent) respond that they would consider starting a business in their country of birth in the future.

However, more important than either brain-drain or brain circulation is the fact that these IIT-trained engineers have been the brokers for making flexible Indian labour available for the IT industry. One of the essential activities in this industry is called 'body shopping'. Contrary to what the focus on IIT-trained engineer–entrepreneurs leads us to believe, software engineering does only partly require high skills. New software development tools like Visual Basic make it possible for labourers with low programming skills to create usable software. One can discern in the software industry a huge labour force of relatively unskilled programmers. The organization of this labour force is in the hands of a number of body shopping companies, including TCS which has 67 offices spanning 18 countries and a global team of 10,000 consultants and NIIT (National Institute of Information Technology) that has a presence in 30 countries. The US is the largest market for software development and H1-B visas are used to bring in software programmers for limited periods. The ethnographic accounts we have of body shopping suggest a strong dependence of workers on the firm, sometimes to the extent of illegal exploitation. In the US Press, one finds with some regularity reports about raids of the police to find undocumented IT workers. By controlling visa and passports as well as housing, finances and actual jobs, the firms can exert a power over workers reminiscent of conditions of indentured labour in the 19th century. The term 'techno-coolies' is therefore perhaps not entirely inappropriate for at least part of this labour force.

The entrepreneurial story is of course a well-known part of the image of the IT industry for especially young Indians. It is the American middle class dream, constantly reinforced by the Indian media. But it is also a dream wonderfully lampooned in the recent movie Guru, where the hero goes to the US to meet his friend who he thinks is rich and drives a Mercedes, but in fact is poor and drives a taxi. That is the reality of techno-coolies. They participate in the hype of dot.com mania, but in fact live a quite limited existence at the margin of the IT industry. The clients make use of body shopping companies to get cheap, flexible labour and pay for that to the company. The company farms the worker out for a particular project, takes a large fee for that (something like half of the salary) and prevents the techno-coolie from being hired on a permanent basis. Even contract workers who have worked for a company for years continue to be 'owned' by the recruitment agent. Especially the Y2K crisis put an enormous pressure on the market for temporary labour during a couple of years at the end of the 1990s. The coolies often live together in a house owned by the company. An important practice is 'sitting on the bench', that is waiting for a client, a period in which one does not earn anything but gets more and more indebted to the company. This kind of practice is central in the body shopping practice, since it creates a large reserve of workers that can be used at any time in a fluctuating market. Mir et al. (2000) report about a 'company apartment' in New York where some of the IT workers were working at minimum wage jobs to buttress a falling income. When they bring their wives, they are granted H-4 visas, which does not allow them to work and reduces them to a very limited existence in the US. While flexibility is a key word in the neoliberal organization of the IT industry, mobility is the key word in describing labour in this context. However, that mobility is very much circumscribed by visa regulations, contracts and debts. In the story of debts, one should also include the investments on education and in obtaining visas. Workers are also willing to pay substantial sums to start with a body shop. All this is based on an optimistic scenario of becoming rich overnight, since one does not hear much about the failures caused by individual collapse or by structural change such as the decline of the dot.com industry. Finally, at the far end of the spectrum, we find South Asian workers at the assembly line as 'low-wage temporary workers' without job security or health insurance. Every computer, printer or electronic device is assembled at production lines. Seventy per cent of the 200,000 labourers working in manufacture in Silicon Valley are of Asian descent (Jayadev, 2001).

Like 19th century indentured labour, body shopping also operates under a legal system that encourages transnational migration. The Indian government is clearly encouraging this sector of activity by providing education to create a labour force and by supporting the jobbers. The system puts people again under limiting legal conditions that are all the more glaring because we are living now under a post-colonial regime. Exploitation of labourers is eminently possible under a system in which they enjoy few rights and can easily be deported.

The general point I want to make is that the IT industry needs all these levels of labour and that these levels are interconnected. The negative effects on Indian economy and society from the emigration of scientists and engineers could, in principle, be reversed by the 'death of distance' inherent in the new information and communication technologies (ICTs,) which makes it possible that the work done by Indian ICT specialists in the US (and elsewhere in the developed world) is also done in India itself. The rise of high-tech sites in Bangalore and Hyderabad, as well as in a number of other cities in India, mainly initiated by US-based or US-returned Indian entrepreneurs may bear that possibility out. This development has great potential which is already clear in the effects of this industry on the Indian economy; but the real issue is whether this industry which is still by-and-large providing low-grade, low-wage, low value-added services to the global market can develop into a more upgraded software industry. Transnational networks of non-resident Indians (NRIs) and their India-based counterparts would be essential for such a development. The social field created by transnational migration would then really grow into a de-centred economic field, enabled by ICTs.

The Nation–State

The system of nation–states with its immigration laws and visa systems forms a central structuring feature of this transnational labour force. Class differences turn out to be crucial not only for the flexibility of labour but also for the flexibility of citizenship. Under what kind of rules does the system of body shopping operate? Immigration laws and regulations in the US and elsewhere remain under constant pressure from the IT industry that demands the raising of the limits. The temporality of contracts and visa is the key to this form of labour migration. By supplying software professionals in time and for the length of time needed, the body shopping firms assist companies to reduce the costs of keeping a permanent labour force (Xiang, 2001). While this is true for all forms of contract labour, organized by recruitment agencies for temporary work, the transnational character of the labour in this case adds the element of temporary immigration to it. This is called flexible labour, but it raises the question of flexible citizenship (Ong, 1999).

Migrant 'legality' and 'illegality' is an epistemological, methodological, and political problem (De Genova, 2002). In the wealthiest nation–states in Europe and America, migration has emerged as the central political issue over the last few years. This has even worsened since 9/11 due to security anxieties in the network society. One cannot separate the Indian labour migration entirely from the more general story of mass migration. Despite the fact that liberalism and also the IT industry itself promises freedom and creativity, it in fact produces at the same time great inequalities and incommensurable interdependencies. In the US, the US Immigration and Nationalisation Service(INS) raids on workplaces have to be announced three days in advance to enable employers to prepare the hiring records. This, obviously, also allows them to fire or temporarily discharge them and to gain complete control over them. Similarly, body shopping firms participate in 'revolving door' policies of keeping migrant labour flexible through a visa system that denies them citizenship entitlements. Policies of citizenship, both in the sending and receiving countries, play a major role in the regulation of migrant labour.

The policies concerning entry visa create huge differences among labourers who are in demand at different levels of the labour market. The highest strata of this system of stratification arrive with legal papers that give in time access to permanent residency and/or full citizenship. The transnational cadre of IIT-trained Indian engineers is firmly middle class. Nevertheless, despite their secure class position, when they want to play a role as 'astronauts', they depend on multiple passports, on green cards, and on special fiscal policies for NRIs as well as on connections with cheap labour. Their entrepreneurial playing field is largely determined by national, international and global arrangements of fiscality and security. Moreover, their activities both in the US and in India depend on the mobilization and immobilization of cheap labour that is dependent on other legal regimes.

India has seen the emergence of a special kind of hyphenated identity: the Non-resident Indian (NRI). The Foreign Exchange Regulations Act of 1973 includes in this category: (*a*) citizens of India living abroad for the purpose of carrying on a business or career, but declaring their intention to stay in India for an indefinite period. (*b*) Persons of Indian origin holding a passport of another country. One is of Indian origin if one has held an Indian passport, or if either of the parents or grandparents was Indian. The wife of a person of Indian origin is held to be of Indian origin too. Neither citizenship nor residence are thus the criteria for deciding who belongs to this category, but 'origin' is and in that sense, it has much in common with the German genealogical definition according to which migrant communities in Eastern Europe belong to the German nation and have the right to return to Germany. One reason for the Indian State to create this category is to raise foreign exchange, since NRIs are allowed to deposit money in Indian banks with competitive, guaranteed rates of interest. However, I would suggest that the main reason is not economic, but political. It is striking that it is not the lower-class migrant labourers in the Gulf region who are the primary targets of this policy, although they are among the migrants by far the most important economic actors in terms of remittances and other effects on the Indian economy. It is also not the older migrant communities of former indentured labourers and their descendants or even the older merchant communities that form the target of this policy. Rather it is the new Hindu middle-class professional and entrepreneurial migrant, in especially the USA, who forms an important focus for Indian politics.

In 1998, the Hindu nationalist party Bharatiya Janata Party (BJP) proposed further changes, such as the introduction of a PIO card (Person of Indian Origin) with a number of benefits attached to it. In 2001, the Indian government, led by the BJP, announced the appointment of an ambassador at large for NRIs and PIOs in the embassy in Washington. It is ironic that a party that derives so much of its political gains from a campaign which stigmatizes the indigenous Muslim community as 'foreign' is so interested in Indians who actually live in foreign lands. Such Indians are primarily perceived as Hindus and Hindu nationalism mobilizes large groups of Hindu migrants all over the world. 'Achievements' like the nuclear explosions of 1998, for example, enhanced enthusiasm under the NRIs in the USA. The announcement of international sanctions against India led to successful fund-raising by the Indian government, under NRIs in the USA. Recent research has also shown how the Rashtriya Swayamsevak Sangh (RSS) and affiliated organizations raise money from the Hindus in the US under the aegis of a purportedly secular charity, the India Relief and Development Fund, which channels its funds to a wide variety of Hindutva-related projects. Even the money raised by this Fund for the victims of the Gujarat Earthquake was discriminately distributed to Hindus only (see www.stopfundinghate.org). Transnational investment, global politics and the cultural capital of 'belonging' go hand in hand here (Kurien, 1999).

As I have argued elsewhere (van der Veer, 1995, 2002), within the politics of space produced by the South Asian diaspora, the articulation of nation and migration forms an important, structural feature. This articulation has been strongly reinforced by religious movements ranging from Arya Samaj missionaries to the Caribbean in the early 20th century, to Sai Baba missions to Fiji in the 1980s and to the

Vishwa Hindu Parishad (VHP) activities today. A striking element of 19th century coolie migration was the importance of religious nationalism in it. Do we find a similar pattern in the current migration of knowledge workers? In his impressive work on the network society, Manuel Castells (1997) argues that while the legitimizing identities of the State are declining in the Information Age, resistance identities and project identities (aiming at total societal transformation) are on the rise. In his view, these identities are produced by social movements which react against three fundamental threats:

> Globalization, which dissolves the autonomy of institutions, organizations and communication systems where people live. Reaction against networking and flexibility, which blur the boundaries of membership and involvement, individualize social relationships of production, and induce the structural instability of work, space and time. And reaction against the crisis of the patriarchal family, at the roots of the transformation of mechanisms of security building, socialization, sexuality, and therefore of personality systems. When the world becomes too large to be controlled, social actors aim at shrinking it back to their size and reach. (Castells, 1997: 66)

Castells' observations are useful, but, at the same time, things look somewhat different when one examines a Hindu movement like the VHP.

What concerns us here are the contradictory faces of the VHP. On the one hand, the VHP is clearly a movement that promotes Hindu nationalism with an anti-secular and anti-Muslim slant and as such it is a movement that continues much of the religious nationalist rhetoric and methods of such movements since the late 19th century. It resists westernization and globalization insofar as they are portrayed as 'foreign' threats to the basic Hindu values of the Indian nation. Muslims as a community signify the 'foreign' as 'the enemy within'. Ideologically, they are portrayed as 'converts', having their allegiance outside of India. The VHP argues that they do not belong to India, but to Pakistan or to Arabia and have thus to be either religiously purified by re-conversion to Hinduism or ethnically cleansed by forced emigration.

On the other hand, the VHP is a movement that is very active globally and one of the prime agents of the globalization of Hinduism. In the USA, it is active since 1974, following sizeable immigration from India. The anti-Muslim politics which is central to its activities in India does not make much sense in the USA. Anti-globalization rhetoric which emphasizes restrictions on foreign capital flowing into Indian companies is conspicuously absent from the VHP propaganda in the USA and rightly so, since its supporters there are strongly in favour of the liberalization and globalization of the economy. As NRIs, they have also direct personal advantage in the free flow of capital. The focus of the VHP in the USA is, as with many religious movements globally, on the family. The great fear of Indian migrants to the USA is perhaps not so much the threat to the patriarchal nature of the Hindu family, since many of these migrants are well-educated professionals and both men and women are income-earners. Rather, it is the struggle to reproduce Hindu culture in a foreign environment in order to socialize their children into the hybridity of Indian-Americans. The fear is often that the children will lose all touch with the culture of the parents and thus, in some sense, be lost to them. Both Internet-chatgroups and youth camps are organized by the VHP to keep Hinduism alive among young Indians in the USA. As Arvind Rajagopal (1997) rightly observes, the VHP needs different tactics, different objectives in different places in order to be able to recruit members. In one VHP camp attended by Rajagopal, the 150 participants were in majority working in the IT sector (Rajagopal, 2000). Instruction included Upanishadic management consultancy.

What is the connection between software engineering and the religious nationalist project of the VHP? First of all, there is no opposition between science and religion. It is important to note that the Hindu nationalism today continues the tradition of 19th century rational religion. The BJP that now dominates the Indian government is very interested in combining the high-level education in techno-science in the IITs with education in Hindu culture. This move to combine Hinduism with techno-science reminds one of similar moves that have taken place in Muslim countries, like Malaysia, where one finds a rise of techno-science education, backed by international corporations, coupled with the internationalization of Islamic higher learning. Very important in this trend is the notion of authenticity (*asala* in Islam) that refers to an acceptance of techno-science while resisting the Western cultural invasion [or Westoxification (*gharbzadegi*), as it is called in Iran].

Science and technology are central to conceptualizations of modernity, both in its secular and its religious variants. The enlightenment notion that there is an opposition between science and religion has been an important element in sociological and commonsense understandings of modernity and secularity, but many religious movements have not accepted such an opposition. It has been an ideological notion in particular secularist movements, especially in Western Europe, but has to be sociologically recognized as such. It is quite striking how provincial the sociological understanding of modernity often has been (van der Veer, 2001). In societies as diverse as India and the USA, 'Enlightenment fundamentalism' (to use Ernest Gellner's term) has only played a very marginal role. There is no reason at all to expect Hindu engineers and scientists to 'lose their religion' and become secular. Hindu modernity includes an ideological valuation of science and technology. The empirical question is rather to explore what religion does for them; what kind of specific needs it produces or addresses. The globalization of production and consumption, including the flexibility and mobility of labour, is addressed by movements like the VHP and is a major element in their nationalist politics of 'belonging'. The idea that 'symbolic analysts' (to use Robert Reich's term) are rootless because they are highly mobile misunderstands the imaginary nature of the roots. To have roots requires a lot of work for the imagination (dream-work). One element of that dream-work is that pride in one's nation of origin is important in the construction of self-esteem in the place of immigration. It gives a different feeling to admit that one is from a country ravaged by famines and floods than to say with pride that one is from a superior civilization that is also very good in high-tech development. The coherence of a Hindu modernity tied to the sovereignty of India's past and territory has gradually given way to a postmodern bricolage of de-territorialized and de-historicized discourses on family values and cyber-spirituality that is very hard to capture. Contrary to what their opponents think, these movements are not 'outside of modernity', they were very much part of it and are now moving beyond it.

Conclusion

In the 19th century, there was a connection between nation, migration and religion. Obviously, the diasporic communities of Indian indentured labourers and their descendants are quite different from the diasporic communities of middle class software engineers. To some extent, this reflects a more general pattern in the global labour market, namely the marked growth of the professional-managerial class that is distinguished by its level of formal training and its position in the service economy. It is this class that is required to show full flexibility even to the extent that their members are required to go overseas to find jobs. It is important to see that this is a global phenomenon in which workers are asked to forget about career-long employment and to learn new skills every so many years. The uncertainties attached to this lifestyle and the required capacity for self-transformation are addressed by a variety of religious movements in Christianity, Islam and Hinduism (Rouse, 1995: 390). In all these religions, these movements carry forward the heritage of 19th century reform movements that connected nationalism and religion and specifically addressed migrants in a missionary fashion. The new movements produce ideas of spiritual rebirth which are central to new versions of globalized spirituality and attuned to the production of flexible subjectivities. This is further connected to ideas of the nation. The hype found among many IT workers that they can and should play an important role in building the nation back home is a crucial part of the building of transnational networks through investments and circular travel (cf. English-Lueck and Saveri, 2000). Nationalist politics of belonging is connected to transnational marriage politics and dowry at the level of the family and to business and investment strategies at the level of the global nexus of capital and labour.

References

Castells, M. 1997. *The Power of Identity* (Vol. 2 of *The Information Age: Economy, Society and Culture*). Massachusetts: Blackwell Publishers.

De Genova, N.P. 2002. 'Migrant Illgality and Deportability in Everyday Life', *Annual Reviews in Anthropology*, 31: 419–47.

English-Lueck, J.A. and A. Saveri. 2000. 'Silicon Missionaries and Identity Evangelists'. Available at http://www.sjsu.edu/depts/anthropology/scvp/iden.html.

Jayadev, R. 2001. 'IT Industry: USA', *Sarai Reader*, I: 169–71.

Khadria, B. 1999. *The Migration of Knowledge Workers: Second Generation Effects of India's Brain Drain*. New Delhi: Sage Publications.

Kurien, P. 1999. 'Gendered Ethnicity: Creating a Hindu Indian Identity in the United States', *American Behavioral Scientist*, 42(4): 648–70.

Mir, A., B. Matthew and R. Mir. 2000. 'The Codes of Migration: Contours of the Global Software Market', *Cultural Dynamics*, 12(1): 5–35.

Ong, A. 1999. *Flexible Citizenship: The Cultural Logics of Transnationality*. Durham: Duke University Press.

Rajagopal, A. 1997. 'Hindu Immigrants in the US: Imagining Different Communities?', *Bulletin of Concerned Asian Scholars*, 51–65.

Rajagopal, A. 2000. 'Hindu Nationalism in the US: Changing Configurations of Political Practice', *Ethnic and Racial Studies*, 23(3): 467–96.

Rouse, R. 1995. 'Thinking through Transnationalism: Notes on the Cultural Politics of Class Relations in the Contemporary United States', *Public Culture*, 7(2): 353–403.

Saxenian, A. 1999. 'Silicon Valley's New Immigrant Entrepreneurs'. Report for Public Policy Institute of California. Available at http://www.ppic.org/publications/PPIC120.PDF.

———. 2002. 'The Silicon Valley Connection: Transnational Networks and Regional Development in Taiwan, China, and India', *Science, Technology & Society*, 7(1): 117–51.

van der Veer, Peter. 1995. *Nation and Migration. The Politics of Space in the South Asian Diaspora*. Philadelphia: University of Pennsylvania Press.

———. 2001. *Imperial Encounters*. Princeton: Princeton University Press.

———. 2002a. 'Transnational Religion: Hindu and Muslim Movements', *Global Networks*, 2(2): 95–111.

———. 2002b. 'Religion in South Asia', *Annual Review of Anthropology*, 31: 173–87.

———. 2003. 'Postmodern India', in Arvind Das and Marcel van der Linden (eds), *Work and Social Change in Asia* (pp. 91–111). Delhi: Manohar.

Xiang, B. 2001. 'Structuration of Indian Information Technology Professionals Migration to Australia: An Ethnographic Story', *International Migration*, 39(5): 73–90.

16

Uncanny Networks: Pirate, Urban and the New Globalization in India

Ravi Sundaram

Introduction

In a recent paper, Partha Chatterjee raised the question: are Indian cities becoming bourgeois at last? His own answer is mixed, expressing scepticism over the legendary South Asian vernacular practices that will inevitably 'corrupt' the forces of urban globalization.[1]

For several decades, the urban's place in the nationalist imagination has remained ambivalent (Prakash, 2002). Contemporary globalization, however, has drawn attention to Indian cities in such ways that could not be foreseen just a decade ago. Indian cities in the 1990s have witnessed a series of concentrated shock experiences: temporal compression, spatial transformation, assaults on older industrial areas, and a vast new mediascape that now envelops cities almost like an all-pervasive skin. Cities have borne the brunt of the new globalization both in transformative and imaginative terms, with changes in infrastructure, social arrangements and constant expansion. Yet, the new focus on cities in India cannot but reveal a paradox. At the very moment that scholarship seems ready to engage with the Indian city, contemporary globalization has in fact slowly but surely eroded the old modernist compact of 'the city'. The technological sublime of the planner imaginary, so central to post-independence India, is giving way to a splintered urban sprawl in the main metropolitan cities. Planning bodies now base their strategies on smaller projects rather than unitary visions, push for privatized de-coupling of infrastructures; transportation design privileges the automobile flyovers and private toll highways to facilitate rapid travel to the suburbs, private builders take over from older, albeit limited concerns with social housing. This splintered urbanism is by no means unique to South Asia; it reflects a larger global process of rapid urban transformation in the contemporary period.[2]

Urbanism in India today has become a significant theatre of elite engagement. Consumption, the 'information' society, spatialized imprints of the media industry like multiplexes/malls, lifestyle and suburbia go hand in hand with the cries of urban decay and pollution, and the restlessness of populations that are increasingly difficult to manage in the new arrangements. Splintering urbanism may in fact suggest a certain strain within older techniques of governmentality, which as Chatterjee has earlier pointed out, was based on the conceptual division between citizens and populations where the latter were empirical categories of people who received administered welfare policies, while citizens were part of a homogeneous national. The relationship between populations and the state was mediated through the domain of political society which was implicated in a series of complex social arrangements and political mobilizations that could not be formulated within the classic state–civil society relationship.

How then do we understand the contemporary globalizing city? Using Delhi's media networks as an example, I want to suggest that new domains of non-legal practices could pose significant problems for classic strategies of incorporation and management within political society. These non-legal domains open up new spaces of disorder and constant conflict in Indian cities, posing a significant threat to the current self-perceptions of the globalizing elite. At the heart of this disorder is a widespread culture of the *copy*, which is implicated in sophisticated local and transnational networks, striking at the heart of the current elites' mantra of protecting the idea of intellectual property.[3]

Perspectives

In recent years, a growing sophisticated global literature has engaged with the new urbanism. Two streams pertain to this essay, the first of which is a geography of globalization that focuses on networks and flows, aided by rapid communication networks, flows of financial capital and transformation of urban spaces. Sasskia Sassen (2001) argues that a new geography of centrality and marginality has emerged globally with financial centres concentrated in certain core cities with a large, increasingly disfranchised low-end workforce helping provide services and back-up. These services—financial, legal and operational—are subject to a high degree of centralization in global cities (typically located in downtowns), managed by an expatriate elite that runs a global network of service sub-contractors and processing firms. Manuel Castells' (1996) network society thesis focuses on how a new space of flows draws producers of information goods everywhere into powerful communication networks. Elite urban enclaves service/house these classes, simultaneously marginalizing other forms of labour in the city. Positioning in the new space of flows becomes part of the strategies of new info-elites. The second stream of literature emanates from what can only crudely be described as a critical phenomenology of urbanism, ranging from the re-discovery of the work of Georg Simmel and Walter Benjamin, and the contemporary engagement with the work of Michel de Certeau and Henri Lefebvre. The latter stream is less explanatory than reflective, working through a series of critical interventions in the urban.

In his remarkable reflection on the contemporary, Zygmunt Bauman poses the idea of a *liquid* modernity as intimating the new epoch. This was in contrast to the earlier modern, which:

> Could be dubbed, for lack of a better name, the era of *hardware*, or *heavy* modernity—the bulk-obsessed modernity ' the larger is better' kind of modernity. (This was) the epoch of weighty and ever more cumbersome machines, of the ever longer factory walls wider factory flows and ingesting ever more factory crews... To conquer space was the supreme goal—to grasp as much of it which one could hold, and to hold to it, marking it all over with tangible tokens of possession and 'No Trespassing' boards. (2000: 113)

The control of space, instrumental rationality and routinized time has been the constant obsessions of heavy modernity. In contrast, says Bauman, light modernity of the software era proclaims (for those with power) the freedom from place and direct engagement:

> Fluid' modernity is the epoch of disengagement, elusiveness, facile escape and hopeless chase. In 'liquid' modernity, it is the most elusive, those free to move without notice, who rule... The disembodied labour of the software

> era no longer ties down capital: it allows capital to be extraterritorial, volatile and fickle…(Capital's) lightness have turned into the paramount source of uncertainty for all the rest. This has become the present-day basis of domination and principal factor of social divisions. (Bauman, 2000: 120–21)

Fluid modernity operates through high-speed networks, which are not linear as in the older forms, but *discontinuous*. The temporal forms have been well documented by Harvey (1989): acceleration, spatial compression, and a perennial speeding-up which produces constant disorientation. Compression is linked to *informationalization* of knowledge; things and ideas die quickly after they are produced. Says Scott Lash:

> Fast-moving consumer goods are also informational in their quick obsolescence, their global flows, their regulation through intellectual property, their largely immaterial nature in which the work of design and branding assumes centrality, while the actual production is outsourced… Power in the manufacturing age was attached to property as the mechanical means of production. In the information age it is attached to intellectual property. It is intellectual property, especially in the form of patent, copyright and trademark, that put a new order in the out-of-control swirls of bits and bytes of information so that they can be valorised to create profit. (2002: 3)

Global network society also produces a range of generic spatial entities: software parks, outsourcing hubs and data parks. These are akin to Mark Auge's 'non places', which have a uniform brand environment worldwide, buttressed by privatized infrastructural, security and cultural networks: multiplexes and carefully controlled shopping areas. Public but non-civil, as Bauman refers to them, these hyper-modern spaces are now part of a global urban sprawl from Bangalore's software city to Gurgaon's call centre zones in India.

I have chosen Bauman's essay as a starting point, since it combines a series of provocations. Both insightful and speculative, Bauman's essay highlights some of the currents in the contemporary global urban environment. In India, they feed directly into the more technocratic refashioning of elite discourses on globalization. The emergence of zones of generic urbanism in India has of course come in the background of a general infrastructural crisis and the widespread perception of urban breakdown. As older systems of urban regulation (state-supported welfare, transport, health and education) get eroded, new practices among middle- and upper-class elites emerge to 'engineer certainty': security agencies in middle-class colonies and flats, CCTV and domestic worker ID cards and demands to register with local police stations. All of this would have been unthinkable a decade or two ago, but this is true of so many of the urban forms that have emerged after the new globalization.

The urge to order, the curse of the planner city and the dream of the new generic suburban design, is something that has evaded the culture of street media practices in contemporary India. While broadcasting has remained the near-monopoly of the media industry, non-legal distribution and production networks have prised open the music and VCD markets to new publics. This form has emerged in the interstices of contemporary urban growth, disorder and fragmentation. Equally, street media culture has spatialized the new urban form in distinct ways. New visibilities, networks-within-networks and conflicts over intellectual property have changed the old world of the planner city. I want to examine this in the following section by looking at the experience of Delhi.

Delhi's Media Urbanism

'The concept city is decaying', wrote Michel de Certeau in 1984. This could also be the story of Delhi's urban landscape for the past 20 years. Urban Planning was operationalized around a series of Master-plans put in motion by the Delhi Development Authority (DDA) from 1957. The DDA sought to organize space through classic modernist urban design principles: enumeration, classification, zoning and slum management. Delhi's the national capital status gave a certain inflationary charge to the rhetorics of urban control and management, something that has continued even after the decline of the planning model. Since the post-Emergency period, this model has been in secular decline,[4] due to a combination of factors: mass migration and urban expansion, the breakdown of old systems of classification and information so crucial to planning models of governmentality. The city has rapidly expanded through the 1980s and 1990s, both in areas of housing and commerce—an expansion that is mediated through a series of non-legal informal arrangements for a range of actors: the urban poor, small business and local markets, affluent house owners wanting to expand private space beyond legal norms, and of course private builders and contractors. This complex system of informal non-legal[5] urban arrangements is by no means unique to Delhi, but takes on a significant edge, given the emergence of neo-liberal and globalizing networks in the region. Small-scale industry, old commodity markets and historic trading communities have been Delhi's strengths and have largely benefited from the decline of the older control mechanisms. Over the last two decades, they have formed new dynamic networks, which have a footprint outside Delhi, often stretching into neighbouring states and Northern India. This expanding commodity culture is using old and new spatial forms: mobile weekly markets, small shops, and has also enabled the entry of hawker networks and street traders from other social groups. There has been, in other words, a production of *urban density*, a domain that enters new zones of conflict/collaboration in the current period.[6] The recent years have seen a concerted effort to reverse these changes with attacks on hawkers and small units, as well as the brutal displacement of the urban poor to the periphery, the long-term results of this process may be significant.[7]

My narrative follows the conventional division of Delhi into the following zones: the old Walled City area, the centre of historic commodity markets and distribution, the New Delhi, NDMC area which is the space of the colonial capital and the current political class, the South which is home for the more affluent and where networks of corporate globalization are stronger than in the other parts of the City.[8] The North and the Western parts of the City took significant parts of the post-Partition population and are mixed areas of both working class settlement as well as middle-class colonies. It is in the East of Delhi, the Trans-Yamuna area, that a significant portion of the city's population lives; the East also houses the various small factories that are crucial to the informal media networks. Two regions, NOIDA in UP and Gurgaon in Haryana, have been prime candidates for the new generic urbanism: an integrated combination of growing global call centres, shopping malls and multiplexes, and private toll road development to service automobile users. This classic secessionary development[9] is the most 'global' spatial form yet in contemporary Delhi.

Media discourses have tended to privilege the lifestyle zones of southern Delhi as representing the future of the city's route while lamenting the crisis of governance, environmental crisis and general urban ruin. The old stories of social conflict have been increasingly replaced by a significant argument about *property*. It is difficult to find a newspaper today that does not on any given day carry police and industry handouts about raids on 'pirate' industries. Along with the figure of the Islamic terrorist, the figure of the pirate is threatening to the emerging regimes of property and control in the media. As we shall see, this has become one of the major sites of everyday conflict around property claims in the contemporary.

As in many Indian cities, the new globalization transformed media networks in Delhi.[10] At the level of the everyday, the old prohibition and regulation on the social life of commodities have proved ineffective. Urban residents are now assaulted with a deluge of cultural products, cassettes, CDs, MP3s, VCDs, cable television, grey market computers, cheap Chinese audio and video players, thousands of cheap print flyers, and signage everywhere. What is remarkable here is that the preponderance of these products comes from the grey or informal sector, outside the effective regulation of the state or large capital. India today has the world's second largest music market, a large film industry with global dreams, a majority grey computer market, hundreds of tiny phone and word processing shops and cybercafes. New media bazaars which supply these networks have emerged, existing in the cusp of legality and non-legality. Everyday, a guerrilla war is being staged between new intellectual property raiders, the police and the unceasing neighbourhood demand for grey ware.

India never saw a print revolution as in early modern Europe, but the cassette revolution of the 1980s transformed popular music culture. As the historian Peter Manuel has shown, cassette culture in the first phase of globalization, largely non-legal, effectively broke the stronghold of the large music companies, by introducing new artists and expanding the market for low cost cassettes that are sold in neighbourhood shops. Long-forgotten 'folk' music was remixed and circulated in the market. This was followed by the cable television and computer expansion of the 1990s. Delhi was a significant site of this transformation as it was the home of T-Series, the first major beneficiary of this phase. Gulshan Kumar, the first proprietor of T-Series, used an opening in the copyright laws to push version recording, an innovative use of less-known artists to sing items sung by well-known singers. In doing so, T-series inaugurated a media form that has developed dynamically all the way to the recent remix culture and also became the 'nodal' form for the development of new music companies. The key to this is the mix of the legal and the non-legal:

- Using a provision in the fair-use clause of the Indian Copyright Act which allows for version recording, T-Series issued thousands of cover versions of GCI's classic film songs, particularly those which HMV itself found to be unfeasible to release. T-Series also changed the rues of distribution by moving into neighbourhoods, shops, grocery shops, *paan wallahs* and tea shops to literally convert the cassette into a bazaar product.

- T-Series was also involved in straightforward copyright infringement in the form of pirate releases of popular hits relying on the loose enforcement of copyright laws.
- Illegally obtaining film scores even before the release of the film to ensure that their recordings were the first to hit the market (Liang, 2003).

The T-Series phenomenon, led to the development of new media markets in the 1980s and the 1990s: Palika Bazaar in Central Delhi for video, Nehru Place for software and hardware and Lajpat Rai Market in the old city for music and hardware for the cable industry. Transnational links with South and East Asia were established for hardware supplies. Through the 1980s, a range of small players emerged in the media markets, developing new networks of distribution and production. Production was concentrated increasingly in the Trans-Yamuna areas and parts of UP and Haryana, while distribution was managed from the media markets linked to neighbourhood entrepreneurs. In the early years of the media boom period, entrepreneurs imported blank media and hardware from East Asia, built satellite dishes and hardware ancillaries, and developed local cable distribution. Music companies emerged catering to a range of tastes all over North India. What is remarkable is that except for T-Series, which is now a mainstream player (committed to intellectual property law!), the bulk of these new enterprises remained small. This was pirate culture in its early phase innovatively built through networks, but still crucially linked to the main media markets.

In an earlier essay, I termed this phenomenon a pirate or recycled modernity (1998), which is dispersed and unconcerned with modernity's classic search for originality, fashioning itself as fluid movements in India's cities and towns. It is a phenomenon that is neither oppositional nor critical in the classical sense, with no charters against the electronic elites or hypermodern spaces. Pirate modernity is part of a culture of insubordination and disorder that marks our time, and is a source of major concern to global and local elites.[11] A number of features that mark this phenomenon need to be pointed out:

First, pirate electronic culture is part of an immanent technological space. In other words, it presumes that classic distinctions between technology and culture, between humans and non-humans have ceased to hold in the contemporary city. The inherent problems of positing a strict human non-human distinction has been pointed out by Bruno Latour (1993). Old style humanist discourses between subject–object, nature–culture, and so on, is rendered fuzzy in the contemporary. In fact media experiences of the 1990s in India can be read as the failed collision between technology and tradition: every aspect of social life in cities has been 'thingified': phones old and new, audio and video systems, electricity legal and illegal, music and a growing mobile network.[12] 'Things' and humans interact and are enmeshed in every possible way, rendering classical distinctions problematic. There cannot be an urban contemporary without the 'technological', something being made possible both by as much by pirate culture as well as the media industry. It is precisely the 'dirty', discontinuous and mobile possibilities of pirate cultures that make this sphere interesting.

Second, pirate electronic networks are part of a 'bleeding' culture,[13] constantly marking and spreading within urban life. Ambient sound and images are now part of all street/neighbourhood life; a crowded pirate aesthetic pervades video culture and local advertizing. This is part of the culture of dispersal, which marks its resilience. In a world where information bleeding is part of the contemporary (SMS, television text scrolling, newspaper inserts, lamp-post stickers, Internet pop-ups, event branding), pirate culture uses the ruses of the city, but immanently, not transcendentally. In doing so it affects the main media industry (in music, version recording/remix is a large and growing market).[14] Today, the pirate video aesthetic, with its informationalized, overcommodified frame, is a compulsory part of any film experience on the local cable network.[15]

Third, pirate culture is a just-in-time culture. The copy arrives on your cable network the weekend the film is released, and the music versions of popular numbers follow almost immediately. Networks in Delhi use a combination of regional and transnational sources (Dubai, Pakistan, East Asia) to ensure the culture of the instant.[16]

Finally and crucially, pirate culture is a culture of the copy.[17] It is part of a world where experience as we know it is increasingly commodified, and informationalized. For the globalizing middle-class in India, this is happening through the more familiar modes of incorporation: credit cards and credit rating agencies, frequent flyers, vacations, niche marketing, ATM cards and monthly billing cycles, corporate consumer campaigns, brand environments, all generating vast amounts of information. This is the more conventional, almost generic world of the new globalization. The networks of pirate culture on the other hand, usually target the urban populations outside this world, but nevertheless are increasingly drawn to the commodified forms of urban experience. Local markets, neighbourhood music/video stores, grayware computer and audio–video assemblers, independent cable operators are usually part of the pirate network of distribution, which also 'bleeds' into other parts of the city. The commodities of the copy are multi-use, recombined/recycled and in constant circulation—moving in and out of new spaces and networks.[18] In Delhi, the media copy exists in a symbiotic relationship with all other commodities and industries: clothes, cosmetics, medicine, household goods, and also car and machine parts. As is evident, copy culture pits pirate modernity right into a global social conflict on definitions of *property*.

A Brief History of the Copy

Historians of print and the pre-print period have shown us complex forms of the reproduction of texts and cultural objects that existed both in the world of Christendom and the Dar-ul-Islam. In the west, medieval monks and notaries toiled away copying books, legal documents and contracts. In particular the medieval notary, played a crucial role in the emerging socio-legal relations of the emerging absolutist state. Says one historian:

> Stenography transforms the spoken word into the written. Copying transforms the One into the Many. Notarizing transforms the private into the public, the transient into the timely, then into the timeless… The notary was a symbol of fixity in a world of flux, yet the making of copies is essentially transformative—if not as the result of generations of inadvertent errors, then as a result of masses of copies whose very copiousness affects the meaning and ambit of action. (Schwartz, 1996: 214–15)

The historian Elizabeth Eisenstein (1980) suggests that with the coming of the print revolution, a 'typographical fixity' was imposed on the word. The sheer volume of the print revolution was incredible; from 1,450 to 1,500 more books had been printed than those copied in the entire history of Islam and Christianity. However, it seems to me that Eisenstien's assertion seems too categorical, for in the first 100 years, errors were rife in printed books, Papal Edicts against 'Faulty bibles' had no effect on the volume of production. Print in fact opened up the floodgates of diversity by the 17th century: historical work on the cultural uses of print in the French revolution shows the proliferation of pornographic, anti-clerical and revolutionary texts. There were deliberate forgeries and the insertion of parodic statements in official texts.

Forgeries, re-interpretations and parodies were common to popular print culture. However, the issues raised by the art forger after the emergence of the modern painting went straight to the heart of authenticity, individuality, uniqueness and historicity as the representational architecture of the bourgeois art work. Discussions of forgeries and copies of art works existed since the Renaissance, but what is interesting for our purposes is the practice of forgery as a cultural act. The Hungarian scholar Sandor Radnoti in his book on the art fake has this to say on the practice of art forgery:

> The forger attacks originality from the point of view of historical authenticity, insofar as his work gives the impression that it *contains* the story that conveys the same historical evidence as the original. However the clock of history is ticking away for the forger's work as well, it too embarks on a life of its own, and it is only a question of quality, good luck, and time that having survived in historical memory sufficiently long, it becomes authentic, a genuine forgery.... (1999: 43)

Forgery, says Radnoti, is a functional art form, which 'interchanges the interchangeable, substitutes the unsubstitutable'. The crisis of authenticity of the cultural object has been present right from the outset of modernity; it intensified rapidly with different forms of mechanical reproduction. This is of course the argument of Benjamin's important and controversial artwork essay, where copies and mechanical reproduction subvert the authenticity of cultural products. This, says Benjamin, subverts the 'here and now' of the artwork, 'its unique existence in the actual place it happens to be'. For Benjamin the aura is the marker of bourgeois art, 'the spritualization of commodity fetishism', something that is destroyed by new techniques of copying. Critics have pounced on the technological innocence of the artwork essay, and the re-appearance of *aura* in the new culture industries, but the value of the artwork essay lies in its synoptic power and a bold insight into the culture of the copy in modernity.

The major transformation of the culture of the copy takes place in the 19th century. From the times of the Renaissance where copying of cultural products was common and legitimate, the 19th century sees the emergence of proprietary regimes of mechanical reproduction, when the culture of the non-legal copy enters a secular period of criminalization and delegitimation.

The Commodity

Benjamin's artwork essay had the merit of posing the key issue: a new form of commodification enables the means of mechanical reproduction in contemporary capitalism. Circulation now emerges not as a 'lack' to the world of production, but as a sphere that enables a range of practices of consumption, re-production and performance. But for most part of the 20th century the radical tradition inveighed against the world of circulation. The generalization of the commodity form is paralleled by the decline of subjectivity and loss, reification, the transformation of the living into the dead. In his important book John Frow uses Guy Debord's *Society of the Spectacle* to open up discussion on the troubled relationship between the radical avant-garde and the commodity (1997: 4).

Debord's essay makes the point of citing Feuerbach's preface to the *Essence of Christianity*, where the present age is one 'which prefers the sign to the thing signified, copy to the original, representation to reality, the appearance to the essence'. For Debord, the modern world presents itself as an accumulation of spectacles. 'All that was once lived directly,' says Debord, 'has become mere accumulation of spectacles'. Detached from life, images become autonomous, producing a reality that is but psuedo-real. The spectacle is the most general form of the commodity conforming to that historical moment when the commodity form completes its colonization of life. Time and space now become abstract and lifeless; the former unity of the world is lost. Debord's essay was widely read and attained a cult status during its time. What is interesting about the essay is an overwhelming sense of loss. The essay has a structure: 'which opposes representation as such to the immediacy and unity of life, which sets the latter pole within a lost past'. For Debord, the spectacle is 'the reigning social organization of a paralyzed history, of a paralyzed memory, of an abandonment of any kind of history founded in historical time. (Thesis No. 158, cited in Frow: 7).

It may be argued that Debord's brilliant but flawed polemic is an easy target. I actually want to use Debord to draw attention to an old tradition in 20th century radical thought, the generalized denunciation of the commodity form as a phenomenon that negates history and memory. The heritage Debord draws upon is a certain version of Marxism, combined with a heady cocktail of critical theory and 1960s counter culture. Marx himself saw the commodity as having a life before capitalism; capitalism differs in that it promotes generalized commodity production, the extension of the principle of exchange and social relationships around it to all spheres of social life. The history of capitalism, therefore, is the progressive extension of the commodity sphere. Immanuel Wallerstein calls it the endless drive to accumulation; there is an accelerative logic to this with the transformation of labour, land and materials into value, to beget more value.

What about the commodification of cultural products? The philosopher Frow argues that this takes place at a number of different semiotic levels:

- In the case of printed texts, we could distinguish between an initial commodification of the material object (the book) virtually coeval with the printing press.
- A second stage of commodification of the information contained within the material object (and conceptualized in legal doctrine as 'the work'), of which the major historical expression is the development in the 18th century of copyright law and the modern system of authorship.
- A third, contemporary moment, developed in relation to electronically stored information, which in addition to the copyright information itself, commodifies access to that information (ibid: 139).

Frow argues that these are stages, 'in the sense that this sequence while not uniform is normally progressive, and refers to the gradual application of property rights over immaterial entities. It is both the restricting of the commodity form as expanding its controlled use (ibid.).

The contemporary struggle in media networks is therefore not about commodification as such; it is about imposing new *property regimes*. Scott Lash (2002) and Jeremy Rifkin (2001) have argued that as contemporary capitalism emancipates itself from spatial restraints, the struggle is not over the factory but over brands and *domains*. As production of global commodities is contracted out globally, the technologies of reproduction become generalized and accessible, brand protection and network control become increasingly central. Generalizing this for all of contemporary capitalism would be questionable, but in the case of the media worlds it takes a dramatic form. Copy costs are low, and distribution mobile.

The extensions of property rights over immaterial objects are key to the informationalization of the world economy, and a significant part of the TRIPS agreements. The emerging global regime of intellectual property legal practice works though pressure on national regimes to change local copyright law, create a global network of enforcement, and generate secure proprietary digital formats for the media[19] (till date an unsuccessful attempt). The key players in this are the US media (film and music) and software industries, but the effort is to build alliances with local media industries. This has been reasonably successful in India with alliances with NASSCOM, the Indian Music Association and Bollywood anti-piracy associations. Says Liang, summing up the whole scenario:

> The information era props up a master plan, similar to that of modernist planning. The institutional imagination of the era relies on the WTO as chief architect and planner, copyright lawyers as the executive managers of this new plan and the only people who retain their jobs from the old city are the executors of the old plan, the police force and the demolition squad. (2003)

The push to crack down on copy culture has led to a number of significant changes in the Copyright Act of 1957, which has increased the penalties. Minimum provisions now provide for a six-month jail term for commercial piracy along with a minimum fine of Rs 50,000, with a maximum limit of Rs 200,000. The period of imprisonment is doubled with a second offence, as is the minimum fine. Illegal 'use' of a proprietary computer programme carries a minimum jail term of seven days, and a fine of Rs 50,000. The International Intellectual Property Alliance's India report says, 'With the exception of the level of fines, which should be increased, these are among the toughest criminal provisions in the world. Unfortunately, they have never been implemented (2003: 14).

Apart from further changes in the Copyright Act to make it TRIPS-compliant, there have been changes in the Cable Television Regulation Act of 1995 that prohibit cable operators from broadcasting programmes without copyright authorization. The act shifts the enforcement to the local District Magistrate and Commissioners of Police, who have been designated 'Authorised Officers' and who can seize the equipment used by the local cable operator.

The enforcement regime in India works at a number of levels. The first is the creation of public discourse: press stories detailing the crimes of piracy to a willing press core. The second is through building networks with policy makers and MPs conducting seminars and workshops on the dangers of copyright violation. Given the current propensity of our political class for making the media/software industry a central part of the country's global brand, the lobbying has had a remarkable degree of success. The third level of intervention operates through workshops for the police, and through building close networks with the Intellectual Property Cells of the Economic Offences Wing. The fourth strategy is legal, generating pre-release injunctions from court orders,[20] and collaborative raids with the police on the copy network. The overall efforts to enforce intellectual property provisions in India are so wide-ranging that it surpasses older US efforts to push modernization discourses in the 1950s. The coalition includes elite legal firms specializing in Intellectual Property Law, corporate lobby groups and local representatives for the US media/software alliances. The Indian media industry is an increasingly crucial player in this equation, with mixed results. In film, for example, loss-making productions have been released in the pirate market in an effort to recover costs.[21]

Ongoing research into copy culture in the urban morphology of Delhi throws up an interesting picture of emerging conflicts on circulation of media after the TRIPS regime. It is increasingly clear that the National Capital Region is one of the main centres of copy culture in the country.[22] Delhi is now the centre of a complex coordinate of media markets, small software and hardware factories, and local shops that interact with customers. Production sites in Haryana and Rajasthan supply just-in-time media to the copy network, in turn linked to factories and routes in Pakistan and Malaysia. Media markets play an important role in distributing to local shops: Nehru Place for computer software and hardware, Palika Bazaar for film media and Lajpat Rai market for low cost hardware and music. Markets typically combine legal and copy nodes, frustrating efforts by the enforcement regime to spatially 'fix' copy culture. Nehru Place is thus one of Asia's biggest computer markets; Lajpat Rai is a distribution centre for music companies along with grey audio hardware, while Palika Bazaar sells clothes and crafts along with media.[23] The rise of new technologies, like the mobile telephone network, low-cost CD-R duplicating machines and forms of digital compression like MP3 for music and VCD for film, has made the copy network more dynamic, with nodes gaining mobility day by day. Until a few months ago in Delhi many neighbourhood stores would keep local CD-R machines where they would make collections of MP3s for customers. Copy media (MP3s, VCDs) arrive though non-legal distributors regularly who liase with media markets and production sites, using the mobile telephone network. In recent months the quality of pirate media has done better and better, suggesting larger players in the field. The old grubby quality of covers for copy media has given way to professional looking design.

Flexibility and network performance mark some of the emerging local companies in music. Research into Nupur (Prasad, 2003b), a small music company in Delhi, shows a world where the company (which works out of a tiny office) operates almost like the new multinational enterprises described in Jeremy Rifkin's book, almost everything: production, studio work, design, and so on—operates through a system of *contract*. Studios and factories that produce the company's music may well also produce a rival's; further, Nupur is an enforcer of Intellectual Property claims in Punjab where its business is strong and relaxed in other parts.

The Intellectual Property enforcement regime in Delhi developed a complex, semi-autonomous architecture to engage with copy networks at the local level. The central place in this architecture is held by the *raid*. The raid is a coordinated act by legal firms, investigation agencies and the local police. The raid is informed by performative violence, staged before the neighbourhood market. Local copy equipment is either destroyed or seized, the software confiscated and an FIR lodged. There were approximately 1,500 raids last year, of which a significant number took place in the National Capital Region. Neighbourhood shops, factories and markets were raided, often leading to significant clashes between the raiding party and local shopkeepers. At the heart of the raid regime is the figure of the 'investigator', who gathers local intelligence on copying, and acts on behalf of a range of clients. These may range from IP legal forms representing large firms, Bollywood film distributors or music companies. The investigators inhabit a murky world of violence, small rewards, and a cynical contempt for their clients. There is universal belief that enforcement will not work, which is remarkable given their profession. In some cases, the investigators may even come from the world they seek to attack: the pirate modern.[24]

The raid is more of an intimidatory and theatrical act at the local level than something that leads to any measure of legal success in the struggle to control copy culture. In actions outside Delhi, the raid sometimes collapses into a comical event. Consider this report in the Indian Express:

> The film industry's attempts to stop video piracy have suffered an embarrassing setback after a raid on a suspected pirate ended with members of the raiding party being arrested by local police and charged with trespass and extortion. It may take more than the intervention of Rajya Sabha MP Shabana Azmi—who took up the issue today—to help it wipe the egg off its face. The story begins with a raid on a house in Jangpura on Sunday morning, where 400 pirated VCDs were recovered. The disclosure pointed to a manufacturing unit in Kundli, Haryana, owned by Mahinder Batla. Owner of a company Lara Music, Batla's two DVD and VCD manufacturing units are worth nearly Rs 10 crores and was setup three years ago. When a raiding party comprising private investigators of the Motion Pictures Association and the Delhi police reached there, they searched the premises for nearly three hours before the local police arrived on the scene. They accused the team of 'planting the pirated tapes' and arrested seven people on grounds of trespass and extortion. Six people were released the next morning; one of the investigators, Vikram Singh, is still under arrest (Jain: 2001).

The Copy and the City

Talking about the second half of the 19th century in Europe, Jean-Louis Comolli said that life was in the grips of what he called the *frenzy of the visible*. This obtained from the constant flow of images and print forms in their millions[25] and the transformation of everyday life. The new globalization in India's cities in the 1990s recalls this 'frenzy' except in more intensive, cross-media forms. At the heart of this extension of the visible has been the production of media commodities outside the legal property regimes of globalization. Copy culture and non-legal distribution networks have been central to the spread of the media, making distinctions between the technological and cultural seem blurred in daily life. A significant section of the urban population derives their media from these networks. Using the tactics of the fragmentary city, the pirate networks have frustrated every effort of the proprietary enforcement regime to control them.

The pirate modern works through and depends on regional–transnational networks.

In his essay on Naples, Walter Benjamin points to the performative openness of the city, 'Porosity is the inexhaustible law of the life of this city, reappearing everywhere…building and action inter-penetrate in the courtyards, arcades and stairways…to become a theatre of new, unforeseen constellations. The stamp of the definitive is avoided.'[26] How do contemporary elites in South Asia's cities deal with the problem of porosity and produce the 'stamp of the definitive'? Older governmental techniques like technologies of visibility and knowledge about populations clearly continue and expand in Indian cities. Along with national elite obsessions like ID cards, computerized crime records, secessionary enclaves and housing societies are setting up CCTV systems, electronic security and control of 'outsiders'. Software companies lobby for national ID cards, which have already been implemented in the border states. Given the absence of any privacy law in India, electronic conversation, both aural and textual, is open for State interception.

In the porosity of the contemporary city, however, the realms of copy culture thrive. This is the sphere almost akin to what Lefebvre calls the 'residual', what is 'left over'. I say almost since Lefebvre would have been deeply uncomfortable with the graphic commodity spheres of the pirate economy, so would an entire generation of radical urbanists who saw critical/redemptive strategies located in the spheres *outside* the commodity.[27] This was the old dream of the transcendence of the everyday through the everyday. The everyday becomes a space/theatre for strategies of defamiliarization, redemption and *detournement*. But as Blanchot points out, 'the everyday escapes. This is its definition' (1987: 5).

Earlier patterns of political society in India allowed non-legal populations and networks to assume visibility and enter networks of welfare and administration. Copy culture and the people who thrive in its networks cannot do so easily; it would violate the fundamental concepts of property in the current global/national regime. However, pirate culture has no strategies of political mediation; it works through immersion and dispersal rather than representation and voice. It is resistant to control as well as radical–critical strategies[28] of intervention, inhabiting networks of disorder that are endemic to contemporary urbanism. This may be its greatest strength and resilience.

Notes

1. Chatterjee, P. (2006).
2. See Graham and Marvin (2001).
3. The intellectual property discourse began with the emergence of the circulation of print; they have become widespread in the contemporary epoch. Originally centred around the romantic figure of the 'author' whose work intellectual property law claimed to protect, the last 300 years have seen the gradual extension of such 'rights' over many commodities through copyright, patent and trademark law. At the heart of the intellectual property rights (IPR) discourse is the control of commodities in circulation by corporate entities holding proprietary authorship/rights. This control is rendered further fragile with new technologies of production and reproduction, as well as critical ideas of the public domain, which have emerged from the open-source software movement. As networks become more dynamic and extensive, so do the stakes to control actors and commodities. For a useful discussion, see Frow (1997), Vaidhyanathan (2001).
4. See Baviskar (2003) for a useful survey of this.
5. Liang (2003) calls this new urban developments since the 1980s a porous legality, which enabled the development of a new media space. Though not specifically about Delhi, Liang's essay draws excellent connections between the growth of non-legal urbanism and the new mediascape.
6. The assault on 'polluting' industries as well as street hawkers are significant examples of efforts to make the city 'ready' for globalization.
7. The discourse of scientific environmentalism is used to justify these transformations, often with court sanction. See Sharan (2002).
8. The South has been in the forefront of the new lifestyle culture, which fills the newspaper supplements: theme restaurants, fashion boutiques, farm-house parties. South Delhi also has the large working class settlement of Daskhinpuri, which is largely invisible in the current discourse.
9. The model fits Gurgaon more than NOIDA, which includes older industrial areas.
10. This section benefits from on-going fieldwork on Delhi's media networks that is part of the Publics and Practices in the History of the Present (PPHP) project at Sarai, Centre for the Study of Developing Societies (CSDS).
11. Local conflicts, in the cable industry are sharp since they involve territorial control. In Delhi large networks have gradually dominated the smaller players who hold approximately 40 per cent distribution in the city. For a fascinating story of a large distributor's defeat of a local cable operator with working class origins, see Sharma (2002).
12. The TRAI posits that India will have roughly 70 million mobile phone users in a few years. Urban infrastructure was always implicated in what Latour calls a 'skein of networks', where networks, places, people are enmeshed constantly producing and performing. For Latour, there are no subjects and objects as in classic Enlightenment thought, but actants. A criticism could be made of Latour that he generalizes network architecture to the extent of ignoring the worlds outside it, but the insights are significant.
13. My colleague Jeebesh Bagchi calls this a 'seepage' culture, using the metaphor of architecture. See also Larkin (2004) for an analysis of the pirate aesthetic in Nigeria.

14. The research by Bhagwati Prasad (2003a) showed that there are at least 37 versions of the explicit tune *Kaanta Lagaa*, ranging from dance to devotional forms.
15. See Larkin (2004) for a fascinating enquiry into the Nigerian experience of pirate video.
16. Just in time culture is used by Scott Lash to describe information society, where info-bits arrive as Virlio says 'at the speed of light'. They depart as easily.
17. Copy culture, in Delhi at least, has pervaded to all forms of consumer commodities. The National Capital region is one of the main centres of copy goods manufacture. See IIPA (2003).
18. See Arjun Appadurai's The Social Life of Things (1986) for an early pioneering attempt to understand the journey's and biographies of the commodity.
19. Every attempt to devise secure formats has been broken by the hacker communities.
20. This was used most dramatically in the Ten Sports case, where the court issued an order preventing alternative telecasts (from South East Asian satellites) of the World Cup soccer matches by cable operators unwilling to accede to Ten Sports' demand for proprietary broadcast fees.
21. Most of our interviews with the locals confirm this. This practice is by no means unique to India; it has also been noticed in the Hong Kong film industry (Wang: 2003).
22. See the IIPA country report detailing the raids and legal proceedings.
23. For Palika Bazaar, see Kumar (2002).
24. Interviews with investigators from two agencies in Delhi.
25. Cited in Friedberg (1993: 15).
26. Cited in Amin and Thrift (2002: 10).
27. Benjamin is an exception. He uses the idea of profane illumination through older, dead commodities/ruins of capitalism, recalled through allegorical strategies, which could profoundly disturb the contemporary.
28. Alternative strategies may act as a critical witness to the pirate modern, while pushing for non-proprietary worlds, and reflecting on the idea of a commons and a new public.

References

Amin, A. and N. Thrift. 2002. *Cities: Reimagining the Urban*. London: Polity.

Appadurai, A. (ed.) 1986. *The Social Life of Things: Commodities in Cultural Perspective*. Cambridge: Cambridge University Press.

Bauman, Z. 2000. *Liquid Modernity*. London: Polity.

Baviskar, A. 2003. 'Between Violence and Desire: Space, Power, and Identity in the Making of Metropolitan Delhi', *International Social Science Journal*, 55(1): 175: 89–98.

Blanchot, M. 1987. 'Everyday Speech', *Yale French Studies*, 73: 12–20.

Castells, M. 1996. *The Networked Society*. Oxford: Blackwell.

Chatterjee, P. 2006. 'Are Indian Cities Becoming Bourgeois at last? Or, if you prefer, we could exclaim: Are Indian cities becoming bourgeois, alas?', in Jodhka.S. Surinder, Mary.E. John and P. Jha (eds), *Contested Transformations: Changing Economies and Identities* (pp. 113–24). New Delhi: Tulika.

Eisenstein, E. 1980. *The Printing Press as an Agent of Change: Communications and Cultural Transformations in Early Modern Europe*. Cambridge: CUP.

Friedberg, A. 1993. *Window Shopping: Cinema and the Postmodern*. Berkeley: UC Press.

Frow, J. 1997. *Time and Commodity Culture*. New York: Oxford University Press.

Graham S. and S. Marvin. 2001. *Splintering Urbanism, Networked Infrastructures, Technological Mobilities and the Urban Condition*. London: Routledge.

Harvey, D. 1989. *The Condition of Postmodernity*. Oxford: Blackwell.

International Intellectual Property Alliance (IIPA). 2003. 'Special 301 Report India'. Available at www.iipa.com.

Jain, S. 2001. 'Video 'Pirates' in Lead Role, Cops Play Villains', *Indian Express*, 19 December 2001.

Kumar, R. 2002. *Palika Seems Tired at 23*. Delhi: PPHP Broadsheet No 2, Sarai.

Larkin, B. 2004. 'Degrading Images, Distorted Sounds: Nigerian Video and the Infrastructure of Piracy', *Public Culture*, 16(2)Spring:289–314.

Lash, S. 2002. *Critique of Information*. London: Sage.

Latour, B. 1993. *We Have Never Been Modern*. Harvester: Wheatsheaf.

Liang, L. 2003. 'Porous Legalities', paper presented at the SSRC Conference, Intellectual Property, Markets, and Cultural Flows, New York, October 2003.

Manuel, P. 1991. *Cassette Culture: Popular Music and Technology in North India*. New Delhi: Oxford University Press.

Prakash, G. 2002. *The Urban Turn*. Delhi and Amsterdam: Sarai Reader 2, The Cities of Everyday Life.

Prasad, B. 2003a. *Piracy: Judte Rishte, Phaylta Bazaar*. Delhi: PPHP Laghupatrika Sarai.

———. 2003b. *Sangeet Udyog ka Chota Khiladi*. Delhi: PPHP Laghupatrika Sarai.

Radnoti, S. 1999. *The Fake: Forgery and its Places in Art*. London: Rowman and Littlefield.

Rifkin, J. 2001. *The Age of Access*. New York: Penguin.

Sassen, S. 2001. *The Global City*. Princeton: Princeton University Press (New edition of 1991 book).

Sharan, A. 2002. *Claims on Cleanliness, Environment and Justice in Contemporary Delhi*. Delhi/Amsterdam: Sarai Reader 02, Cities of Everyday Life.

Sharma, L. 2002. *A Brief Biography of a Cable Operator*. Delhi: PPHP Broadsheet No 2, Sarai.

Schwartz, H. 1996. *The Culture of the Copy*. New York: Zone Books.

Sundaram, R. 1998. Pirate Electronic Cultures in India, Third Text.

Vaidhyanathan, S. 2001. *Copyrights and Copywrongs: The Rise of Intellectual Property and How it Threatens Creativity*. New York: New York University Press.

Wang, S. 2003. *Framing Piracy: Globalization and Film Distribution in Greater China*. Boulder, Colo: Rowman & Littlefield.

About the Editors and Contributors

About the Editors

Ashwani Saith is Professor of rural economics at the Institute of Social Studies, The Hague, and Visiting Professor at the London School of Economics & Political Science, and at the Institute for Human Development, New Delhi. He has taught and researched on a range of themes in development studies, with a focus on processes and politics concerning poverty and socio-economic vulnerability, labour and work in unorganized/informal economies, rural development, rural industrialization and non-farm economies, migration, information technology, reform and transitions, and globalization processes and patterns. Regionally, his work focuses on Asia, especially India and China.

Dr Saith has served as a research analyst and policy advisor with several international development agencies, including ILO, FAO, IFAD and UNDP. He has been associated in an editorial capacity with several leading academic journals in the field of development, including *Development and Change*, *Journal of Agrarian Change*, *Journal of Peasant Studies*, *Journal of Development Studies*, *Indian Journal of Labour Economics*, *Indian Journal of Human Development* and *Labour and Development*. He has published extensively in international journals, and also authored and edited several books.

M. Vijayabaskar is a faculty member at the Madras Institute of Development Studies, Chennai, India. He has worked and published extensively on labour issues in India, particularly in globalizing sectors like garments and information technology. He is currently working on the process and implications of urban reforms in India. He is also co-editor of the companion volume *ICTs and Indian Economic Development: Economy, Work, Regulation*.

V. Gayathri is a Visiting Fellow, Institute for Human Development. Until recently, she was a Project Officer in ILO, Delhi, looking at child labour issues in India. She has written a number of papers on labour and other related issues, and is the Co-editor of *Social and Economic Security in India* (Institute for Human Development and Manohar Publishers), New Delhi, 2001.

About the Contributors

Reuben Abraham is with Cornell University's Center for Sustainable Global Enterprise at the Johnson School of Management and a Visiting Faculty at the Indian School of Business (ISB) in Hyderabad. He is the Director of the Base of the Pyramid Learning Lab in India, a joint venture between ISB and Cornell. He serves on the global board of directors of George Soros' Economic Development Fund (SEDF), a $155 million fund which aims to catalyze growth in emerging markets. He was also a TED Global Fellow for 2007 and a member of the Clinton Global Initiative. Reuben received his M.A., M.Phil and Ph.D from Columbia University in New York.

Piush Antony, a sociologist, is an Assistant Project Officer with UNICEF, Hyderabad. She has been a Visiting Fellow at the Institute for Human Development, Delhi, and has co-edited *Social and Economic Security in India* (Institute for Human Development and Manohar Publishers), New Delhi, 2001.

Xiang Biao is a RCUK Academic Fellow at the Institute of Social and Cultural Anthropology and a Research Officer at the ESRC Centre on Migration, Policy and Society, University of Oxford. He is the author of *Transcending Boundaries* (Chinese by Sanlian Press, 2000; English by Brill Academic Publishers, 2005), *Global 'Body Shopping'* (Princeton University Press, 2007), and over 30 papers on migration and social change in Chinese and English. He is currently a Visiting Fellow at Asia Research Institute, National University of Singapore.

Atanu Dey is an Economist working at Netcore Solutions. Previously he worked at Hewlett-Packard in California. He received his Ph.D in economics from UC Berkeley, and holds graduation degree in computer science and an undergraduate degree in mechanical engineering. He was a Reuters Fellow at Stanford University. His interests include education and rural development.

Rajesh Jain is a serial entrepreneur. He had founded IndiaWorld, which grew to be India's largest family of Internet portals. It was acquired by Sify in 1999. He currently heads Netcore Solutions, and has a number of investments in early-stage companies in the mobile and broadband space in India.

Ashok Jhunjhunwala is Professor of the Department of Electrical Engineering, Indian Institute of Technology, Chennai. He also leads the Telecommunications and Computer Networks group (TeNeT) at IIT Madras. This group works closely with industry in the development of a number of telecommunications and computer network systems. TeNeT group has incubated a number of technology companies, which work in partnership with TeNeT group to develop world class telecom and banking products for rural markets. He is the recipient of the Padma Shri in 2002 and the Dr Vikram Sarabhai Research Award for the year 1997, among many others. He is a Director in the Board of SBI and on the board of several companies including TTML, BEL, Polaris, 3i Infotech, Sasken, Tejas, NRDC, and IDRBT. He was also member of Prime Minister's Scientific Advisory Committee.

Shirin Madon has been researching the impact of ICTs on planning and administration in India for the last 15 years. Initially, this broad research theme was directed at conceptualizing the context of development administration and planning in India, and tracing how this context shapes and is shaped by the process of technological intervention. More recently, her research has centred around on e-governance reform in India and Tanzania, with a focus on telecentre projects and health information systems. Her publications include *The Digital Challenge: Information Technology in the Development Context*, Ashgate Publishing Limited, Hampshire (co-edited with S. Krishna).

Amit Mitra currently lectures on information systems at the Department of Information Systems, Cranfield University, UK. Over the last decade, he has been an active researcher in areas of theoretical and applied computing. Whilst pursuing doctoral research, Amit also simultaneously worked on a variety of assignments like evaluating MIS implementation in the Philippines for a UN agency. Ever since 1998, Amit Mitra has been lecturing on topics that have included inter alia knowledge management, research methods in information systems and public sector informatics within university-based academic departments. Amit Mitra's research has also been published in peerreviewed national and international journals.

Vivek Padmanabhan is the Senior Vice President for Global Shared Services at eTelecare Global Solutions (NASDAQ: ETEL), a business process outsourcing company. He joined eTelecare Global Solutions from McKinsey & Co's Berlin office. His international business experience also includes work as a consultant in Asia. He holds an MBA from the University of Chicago, a Masters in Computational Mechanics from Duke University and a Bachelor of Technology in Civil Engineering from IIT Madras.

Vinod Pavarala is Associate Professor of communication at the University of Hyderabad. He has consistently supported and initiated studies on community radio. He has been actively involved in the national campaign for legalizing community radio in India. His other research interests include analyses of popular media, and mapping of discursive construction of corruption as a social problem in India. He has recently co-authored the book *Other Voices: Struggle for Community Radio in India*, along with Kanchan K. Malik, which has been published by SAGE.

Anuradha Ramachandran is a Financial Consultant with Ventureast Fund Advisors India Ltd, which advises venture capital funds in India, and has launched Ventureast-TeNet Fund.

Sundar Sarukkai is Professor of Philosophy at the National Institute of Advanced Studies, Bangalore. He is the author of the following books: *Translating the World: Science and Language*, *Philosophy of Symmetry* and *Indian Philosophy and Philosophy of Science*.

Alwyn Didar Singh is a Senior Civil Servant from India, presently posted as Member (Finance) at the National Highways Authority of India in the Ministry of Shipping, Road Transport and Highways where he is leading e-Government initiatives. Earlier, he was also CEO of two IT companies in Punjab, India. Dr Singh is a Strategy Consultant who has worked with several international agencies, including International Trade Centre, UNCTAD, World Health Organization and the ILO. He has several publications including two books in the areas of e-Commerce, e-Government and Trade.

Nirvikar Singh is Professor of Economics and Co-Director of the Center for Global, International and Regional Studies at the University of California, Santa Cruz. He was also a Co-founder of the Santa Cruz Center for International Economics. In 2007–08, he will serve as Special Advisor to the UCSC Chancellor, working on developing professional education in the Silicon Valley. He received his Ph.D from the University of California, Berkeley. His current research topics include electronic commerce, business strategy, information technology and development, federalism and political economy, and economic reform in India. He has authored over 100 research papers and his book, *The Political Economy of Indian Federalism*, co-authored with M. Govinda Rao, has recently been published by Oxford University Press.

T.T. Sreekumar is Assistant Professor at the National University of Singapore. Sreekumar holds a Ph.D in the STS (Science, Technology & Society) stream from Hong Kong University of Science and Technology, Hong Kong. His research interests include the impact of the Internet in developing countries, with a focus on Community Informatics and Digital Divide, ICTs and Innovation Systems in Asia, ICTs and Civil Society in South Asia, ICTs and Political Economy of Development, IT, Global Civil Society and New Social Movements. He has written and published extensively on these issues.

Ravi Sundaram is a Fellow at the Centre for the Study of Developing Societies, Delhi and Co-Initiator of the Sarai programme on media and urban culture (www.sarai.net). His work deals with the intersection of the city and contemporary electronic cultures, issues of legality and non-legality, and new conflicts around property and the electronic commodity. He has co-edited the critically acclaimed series: *The Sarai Readers: The Public Domain* (2001), *The Cities of Everyday Life* (2002), *Shaping Technologies* (2003), *Crisis Media* (2004) and the new *Turbulence* (2006). His *After Media: Pirate Life and Urban Culture* is forthcoming from Routledge (2008).

Gopakumar Krishnan Thampi is Director of Public Affairs Centre (PAC), Bangalore. He was also formerly an Executive Director at the Transparency International's Secretariat in Berlin. He holds a doctoral degree in Entrepreneurial Studies and postgraduate qualifications in economics, journalism and mass communication, and conflict resolution. He writes frequently on themes related to politics, civil society, governance and international affairs. He is an Advisory Member of Transparency International's Steering Group on Surveys and Indices, and also the Editor of *Public Eye*—a quarterly publication from PAC.

Peter van der Veer is Professor of Comparative Religion, at Utrecht University. He has authored several papers and books on the role of religion in modern society, and the production of communal identities in modern spaces. His well-known book, *Imperial Encounters: Religion and Modernity in India and Britain*, (Princeton University Press, 2001), for instance, maps the simultaneous co-production of religious and secular spaces in the UK and colonial India. He has also worked on identity formation among several migrant communities.

INDEX